# PHYSICAL METHODS

IN

# DETERMINATIVE
# MINERALOGY

# PHYSICAL METHODS
# IN
# DETERMINATIVE
# MINERALOGY

*Edited by*

## J. ZUSSMAN

*Department of Geology and Mineralogy, University of Oxford, England*

## ACADEMIC PRESS

LONDON and NEW YORK · 1967

ACADEMIC PRESS INC. (LONDON) LTD

Berkeley Square House

Berkeley Square

London W.1


*U.S. Edition published by*

ACADEMIC PRESS INC.

111 Fifth Avenue

New York, New York 10003

*Library of Congress Catalog Card Number:* 67–24319


PRINTED IN GREAT BRITAIN BY
W. S. COWELL LTD, BUTTER MARKET, IPSWICH

187114

# List of Contributors

S. H. U. BOWIE, *Atomic Energy Division, Institute of Geological Sciences, 15 & 17 Young Street, London, England* (pp. 103, 467)

B. W. CHAPPELL, *Department of Geology, Australian National University, Canberra* (p. 161)

J. V. P. LONG, *Department of Mineralogy and Petrology, University of Cambridge, England* (p. 215)

R. J. P. LYON, *Department of Geophysics, Stanford University, California, U.S.A.* (p. 371)

J. D. C. McCONNELL, *Department of Mineralogy and Petrology, University of Cambridge, England* (p. 335)

R. J. W. McLAUGHLIN, *Department of Geology and Mineralogy, University of Melbourne, Australia* (pp. 405, 475)

I. D. MUIR, *Department of Mineralogy and Petrology, University of Cambridge, England* (p. 31)

L. D. MULLER, *Warren Spring Laboratory, Stevenage, Hertfordshire, England* (pp. 1, 459)

G. D. NICHOLLS, *Department of Geology, University of Manchester, England* (p. 445)

K. NORRISH, *C.S.I.R.O., Division of Soils, Adelaide, Australia* (p. 161)

J. ZUSSMAN, *Department of Geology and Mineralogy, University of Oxford, England* (p. 261)

## Acknowledgments

For permission to use in whole or part, certain diagrams and tables, we are indebted to the following:

American Geophysical Union; American Society for Testing and Materials; Butterworths; Cambridge Instrument Co., Cambridge, England; Centrex Publishing Co., Eindhoven; "Cameca" Company, Paris; Edward Arnold Ltd; Elsevier Publishing Co.; H.M. Stationery Office; Holt, Reinhart and Winston, Inc.; Institute of Physics and the Physical Society; John Wiley and Sons, Ltd; Journal of Earth Sciences, Nagoya University; Macmillan and Co. Ltd; Macmillan Journals Ltd; McGraw-Hill Publ. Co. Ltd; Mineralogical Society, London; Munksgaard International Booksellers Ltd; Prof. P. R. J. Naidu, Mysore; Philips Electronic Instruments; Taylor and Francis Ltd; Vickers Instruments Ltd.

Appropriate acknowledgments are given in the captions to illustrations.

# Preface

The physical methods described in this book are those most commonly in use for the study of rocks and minerals, but many of them are also applicable to a much wider range of materials, including metals, ceramics, refractories, glasses and cements. Thus it should be useful not only to the geologist, mineralogist, geochemist and geophysicist, but also to the wider range of researchers concerned with "materials science".

The book is written with the laboratory research worker primarily in mind, but the treatments in many chapters are such as to be of value also to the advanced student, or postgraduate student, who has not yet embarked upon research.

A mineral can be said to be "determined" when we have identified the mineral phase and have determined its chemical composition. These two kinds of information about a mineral may be sought and obtained with varying degrees of thoroughness. The composition may be determined for major elements alone, but the determination may also embrace minor and trace element concentrations down to the lowest possible limits of detection. Knowledge of the valency state of certain elements is often required, and for some elements the isotopic constitution may be of interest. Structural as well as chemical details may be investigated in either more or less detail. For example, the identification of a mineral as, say, muscovite, might suffice, but one can also specify the sequence of layer stacking in its structure, and the extent to which it is ordered or disordered. A zircon can be either normal or metamict, an albite may be in the high-temperature or low-temperature form, and two amphiboles of very similar composition could conceivably differ by the way the iron and magnesium ions are distributed in the crystal structure. Although methods for making the latter, more subtle, distinctions between specimens are given some consideration, the major part of this book is concerned with physical methods for identifying a phase and determining its composition in the broad sense.

The physical properties of minerals (e.g. density, refractive index) are influenced principally by chemical composition and crystal structure, and so the quantitative measurement of such physical properties can often be used to infer the fundamental character of the mineral. The properties which are made use of in this way are those which can be measured most easily and with most accuracy, and those which are more sensitive to small variations in the mineral. Another class of physical method that is useful is that which involves the recording of a characteristic electromagnetic spectrum, as in emission, atomic-absorption or infrared spectroscopy and in X-ray fluorescence or electron-probe analysis. A third important kind of examination of materials is that which uses diffraction, e.g. of an electron, X-ray or neutron beam.

The physical methods dealt with in this text are in general those for which many geological research laboratories are, or can reasonably be, equipped with the necessary apparatus and staff at the present time. Thus methods such as neutron diffraction, nuclear magnetic resonance, electron spin resonance, mass spectrometry and Mössbauer spectroscopy are omitted. Spectrophotometry and colorimetry are not dealt with because they are more generally linked closely to chemical analytical procedures and are rather specific for particular elements. Atomic absorption (Chapter 13), however, is included, since it is a newly developing method that may have more general application. Although the thermal techniques described in Chapter 9 could be regarded as chemical rather than physical, they are included here since chemical reagents (apart from the gaseous atmosphere) are not involved.

The various methods included here require and are given different kinds of treatment by the various contributors. For those which are well established (e.g. transmitted-light microscopy and X-ray diffraction) there is only brief discussion of fundamental principles and more on techniques and applications, whereas for those methods that are newer or are developing fast (e.g. X-ray fluorescence analysis, electron-probe microanalysis), there is a more comprehensive treatment and some discussion of recent developments and future possibilities.

Although an increasing amount of information can be obtained about a mineral without necessarily separating it from the rock (e.g. by X-ray diffraction, electron diffraction, electron-probe microanalysis) many determinative techniques require pure mineral specimens. Chapter 1 therefore deals with mineral separation, although it is not in itself a determinative method. Some other chapters also require special editorial comment.

The method of reflected-light microscopy is in principal an old one, but only comparatively recently has it become really quantitative. Chapter 3 gives it detailed treatment and demonstrates its usefulness particularly in conjunction with quantitative hardness measurement.

The method of X-ray fluorescence analysis (Chapter 4) is comparatively new, and an abundance of techniques are described in the literature. Some are good and some not so good, some specific to a particular application and some general. Only now is there emerging something like a consensus of opinion as to a comparatively few recommended techniques for the main kinds of analytical problem presented by minerals.

Electron-probe microanalysis (Chapter 5) is essential for studying fine-textured specimens where separation and subsequent analysis of very small amounts of the desired material is either very difficult or impossible. As the accuracy, ease of operation, and the range of elements covered by the method all increase, the method might also be found to be preferable to others even for coarse specimens for which separation presents no problem.

Electron microscopy (Chapter 7) has for some time been used for the study of fine-grained materials, such as clay minerals, but insofar as observations were merely of grain size and shape, the method was considerably underexploited. Electron-diffraction patterns can be used to identify and characterize mineral grains, and are particularly sensitive to structural variations. More versatile specimen handling facilities are becoming available in electron microscopes and will increase the scope of this method. Furthermore, diffraction phenomena need also to be considered carefully in making a true interpretation of electron micrographs. These aspects are emphasized in the present text.

The use of infrared spectra empirically, for identifying and possibly further characterizing minerals, is reasonably well established and is dealt with and fully documented here (Chapter 8). The fine details of the infrared, visible and ultraviolet absorption spectra are, of course, related to specific chemical and structural features in the specimen. The understanding of details of spectra from substances as complex in chemistry and structure as are the silicate minerals, however, has begun to emerge so recently and is still so limited that this can barely be mentioned in the present text.

Although the principal application of emission spectrography to geology is in the determination of elements in trace or minor concentrations, there remain, nevertheless, certain aspects of emission spectrography (Chapter 10) that are usefully dealt with in the present context of determinative mineralogy. Brief mention is also made of the technique of spark-source mass-spectrography.

Each of the methods dealt with has its advantages, disadvantages and limitations, and each has its specialist devotees. It is hoped that the juxtaposition in this book of a number of different methods for determinative mineralogy will serve as a warning against the danger of too narrow an approach to analytical problems, and encourage the specialist in one technique to try, or at least to be aware of, several others that may be useful.

J. ZUSSMAN

*Oxford*
*March, 1967*

# Contents

# CHAPTER 1

# Laboratory Methods of Mineral Separation

## L. D. MULLER

*Warren Spring Laboratory, Stevenage, Hertfordshire, England*

## I. Introduction

The separation and recovery of specific constituents from heterogeneous assemblages of minerals, such as occur in both coherent and incoherent crystalline rocks and ores, frequently forms an essential, and often time-consuming, preliminary stage to subsequent detailed studies. Many methods of mineral separation are available, none are universally applicable and many are primarily designed for large-scale commercial operation.

In this chapter only laboratory bench-scale methods of separation are discussed, the more important and widely applicable of these being described in some detail, and brief reference only being made to those methods which may have occasional specific application. In general it is assumed that the amount of material to be processed does not exceed a maximum of 1–2 kg, but may be considerably less, and that the weight of end product(s) required is unlikely to be greater than about 25 g. In addition, emphasis is normally given to the preparation of a pure or high-grade product and not to maximum recovery. As the quantitative aspects of mineral separation are not paramount, the need for working with representative samples is not stressed. If, however, it is proposed to separate and recover minerals on a quantitative basis, the preliminary collection and subsequent sampling of bulk materials to provide representative working samples is of importance and must be considered. Both Brown (1960) and Krumbein and Pettijohn (1938), discuss the problems and methods of collection and sampling, while Gy (1954, 1956) has developed a method of determining the minimum weight at a given particle size that must be cut from bulk material to ensure a representative sample with defined accuracy; see also Ottley (1966).

In the separation and recovery of any one specific mineral constituent it is usually simpler and quicker to exploit more than one of its physical properties by employing a combination of two or more separating methods rather than to rely on the repeated use of one technique. Similarly it is invariably disadvantageous to attempt to achieve maximum grade (i.e. purity) and maximum recovery in one operation. It is almost always preferable to make a preliminary bulk separation (that is, to obtain a maximum recovery at a relatively low grade) of the sought-after constituent, which may then be further purified by one or more additional separatory techniques. Wherever possible the maximum quantity of material available for processing should normally be used as this usually facilitates the preparation of high-grade products; it becomes essential when the required mineral is present as a minor constituent. When only a small amount of material is available, considerable care must be exercised in carrying out the selected methods of separation and it may, at times, be necessary to sacrifice grade for recovery.

Comminuted material with a wide size-range may be processed by most methods of separation, but the efficiency of such separations is normally considerably improved if a closely sized fraction of the material is used. It is therefore preferable to sieve or screen the crushed material into two or three size fractions as a preliminary to separation; each fraction, if required, can be subsequently individually treated. In most instances, where the size of liberation of the required constituent permits, the minus 300 mesh B.S. ($-53$ $\mu$m) material should be discarded at the outset, for although it is often possible to process such particles they are usually difficult to separate and recover, and particularly so when it is proposed to use hand-picking techniques in the final stages of purification. With panning and tabling methods of gravity separation optimum performance is obtained with materials which have been hydraulically classified to give two or more sized products, the minus 300 mesh product once again being rejected where possible.

The ease of separation and recovery of minerals is determined by maximum differences in their physical properties, their initial grain size and percentage contents. When froth flotation is used as a means of separating minerals their physico-chemical surface properties should also be taken into account. It is essential, therefore, as a preliminary to separation, to establish the mineral suite as completely as possible and to assess the composition level and average grain size of the required constituent(s). A detailed knowledge of the mineral suite leads to the most suitable choice of separatory methods, determines

FIG. 1. Manually operated hydraulic rock splitting and crushing machine. (By courtesy of Cutrock Engineering Co. Ltd.)

their order of application and provides an understanding of any problems that may arise. Composition levels and grain size govern the amount of material that should initially be processed and the degree of comminution required. This latter factor is of particular importance when only small amounts of material are available as, in such instances, over-grinding, leading to the production of large amounts of very fine material, must be avoided.

## II. Preparation of Material

### A. COMMINUTION

For purposes of mineral separation, comminution (size-reduction) of solid materials by crushing or grinding is undertaken only to the extent necessary to liberate the sought after constituents. As all forms of comminution inevitably produce a proportion of fine material which is difficult to treat, this should be kept to a minimum by periodic sieving during comminution to remove the undersize before further reduction of the oversize material. The method of comminution and the type of equipment used will usually be dictated by the amount and nature of the material to be reduced.

FIG. 2.   Steel percussion mortar and pestle. Diameter of the larger mortar is about 4 in. (Crown copyright; reproduced by permission of Controller of H.M.S.O.)

For small samples, chips of suitable size may be flaked off a specimen with a geological hammer or by hammering on a small anvil. Alternatively, the material may be crushed in a manually operated hydraulic splitting machine (Fig. 1). The small pieces of rock thereby obtained may then be reduced to the required liberation size in an iron or ceramic mortar and pestle, or in a percussion mortar of the type shown in Fig. 2.

With somewhat larger amounts, the laboratory roller-crusher shown in Fig. 3 may be used to reduce 1–2 cm material to the required size. This machine has two contra-rotating motor-driven 7-in rollers, each of which is grooved for half its length, the other half being smooth. The gap between the rollers is adjustable from $\frac{1}{2}$ inch to contact. In operation, the material is fed from a hopper on to the grooved portion of the rollers and then, after collection, repassed several times through the smooth portion, the gap being progressively adjusted to give the desired final particle size. To facilitate cleaning, the rollers may be independently removed. A 4- or 6-in high-speed hammer mill provides an alternative means of size-reduction. The action of the mill is through four small swing hammers, each pivoted on a disc rotating at high speed. As the material is pulverized, the fine particles are

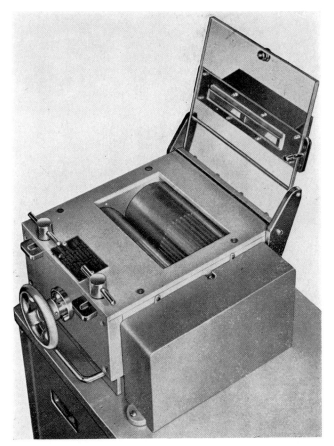

FIG. 3. Laboratory scale mechanically operated rock crusher. (By courtesy of Chas. W. Cook and Sons Ltd.)

forced through a grate and into a collection receptacle. A choice of grate sizes assists minimum production of fine material. Such a mill may reduce minus $\frac{1}{4}$-in material through 72 mesh B.S. at a throughput of approximately 2 kg/h.

For crushing greater weights of material at initially larger sizes, some form of laboratory jaw-crusher (Fig. 4) is preferable. This type of jaw-crusher may have a capacity of from 100 to 600 lb/h and can accept pieces of rock from about 2 to 6 in, depending on the size of the crusher. Simple adjustment of the jaws regulates the size of the crushed product to a minimum of approximately $\frac{1}{8}$ in.

Fig. 4.   The Roll jaw crusher, shown in the open position to facilitate cleaning. (By courtesy of Sturtevant Engineering Co. Ltd.)

With carbonate rocks, calcination may at times be successfully employed in place of the more usual methods of comminution. In this instance the lumps of rock are calcined in a suitable furnace at a temperature of approximately 1000°C. The calcine is then quenched in water and the resulting very fine slaked lime decanted to leave a heavy minerals residue, usually liberated at its natural size (Jones and Fullard, 1966).

### B. SIEVING AND CLASSIFICATION

In the context of mineral separation sieving is usually carried out either as part of the comminution cycle to remove the finer fractions and so reduce the incidence of over-grinding, or to prepare closely sized products to increase the efficiency of subsequent separatory methods. At times, when an adequate size differential exists between the wanted and unwanted constituents, e.g. due to shape differences, sieving may give an initial worthwhile concentration.

Sieves (Endecotts, 1963; Pryor, 1965) normally comprise a metal screen cloth retained in a circular frame, the most usual size being 8 inches in diameter, though for special purposes both larger and smaller diameter frames are available. The two fundamental dimensions of a screen cloth are the thickness (diameter) and spacing of the woven wires; the spacing, or aperture width, gives a measure of the particle size. Sieves are usually identified by a mesh number, which is defined as the number of wires per lineal inch. A number of sieve scales are in use, the two most widely met with being the British Standard Fine Mesh Series (B.S.) and the American equivalent (A.S.T.M.). The apertures of the B.S. fine series extends from 4000 to 45 $\mu$m and the A.S.T.M. series from 5660 to 37 $\mu$m. Particle size or sieve fractions are generally expressed in terms of mesh size; e.g. $-72 + 100$ mesh B.S., signifying that the material has passed through a 72 mesh (210 $\mu$m) sieve and has been retained on, and is therefore coarser than, the finer 100 mesh (150 $\mu$m) sieve.

Hand-sieving of dry materials is usually restricted to a nest of from one to four sieves, the coarsest at the top and the finest mounted into a bottom collecting pan. The material is placed on the top sieve and the nest shaken with a rotating and intermittent tapping action until screening is judged to be complete. Manual operation may be replaced by machine sieving, in which a greater number of sieves may be stacked into some form of mechanical shaker. After use, all sieves should be cleaned with either a soft brass-wire brush or hair-brush (for sieves finer than 100 mesh). Immersion in an ultrasonic bath is an alternative and efficient method of cleaning.

Wet sieving may at times be practised with advantage, particularly when large amounts of fine material are present. With this type of sieving a single sieve is used and is gently swirled and rocked in a bowl of water (to retain the under-size products) until the fine material no longer passes through the screen cloth. Additional water, as a spray, is frequently used to assist the screening process.

When the possibility of screen contamination cannot be accepted, cloth sieves may be employed. The cloth, of required mesh, is cut to the desired size and tightly stretched across plastic sieve holders; after use the cloth is discarded. With all forms of sieving, overloading of individual screens should be avoided.

Though special sieves finer than 350 mesh B.S. (45 $\mu$m) are now available for sizing materials in the sub-sieve range, such sizing is more usually carried out in the laboratory by some form of beaker-decantation. Unlike normal sieving, this method is based on the settling velocities of particles in a liquid, usually water, and is fully described by Pryor et al. (1953).

Hydraulic classification (Pryor, 1965, pp. 210–213) though not strictly a form of sieving, sorts the mineral constituents into size fractions on the basis of equal settling particles. Classification may be carried out in an elutriator of suitable size (see page 26) at high pulp densities to give hindered settling conditions, the classified products being taken off at increasing flow-rates. The use of hydraulically classified material is normally restricted to panning and tabling methods of separation.

## C. CLEANING

Incoherent rocks and coated grains may frequently be adequately dis-integrated and/or cleaned by scrubbing the material at a high pulp density in a suitable container with some form of mechanical stirrer. The addition of small amounts of a dispersant such as Calgon (sodium hexametaphosphate) or sodium silicate is generally essential for efficient cleaning. After scrubbing, the fine suspension is decanted and the sands further washed to give a clean product. More recently, the use of ultrasonic cleaning has been proposed by Revnivtzev and Dmitriev (1964).

## III. Separations by Gravity

All methods of separation by gravity seek to exploit differences in density between the required minerals and the associated, unwanted, constituents. For relatively small-scale laboratory separations, heavy liquids are widely used and provide the most rapid and efficient methods. Other methods in-clude the use of panners and the shaking-table, the latter being of particular value when handling greater weights of material.

### A. HEAVY LIQUIDS

Mineral separations using heavy liquids usually require that the density of the liquid is such that the wanted mineral either sinks or floats in the liquid while the unwanted minerals react in the reverse manner. The use of heavy liquids has wide application and is capable of considerable refinement particularly through the use of the heavy liquid density gradient which enables many minerals with very close densities to be individually recovered (Muller, 1960; Muller and Burton, 1965).

Separations may be made at any particle size (subject to adequate libera-tion) and down to a limiting size of about 10 $\mu$m. Below 100 mesh B.S. (150 $\mu$m) a centrifuge normally becomes necessary to accelerate the settling rates of the fine particles and to prevent mineral entrainment and/or agglo-meration. The weight of material processed may vary from a few milligrams to 200–300 g, though Fairbairn (1955) describes methods of handling kilo-gram quantities.

The range of densities practically available when using true liquids or solutions of salts extends from less than unity to 4·28 g/ml (Clerici solution at room temperature). For higher densities up to about 5·0 g/ml, melts of salts of low melting-point and adequate transparency may be used while relatively low-melting-point alloys increase the density range to over 10 g/ml, though such liquids are opaque. Mercury dispersions in bromoform to give *suspen-sions* with densities varying from about 3·5–7 g/ml have recently been described by Desnoes (1963), while Browning (1961) refers to the Cargille heavy liquids which are also suspensions of metals in organic liquids having the same density range.

A comprehensive list of heavy liquids, solutions and low-melting-point solids are given by Milner (1962) and Krumbein and Pettijohn (1938). Engineered Materials of New York are suppliers of suitable metal alloys.

The three organic heavy liquids most widely used in the laboratory for mineral separations are:

|  | Density*; g/ml at 20°C |
|---|---|
| Bromoform | 2·89 |
| Tetrabromoethane (acetylene tetrabromide) | 2·964 |
| Methylene iodide (di-iodomethane) | 3·325 |

* This is numerically equivalent to specific gravity at 20°C referred to water at 4°C.

These liquids may be diluted to lower densities with a number of miscible organic solvents, which may include benzene, carbon tetrachloride, acetone and alcohols; the use of *NN*-dimethylformamide has been proposed as a diluent for methylene iodide and bromoform respectively by Meyrowitz *et al.* (1960) and Hickling *et al.* (1961). This diluent, like acetone, is completely miscible with water. All three heavy liquids are light-sensitive, the break-down products producing marked discoloration; they should therefore be stored in dark-coloured containers.

Clerici solution (Clerici, 1907; Vassar, 1925) provides a natural extension to the density range of the organic liquids and is of great value in mineral separation. It comprises a saturated solution of equal weights of thallous formate and thallous malonate in distilled water and is commercially available or may be simply prepared. Details of the method of preparation are concisely given by Jahns (1939); 300 g of each salt with 40 g of water giving approximately 135 ml of solution with a density of 4·28 g/ml at room temperature. With increasing temperature and concentration of salts, densities of 4·65 at 50°C and about 5·0 at 90°C may be obtained. The solution is readily diluted with distilled water or reconcentrated on a water-bath; it is not light-sensitive.

When quantities of heavy liquids greater than normal are required, lead sulphamate solutions may be considered. Weavind and Linari-Linholm (1958) have described the preparation and use of this solution, which has a density of 3·64 g/ml at 80°C.

The halogenated heavy liquids, as with other organic liquids and solvents, all exhibit a certain degree of toxicity and although commonly used in the laboratory, require normal handling precautions. These should at least

B*

ensure that all operations are carried out in a well ventilated room or, preferably, in a vented fume-cupboard; precautions to prevent ingestion and contact with the skin or eyes should also be taken. With Clerici solutions, the thallium content makes these liquids extremely poisonous and similar precautions must be rigidly observed. O'Connell (1963) discusses in detail the properties, hazards and the precautions necessary when handling these liquids. Reference may also be made to Sax (1963) and Patty (1962). Bloom (1963) discusses the toxic properties of several commonly used organic diluents. Weavind and Linari-Linholm (1958) do not comment on the toxicity, or otherwise, of lead sulphamate.

Depending on the diluent used, the organic heavy liquids may be recovered from diluted mixtures or washings by fractional distillation (Milner, 1962) at reduced pressure in the case of methylene iodide, or more simply by evaporation of the diluent in the open. Dilute Clerici solutions may be reconcentrated or recovered from washings by careful evaporation on a sand or water-bath. Considerable care must be taken to avoid overheating as a dark-coloured residue may be formed. O'Meara and Clemmers (1928) and more recently Browning (1961) discuss various methods for the recovery of the organic liquids and Clerici solutions, while Fairbairn (1955) gives similar information for the recovery of relatively large volumes of bromoform, tetrabromoethane and methylene iodide. When acetone is used as the diluent it may, owing to its high solubility in water, be completely removed by streaming water through the diluted heavy liquid until a suitable density marker or mineral fragment placed in the liquid floats. The purified liquid is then separated from the excess water and finally dehydrated by shaking with calcium chloride followed by filtration.

Heavy liquids may darken considerably with repeated use and so become difficult to use. A number of methods have been proposed for decolorizing these liquids; both O'Meara and Clemmers (1928) and Browning (1961) detail several chemical methods for both organic heavy liquids and Clerici solutions while Griffitts and Marranzino (1960) advocate the use of small quantities of Fuller's earth, the earth being vigorously shaken or stirred with the liquid and then filtered. Fairbairn (1955) recommends dilution of the liquid with acetone (at least 1:5) followed by gentle warming and the addition of finely ground bone charcoal. When most of the acetone has evaporated the liquid is filtered and collected for final recovery with water. Clerici solutions may also be decolorized with bone charcoal (see Rankama (1936), who also outlines two chemical methods for the purification of these solutions containing excess amounts of dissolved impurities).

For checking the density of the various liquids, a set of small glass blocks* of known densities, or selected mineral fragments are frequently used.

* Commercially available from the Gemmological Instruments Ltd.

Dolomite (sp.gr. 2·85), aragonite (sp.gr. 2·94), andalusite or sillimanite (sp.gr. 3·20) and willemite or rutile (sp.gr. 4·2) are suitable indicators for bromoform, tetrabromoethane, methylene iodide and Clerici solution, respectively. The density of the liquid will be precisely given when the indicator is in hydrostatic balance with the liquid.

When accurate density measurement is needed, a Westphal balance may be used (Holmes, 1930). The determination of the refractive index of a liquid with an Abbé refractometer provides a further means of accurate density checking. As a graph of refractive index against density is usually substantially linear this method enables the density of a liquid, diluted to any value, to be determined rapidly once the appropriate curve has been established. As both refractive index and density are temperature dependent, this factor must be considered when accurate values are sought. Methods of obtaining appropriate curves and of applying temperature control are given by Muller and Burton (1965).

To obtain a liquid of preferred intermediate density by dilution the following equation:

$$V_b = \frac{V_a(\rho_a - \rho_m)}{(\rho_m - \rho_b)}$$

may be used to calculate the volume ($V_b$) of a diluent of density ($\rho_b$) which has to be added to a known volume ($V_a$) of heavy liquid of density ($\rho_a$) to give the liquid of preferred intermediate density ($\rho_m$). Sclar and Weissberg (1961) have solved this equation graphically and give a density chart from which it is possible to read off the appropriate volume of diluent directly to give the required intermediate density for any pair of miscible liquids. Alternatively a suitable glass or mineral indicator may be used when preparing an intermediate liquid. The appropriate indicator is placed in the heavy liquid, and the diluent slowly added with careful stirring until the indicator neither sinks nor floats. A Westphal balance may be similarly used. In this instance the balance is set to the required density and the diluent carefully mixed into the pure liquid until the balance reaches the set value.

Separation of the required mineral(s) from coarse powders may be simply carried out in a suitable beaker filled with the appropriate heavy liquid. The mineral grains are stirred into the liquid to ensure complete wetting and those with densities greater than that of the liquid will settle to form a "sinks" product. After separation, the "floats" product remaining on the surface may be skimmed off with a fine wire mesh scoop. Following removal of the two products each is separately filtered to recover the heavy liquid and then washed with a suitable solvent to remove the final traces of liquid. Another commonly used and simple method of separation requires a glass filter funnel mounted vertically in a clamp and fitted with a length of rubber tubing provided with a pinch-clip. The grains are again gently stirred into the liquid in the funnel and on separation the heavy minerals collect into the rubber tubing from which they may be discharged into a filter by manipulation of the pinch-clip. Browning (1961) replaces the filter funnel by a 250 ml pear-shaped separatory funnel. Muller and Burton (1965) describe a separator in which the normal stop-cock is modified to provide a small container to receive the heavy mineral crop. This has the advantage of readily eliminating any cross-contamination of the floats and sinks products during their recovery. Jones (1965) gives details of a continuous method of separation suitable for $-8$ mesh $+150$ mesh material, while Sarkar and Manchanda (1962) illustrate an enclosed apparatus for the sink/float separation of coals finer than $\frac{1}{8}$ inch. For handling minute amounts of mineral grains, both Foster (1947) and Rodda (1951) have developed microseparation methods using drops of liquids on microscope slides.

For treatment of relatively large quantities in the $-40$ $+200$ mesh size-range Fairbairn (1955) has described a simple and larger version of the filter funnel capable of processing about 1·5 kg. For amounts up to about 15 kg a semi-continuous method of beaker separation is described. In this instance the material is contained in a 9-quart stainless-steel vessel with tetrabromoethane (a 2:1 mixture is recommended). This pulp is kept continuously stirred and is slowly discharged to a 4-quart beaker filled with the same liquid. The beaker has an internal collar which is jigged to displace the floats into a collection launder surrounding the beaker.

In general, minerals finer than about 100 mesh are not efficiently separated by these methods as the settling velocities through the liquids are sufficiently slow to allow entrainment and/or agglomeration to take place. To obtain clean separations the heavy liquid–mineral suspension should be centrifuged; with such separations the chief difficulty arises in making a complete and separate recovery of the two products, particularly from a normal centrifuge tube. An early method for recovery from such a tube entailed the rapid freezing of its contents followed by halving the resulting solid plug and final remelting to recover the floats and sinks. Fessenden (1959) discusses the disadvantages of this method and proposes the controlled freezing of that portion of the tube containing the heavy minerals. Scull (1960) and Pollack (1962) give improvements on Fessenden's technique. Barsdate (1962) recommends the use of a hypodermic needle to recover the sinks product while Nickel (1955), Cheeseman (1957) and Brown (1960) illustrate centrifuge tubes with various designs of inner tubes which may be sealed to isolate the two products before their removal. Muller and Burton (1965) have described a further method in which a thin-walled, flexible, polythene tube is used to sleeve an ordinary centrifuge tube. After centrifuging, the polythene tube is withdrawn and the two contained separates sealed off from each other by closure at the middle of the tube with a modified crocodile clip.

When using the methods already described to separate constituents with density differences of less than about 0·1, practical difficulties arise in attempting to adjust and maintain the separating liquid accurately at the required intermediate density. The heavy-liquid density gradient column, first proposed by Sollas (1886, 1891) and described more fully by Holmes (1930)

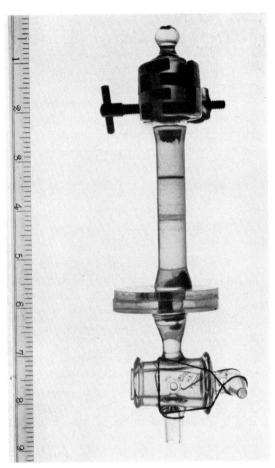

FIG. 5.   A heavy-liquid density gradient column with separated mineral layers. The mineral densities range from 2·39 to 2·69 g/ml. (Crown copyright; reproduced by permission of Controller of H.M.S.O.)

overcomes this problem and enables this type of separation to be undertaken. In such a column the density of the liquid increases *uniformly* from top to bottom so that when a mixture of liberated minerals is added, individual grains settle to their appropriate density levels to form clearly defined mineral layers (Fig. 5). The change in density throughout the column can be made very gradual so that constituents with similar densities (ca 0·005 g/ml) can be separated. Any of the normally accepted heavy liquids and Clerici solutions can be used to establish a gradient column. Such columns are quite stable, remain almost unaltered over a period of several days and may be centrifuged without disturbing the gradient. They are therefore suitable for the separation of materials down to a particle size of about 10 $\mu$m and may sometimes be successfully applied to the separation of clay constituents. Rapid methods for setting up pre-selected density gradients and the equipment and techniques for separately removing the individual mineral layers from a column have recently been described by Jones (1961) and Muller and Burton (1965), while Woo (1964) advocates the use of *stepped* gradient columns. Beevers (1961) and British Standard 3715 (1964) detail alternative methods of setting up density gradients. Weights of material to about 70 g may be processed depending on the method and type of separator used.

A novel application of the use of heavy liquids to recover montmorillonite and montmorillonite and halloysite has been described respectively by Rodda (1952) and Loughnan (1957). In these instances the specific gravities of the minerals were modified by absorption of certain of the heavy liquid constituents.

## B. PANNERS AND THE SHAKING-TABLE

Panners—which may include the miner's pan and Vanning plaque, the superpanner and micropanner—and the laboratory-size shaking-table, are all capable of separating materials with densities beyond the scope of normally available heavy liquids. In addition, some are designed to handle large amounts of material simply and rapidly.

The basic principle effecting separation with these types of concentrators is that of flowing film concentration wherein a liquid film (normally water) in laminar flow across a flat inclined surface has zero velocity at that surface and a maximum velocity a little below the liquid–air interface. In such a film mineral grains tend to sort and stratify themselves through the liquid so that the heavier grains lie in the regions of lower velocity while the lighter ones move into those of higher velocity, thus the lighter grains travel faster down the inclined surface and the minerals tend to separate. The inclined surface (deck) may be additionally subject to various forms of bumping or reciprocating motions to increase efficiency of separation. The theory and application of flowing films to table separation are discussed in detail by Pryor (1965) and Gaudin (1939), while Muller (1958) summarizes some of the principles underlying operation of the superpanner and micropanner.

When separating by panning or tabling, a minimum density difference of about 1·0 is usually necessary between the constituents for the method to be effective. A more precise indication of possible separation is given by the "concentration criterion" ratio (Taggart, 1945) which may be expressed as:

$$\text{concn crit.} = \frac{\rho_H - \rho_M}{\rho_L - \rho_M}$$

where $\rho_H$ = density of heavy constituent, $\rho_L$ = density of light constituent, $\rho_M$ = density of separating medium (normally water).

Generally, with a value of the ratio greater than 2·5 separation is possible at all sizes, while below 1·25 it is normally not possible; with intermediate values the efficiency of separation is somewhat dependent upon particle size. Though a wide size-range can be accepted, separation performance is considerably improved when closely screen sized or, preferably, hydraulically classified material is used. Separations below 300 mesh B.S. (53 $\mu$m) are possible, but difficult to achieve.

The miner's pan consists of a shallow metal dish with flared rim, about 18 inches diameter and 4 inches deep. It can take up to 25 lb of material such as gravels or sands. The pan is initially held in a horizontal position below the surface of a pool of water and the contents swirled to remove very fine material, such as clays, to leave a clean sand. The pan is then tilted forward and held in contact with the water surface and the lighter materials panned out using a combined circulatory and to-and-fro motion until only the heavy minerals remain. Both skill and experience are required to prevent too great a loss of the finer fractions of the concentrate. Raeburn and Milner (1927), Krumbein and Pettijohn (1938) and Pryor (1965) all describe the use of the miner's pan. The Vanning plaque provides a simple method of laboratory panning and is similarly operated. It comprises a shallow enamelled iron saucer-shaped dish, 11 inches in diameter. A maximum of 50 g of material is used; it is not suitable for coarse sizes.

The superpanner (Haultain, 1937; Taggart, 1945) (Fig. 6) is essentially a mechanized version of the miner's pan and is one of the most valuable laboratory tools available for gravity separation. It consists of a long, shallow V-shaped deck mounted on a 3-point suspension and reciprocated by a cam which gives a bump during each cycle. The deck can be adjustably sloped and also has a differential side-shake applied to it. Wash-water is fed to the top end and is removed at the opposite end by a suction device. The operating variables are:

1. Slope of deck; 2. Intensity of bump; 3. Length of stroke; 4. Speed of stroke; 5. Amplitude of side-shake at each end of deck; 6. Amount of wash-water; 7. Depth of pool.

From about 1–30 g of material is normally placed in a shallow pool formed at the bottom end of the sloping deck. On operating the superpanner the bumping and side-shake motions combine to dilate the settled pulp and

allow the heavier constituents to stratify to the deck surface. Continued bumping causes this material to move up along the deck to form a concentrate
while the counter-current flowing film of water washes the lighter fractions
into the pool where they may be continuously removed by the suction device;
an intermediate trap in the suction line retains this material, if required. After

FIG. 6.    The Haultain Superpanner for the gravity concentration of small amounts of
granular material. (Crown copyright; reproduced by permission of Controller of H.M.S.O.)

preliminary setting of the operating variables it is normally only necessary to
adjust the deck slope, amount of wash-water and speed of bumping to obtain
the required separation, the heavy minerals being distributed up the length of the
deck in order of increasing density. All variables are adjustable during operation.

The micropanner (Muller, 1958) is a small-scale version of the superpanner
and accepts from a few milligrams to about 5 g of material. It has the same
operating variables and is used in the same manner. As an alternative to water,
heavy liquids may be used to enhance the concentration criterion ratio for
minerals with close densities. Similarly, the alternative use of light liquids
(acetone, benzene, etc.) with their lower densities and viscosities increases the
apparent density and settling velocity of the minerals and so enables finer
materials to be handled.

Unlike the panners, which are all "batch" separating devices, the laboratory-
size shaking-table (Taggart, 1945) (Fig. 7) operates continuously at an

approximate feed-rate of about 25 kg/h, depending on the size-range of the material. Such a table is particularly useful when a preliminary bulk separation of minor heavy constituents entails the processing of large amounts of material; Harris *et al.* (1967) have demonstrated its application in the recovery of minor zircon occurring in a granite. The shaking-table consists of

FIG. 7. A shaking-table for continuous separation of large amounts of material. (Crown copyright; reproduced by permission of Controller of H.M.S.O.)

a small (39 × 20 in) lino-covered rectangular deck fitted with a number of tapered wood riffles and a feed-box. The deck, which may be adjustably sloped across its length, is asymmetrically moved backwards and forwards at about 280 strokes/min by a head-motion. Wash-water is also provided along one side of the deck. In operation, the material is fed at a suitable feed-rate through the feed-box to the table deck at a pulp-density of between 20 and 35% solids by weight. The pulp distributes itself across the deck, the heavy minerals become stratified and are trapped by the riffles so that they are gradually moved towards the end of the deck by the asymmetric motion to form a concentrate band of heavy minerals. The lighter mineral fractions are successively washed over the riffles and across the deck by the combined movement of the pulp and wash-water to form wider mineral bands of decreasing density. Suitably placed splitters and launders around the deck collect the various products. Pryor (1965), Gaudin (1939) and Taggart (1945) all discuss the underlying principles of table separation and describe their method of operation and applications.

## IV. Magnetic Separation

Minerals are termed paramagnetic when attracted toward a magnetic field and diamagnetic when repelled; ferromagnetic minerals are those (such as magnetite) which carry residual magnetism after removal of the field. For purposes of separation, minerals may be classed as highly magnetic (magnetite, pyrrhotine), moderately magnetic (ilmenite, chromite, almandine), weakly magnetic (monazite, tourmaline) and almost non-magnetic (quartz, zircon). The magnetic susceptibility of individual mineral species will vary, however, due to compositional variations or to the presence of minute inclusions of differing susceptibility.

Highly magnetic minerals are readily separated by using any suitable permanent magnet of the bar or horse-shoe type, the poles being covered by thin paper to facilitate subsequent release of the mineral grains. Alternatively a simple electromagnet of the type illustrated by Krumbein and Pettijohn (1938) may be used. Adjustable curved pole-pieces fitted to these types of

FIG. 8.   Frantz isodynamic magnetic separator. (By courtesy of the S. G. Frantz Co. Inc.)

magnet increase the sensitivity of separation. After separation, highly magnetic mineral grains may form stringers or aggregates due to their residual magnetism. If necessary, they may be demagnetized by first placing the grains, suitably contained, in an a.c. solenoid and then smoothly withdrawing the container.

For the very much larger group of less magnetic minerals, the iso-dynamic type of electromagnetic separator (Fig. 8), originally due to S. G. Frantz, is now widely used and has almost completely replaced earlier types such as those described by Hallimond (1930) and Officer (1947). The separator consists of an electromagnet with two specially shaped pole-pieces having a long narrow air gap. A flat vibrated chute, which divides into two at one end, is located in the air gap and enables material to be fed along the length of the electromagnet. The configuration of the pole-pieces is such that a constant magnetic force (perpendicular to the length of the chute) is applied to a grain of given susceptibility during its travel along the chute; thus ensuring maximum response to the forces of separation. The electromagnet is universally mounted and may be tilted to any position both transversely and longitudinally, the latter to assist material transport along the chute. The d.c. current controlling the magnetic flux can be sensitively varied by a two-stage rheostat system. In operation, the dry material is fed from a small hopper to the chute whose transverse slope is adjusted so that the gravitational and magnetic forces are opposed. By suitably balancing these forces separation can be obtained between minerals with very close susceptibilities, the more magnetic being moved against the chute slope and eventually discharging into one of two small collector buckets. By alteration of the transverse slope, both para- and diamagnetic minerals may be separated. As with most types of magnetic separators, performance is improved by using closely sized fractions of the material.

When used with the electromagnet in the inclined position, the separator can operate successfully with dry materials as fine as about 300 mesh (53 $\mu$m) and up to a maximum rate of 250 g/h, depending on the size and nature of the material. With the electromagnet in the vertical position and without the chute, rapid but less sensitive separations are made at high through-put rates. In either inclined or vertical positions, ferromagnetic minerals must first be removed as these rapidly clog the air gap.

Hess (1956) gives practical operating data and instrument settings for several types of separation while Gaudin and Spedden (1943) discuss the separation of sulphide minerals; this paper includes an appendix by Frantz on the construction and operation of the machine. McAndrew (1957) has used the separator to determine rapidly the magnetic susceptibility of minerals and composite grains.

The high-intensity, induced-roll magnetic separator (Fig. 9) is a suitable alternative to the Frantz type machine when relatively large amounts of material have to be treated. It is a continuous-flow machine and can operate at all feed-rates up to approximately 1 kg/min. It accepts dry feed in the size-range from about 36–300 mesh, depending on the nature of the material.

The separator has a laminated metal cylinder located between the shaped pole-pieces of an electromagnet. A thin curtain of material is fed onto the rotating cylinder from a small triangular-shaped hopper having an adjustable gate. The non-magnetic grains are thrown off by centrifugal forces and collected into a chute, while the magnetic minerals adhere to the cylinder and are carried round and brushed off into a second chute. The operating variables include rate of feed, speed of rotation of the cylinder, variation of magnetic flux and adjustment of the air gap between the rotating field and pole-pieces.

FIG. 9.   High-intensity, induced-roll magnetic separator. (By courtesy of "Carpco".)

Dry magnetic separation of fine material is normally difficult as particles tend to aggregate due to surface forces. This type of material can be treated in a Dings-Davis tube, a wet type magnetic separator. It has an inclined glass tube, filled with water and placed between the poles of an electromagnet. The tube is mounted on an adjustable carriage which gives the tube a combined oscillatory and twisting motion. The material is fed as a slurry and settles through the tube, the magnetic particles adhering to the tube walls adjacent to the magnet; the complex motion of the tube releases any entrained non-magnetic particles. The latter are collected into a small flask and, after removal, the magnetic product is separately recovered after switching-off the magnet current. Adjustment of feed-rate, magnetic flux, and inclination and speed of oscillation of the tube lead to selective separation. The machine has a relatively low intensity and is usually only used to separate the more highly magnetic minerals.

For separating small quantities of very fine materials du Fresne and Anders (1962) have recently described the design and operation of a wet magnetic separator which is mounted on the stage of a microscope. With it they have prepared small homogeneous samples with differing magnetic susceptibilities.

## V. Electrostatic, High Tension and Dielectric Methods

These methods may frequently be of value when more conventional methods of separation prove inadequate. Electrostatic and high-tension separations depend on the constituents of a mineral mixture having different values of surface resistance. Good conductors have a low surface resistance and lose their electrostatic charge rapidly, while poor conductors, with higher values, retain their charge for longer periods. The terms electrostatic and high tension (electrodynamic) are not entirely synonymous and may be related to the method of charging. The former term implies that no current is flowing during the process of charging, while high tension applies when either a slight current flow (spray discharge) or a high rate of flow (corona discharge) is employed. These methods all operate in dry conditions and humidity has an important effect. Conversely, dielectric separation is a wet method, the dielectric constant of the liquid chosen being either intermediate to, or greater than those of the minerals being separated.

With electrostatic and dielectric methods the forces available for separation are relatively small and in general the size range of the material lies between about 36 and 300 mesh.

Ralston (1961) and Fraas (1962) both discuss the fundamentals of these methods of separation together with the various types of equipment and their applications. For laboratory use, Holmes (1930) has described some earlier forms of simple electrostatic separators. More elaborate continuous-flow high-tension separators suitable for handling rather larger amounts of material are marketed by both Rapide Ltd. and Carpco Manufacturing Inc. These separators are of the roll-type in which the material is fed as a thin curtain on to an earthed rotating metal roll. A fine-wire-beam electrode, backed by a larger diameter electrode, is mounted adjacent and parallel to the axis of the roll and provides a very strong and variable corona discharge. The good-conductor minerals in contact with the roll discharge rapidly and are flung off the rotating roll while the poor conductors adhere until removed by a suitably placed brush. For processing small amounts of material without loss, Woodley and Duffell (1964) describe a high-tension separator in which the roll has been replaced by a stainless-steel disc rotating at 80 rev/min. Mineral grains are fed to the disc to pass under a fine-wire-corona source. As the grains lose their charge by leakage to the disc they are progressively thrown off into a series of small collecting trays surrounding the disc. The separator is totally enclosed and its humidity may be controlled.

Hatfield (1924), Holman and Shepherd (1924) and Berg (1936) describe various small batch-type dielectric separators and the appropriate liquids. More recently Pohl and Plymale (1960) have given details of an experimental continuous-flow separator, the isomotive cell; Verschure and Ijlst (1966) have described a similar separator. Rosenholtz and Smith (1936) give a comprehensive list of minerals and their dielectric constants while reference may be made to the International Critical Tables (1929) for the constants of various liquids. It must be remembered that the values vary with the methods of measurement employed and with some, different figures are given by various authorities.

## VI. Miscellaneous Methods

### A. FROTH FLOTATION

Almost all minerals are hydrophilic (water wetted or adherent) but may be made hydrophobic (water repellent or air adherent) by alteration of the nature of the mineral surface using certain organic compounds (collectors). These collectors, which may be either anionic or cationic compounds, are heteropolar; the polar grouping attaching to the mineral surface while the hydrophobic non-polar group is free to adhere to an air–water interface. The ability of such collectors *selectively* to coat, or condition, the surface of *specific* mineral particles, so that they can adhere to air bubbles, forms the basis of froth flotation.

This method of mineral separation, which is used extensively in commercial practice for the recovery of a wide range of minerals, can be successfully applied in the laboratory to handle from about 3 g to 1 kg of material. It is, therefore, a separatory method of considerable importance and can often be used with time-saving advantage to obtain a medium to high-grade bulk concentrate of the required mineral(s) as a preliminary to the preparation of a high-purity product by other more conventional methods of laboratory separation.

In the laboratory, the crushed rock, in the form of a water-dispersed pulp with a pulp density of 20–30% by weight, is normally conditioned in some type of flotation cell by the addition of a collector appropriate to the specific mineral required. The cell is usually provided with an impeller or other means of maintaining the pulp in suspension and also of generating a stream of air-bubbles to rise through the stirred pulp. A suitable frothing agent, which may include pine oil, cresylic acid or one of the higher alcohols, may also be added to the pulp to ensure that the air-bubbles form a stable froth at the surface. On floating, the selectively conditioned mineral particles attach themselves to the air–water interface of the air bubbles and are lifted into the surface froth. The mineralized froth is continuously skimmed off to give the desired mineral concentrate.

To increase selectivity of flotation, activators and/or depressants may be used. Activators increase the flotability of the required mineral(s) while depressants prevent flotation of the unwanted or gangue mineral particles. Many reagents used, including collectors and frothers, are water soluble. Others may be mechanically dispersed into the pulp or added as neutral solutions or as emulsions.

Sutherland and Wark (1955) list thirty-two flotation variables, ten of which are characteristic of the rock or ore and cannot be controlled. Of the remaining twenty-two variables the more important may be briefly mentioned. The measurement of the pH value of the pulp and its control, by the addition to the pulp of such chemicals as sulphuric acid, lime or sodium carbonate, is of paramount importance as the pH largely regulates the behaviour of the collector used. The presence of slimes ($-10\ \mu$m particles) in a pulp should be avoided as these may lead to excessive reagent consumption (particularly of cationic collectors) and/or inhibit flotation by coating the surface of the larger mineral particles (normally 350 to $10\ \mu$m). Mineral surfaces should normally be fresh and should not be exposed to air or other sources of contamination for long periods after crushing or scrubbing. Contaminated surfaces may cause flotation of unwanted minerals or alternatively prevent the selective conditioning of the wanted mineral(s). Conditioning times may vary from a few seconds to 20 min, while the pulp density and degree of aeration may also be factors of importance. Reagent quantities are generally small and are normally expressed as lbs of reagent per ton of material treated, e.g. 0·1 lb/ton, a frequently used quantity.

Many types and sizes of laboratory flotation machines (cells) are commercially available and others may be simply constructed. Fuerstenau *et al.* (1957) illustrate and describe the use of the modified Hallimond tube (a small glass flotation cell suitable for a 2–3 g charge), while Taggart (1945) describes the Mayeda cell, which accepts a 10 g charge. The Pryor and Liou (1948) pneumatic cell also utilizes this weight of material. For charges up to 50 g a straight-sided, 4-in diameter Buchner funnel with glass frit (porosity 3) may be used (see Fig. 10). The funnel should be extended to give a volume of about 500 ml; this cell may be stirred mechanically or by hand and is aerated by connection to an air-supply. For use with greater weights of

FIG. 10.   A simple pneumatic glass flotation cell. (Crown copyright; reproduced by permission of Controller of H.M.S.O.)

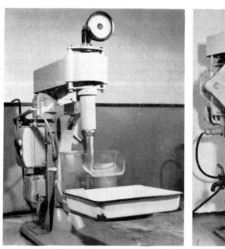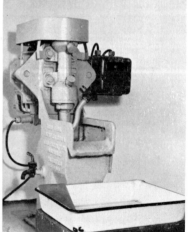

Fagergren cell                                    Denver cell

FIG. 11.   Laboratory flotation cells: charges of 500 g to 1 kg may be used with these cells. (Crown copyright; reproduced by permission of Controller of H.M.S.O.)

charge (and these are to be preferred to facilitate subsequent clean-up stages) several types of 500 g and 1 kg cells are marketed. Two of the most widely used are shown in Fig. 11.

For the theory and practice of froth flotation reference may be made to Pryor (1965), Sutherland and Wark (1955), Gaudin (1939, 1957) and Fuerstenau (1962). Pryor, Sutherland and Wark and the Denver, American Cyanamid and Float-Ore handbooks all give useful tables of floatable minerals together with the most suitable reagents, pH values and other data relevant to their flotation. Harris *et al.* (1967) describe in detail the laboratory flotation conditions necessary for the recovery of a number of minerals from typical igneous and metamorphic rocks, such as granites, dolerites and schists. The species discussed include many of the major and accessory rock-forming minerals.

### B. ASYMMETRIC VIBRATION

This method of dry separation makes use of differences in the shape factor of certain minerals and may be applied to the separation of rounded or equant grains from those of a platy or otherwise well-cleaved nature, for example quartz from micas. In its simplest form the separator may consist of an inclined glazed sheet of cardboard which is rapidly tapped. Faul and Davis (1959) have described a simple mechanized mica separator, similar in principle to the glazed card, which will separate a number of minerals on the basis of differences in shape and cleavage; the difficult biotite–chlorite separation is also stated to be feasible.

### C. ELECTROCHEMICAL SEPARATION

Vincent (1951) has proposed a novel method of separation using differences in the electrical conductivity of minerals. It is based on the selective deposition of metallic iron on conductor minerals enabling them subsequently to be separated from the non-conductors by magnetic methods. Shcherbak (1957) has developed a variant of this method and has applied it to a mixture of non-conducting, semi-conducting, and conductor minerals which are distributed on a mercury surface in a suitable container and covered by a $FeCl_2$–$CaCl_2$ solution. Passage of current through the electrolyte deposits an iron film on the surface of the conductor minerals and forms an iron–mercury amalgam. The coated minerals become wetted by the amalgam and sink below the surface. The non- and semi-conductor minerals remain on the mercury surface and may be washed off before recovery of the submerged conductor minerals.

The equipment and methods of operation are given in detail and the technique may process 1–2 g of material in the size-range 98 to 20 $\mu$m. The minerals discussed include pyrite, galena, chalcopyrite, cerussite, cassiterite, wolframite, malachite, chrysocolla, zircon, barytes, siderite and dolomite. The possibility of separating a mixture of sulphides is also discussed.

## D. ELUTRIATION

Elutriation of closely sized material in a rising current of fluid (normally water) may at times provide an adequate method of separation. Frost (1959) describes an elutriating tube for the separation of minerals. In such tubes the closely sized material is charged into an upward flow of water delivered from some form of constant head device; a screw-clip or small needle-valve controls the desired flow-rate. The flow-rate is adjusted so that the higher density mineral settles slowly into the collecting jar against the ascending current, the remaining minerals being collected into the overflow. The flow-rate used may be obtained empirically or may be determined by calculating the maximum settling velocity of the higher density mineral particles using Stokes Law, the required flow-rate being equal to the settling velocity. Stokes Law is normally expressed as:

$$V_m = \frac{2}{9} \times \frac{(\rho - \rho_1)r^2 \, g}{\mu}$$

where $V_m$ = max. settling velocity of mineral particle; $\rho$ = sp.gr. of particle; $\rho_1$ = sp.gr. of fluid (water); $r$ = half mean particle size; $g$ = acceleration due to gravity (981 cm/sec/sec); $\mu$ = viscosity of fluid. (All expressed in c.g.s. units.)

Settling velocity is, therefore, proportional to the *square* of particle size and to the apparent sp.gr. of the settling particles. Efficiency of separation increases with increase in the density difference between the minerals being elutriated, but decreases rapidly if the size-range of the minerals is increased.

For purposes of elutriation Stokes Law is normally applicable to particles in the size-range 5 to about 200 $\mu$m.

## E. HAND-PICKING

As a means of separation this method is both slow and tedious and should only be used as a last resort or as a means of removing the remaining traces of a contaminant (normally composite grains) to obtain an absolutely pure product. The material handled should be as coarse as possible compatible with adequate liberation; for practical purposes the lower size limit may be set at 200 mesh B.S. (75 $\mu$m). The simplest method of hand-picking is to disperse the dry mineral grains sparsely in a Petri dish and, viewing through a stereoscopic binocular microscope, to push unwanted grains to one side with a suitable wooden probe. Alternatively the grains may be individually recovered and transferred to a small watch-glass in the Petri dish. Hilder-brand (1952) has proposed the use of a steel needle coated with an aerosol which forms an even thin adhesive layer on the needle. When picking grains from a liquid, Partridge (1934) advocates the use of a greased wire.

Such methods are extremely slow and the wanted grains may be recovered much more rapidly by using a finely tapered glass tube connected, through a collection bottle, to a water suction pump. A small hole in the tube, suitably sited for finger-tip closure, provides a simple means of controlling the suction. Murthy (1957) and Savolanti and Tyni (1960) have both described useful variants of this type of suction picker. To assist in the picking of larger quantities of material, Senftle (1951) has described a small motor-driven turn-table mounted on the stage of a low-power microscope. A small rotating drum-feeder discharges a thin stream of mineral grains on to the turn-table from whence the required mineral constituent is removed with a suction picker, the remaining material being swept off by a micro air-jet. Muller (1960) gives details of a similar apparatus in which the turn-table is replaced by a small endless polythene conveyor belt.

### REFERENCES

American Cyanamid Co. (1954). Mineral Dressing Notes, No. 20. New York.

Barsdate, R. J. (1962). Rapid heavy mineral separation. *J. sedim. Petrol.* **32**, 608.

Beevers, A. H. (1961). Preparation of sensitive linear density gradients. *Proc. Soil Sci. Soc. Am.* **25**, 357.

Berg, G. A. (1936). Notes on the di-electric separation of mineral grains. *J. sedim. Petrol.* **6**, 23.

Bloom, H. (1963). Toxic properties of several organic solvents used in geochemical exploration. *Am. Miner.* **58**, 1000.

British Standards Institution (1964). Specification for concentration gradient density columns. B.S. 3715, London.

Brown, G. M. (1960). *In* "Methods in Geochemistry," (A. A. Smales and L. R. Wager, eds.), pp. 4–32. Interscience, New York.

Browning, J. S. (1961). Heavy liquids and procedures for laboratory separation of minerals. U.S. Department of the Interior, Bureau of Mines, Information Circular 8007, 1.

Cheeseman, D. R. (1957). A new technique in centrifugal mineral separation. *Can. Mineralogist* **6**, 153.

Clerici, E. (1907). Preparazione di liquidi per la separazione di minerali. *Atti Acad. naz. Lincei Rc.* **16**, 187.

Denver Equipment Co. (1962). Mineral Processing Flow-sheets. Denver.

Desnoes, A. (1963). Utilisation des suspensions denses de mercure dans la bromo-forme au laboratoire. Bureau de Recherches Géologiques et Minières (Minérais et Metaux), RE-868 (2), 1.

Endecotts (Filters) Ltd. (1963). Test Sieving Manual, London.

Fairbairn, H. W. (1955). Concentration of heavy accessories from large rock samples. *Am. Miner.* **40**, 458.

Faul, H. and Davis, G. L. (1959). Mineral separation with asymmetric vibrators. *Am. Miner.* **44**, 1076.

Fessenden, F. W. (1959). Removal of heavy liquid separates from glass centrifuge tubes. *J. sedim. Petrol.* **29**, 621.

Float-Ore Ltd. "Flotation Reagents." London.

Foster, W. R. (1947). Gravity separation in powder mounts as an aid to the petrographer. *Am. Miner.* **32**, 462.

Fraas, F. (1962). Electrostatic separation of granular materials. U.S. Department of the Interior, Bureau of Mines, Bulletin 603, 1.

du Fresne, E. R. and Anders, E. (1962). On the chemical evolution of carbonaceous chondrites. *Geochim. cosmochim. Acta* **26**, 1085.

Frost, I. C. (1959). An elutriating tube for the S.G. separation of minerals. *Am. Miner.* **44**, 886.

Fuerstenau, D. W. (Ed.) (1962). "Froth Flotation." 50th Anniversary volume. The American Institute of Mining, Metallurgical and Petroleum Engineers, Inc., New York.

Fuerstenau, D. W., Metzger, P. H. and Seele, G. D. (1957). How to use the modified Hallimond tube. *Engng. Min. J.* **158**, 93.

Gaudin, A. M. (1939). "Principles of Mineral Processing." McGraw-Hill, New York.

Gaudin, A. M. (1957). "Flotation," (2nd Edn.). McGraw-Hill, New York.

Gaudin, A. M. and Spedden, H. R. (1943). Magnetic separation of sulphide minerals. *Trans. Am. Instn Min. Metall. Engrs.* **153**, 563.

Griffitts, W. R. and Marranzino, A. P. (1960). Fullers earth as an agent for purifying heavy organic liquids. *Am. Miner.* **45**, 739.

Gy, P. (1954). L'échantillonage des minérais—Erreur commise dans le prelèvement d'un échantillon sur un lot de minérai. *Revue Ind. miner.* **35**, 311.

Gy, P. (1956). L'échantillonage des minérais—Pois à donner à un échantillon, Abaques d'échantillonage. *Revue Ind. miner.* **38**, 53.

Hallimond, A. F. (1930). An electromagnetic separator for mineral powders. *Mineralog. Mag.* **22**, 377.

Harris, P. M., Hollick, C. T. and Wright, R. (1967). Mineral separation for age-determination. (In preparation.)

Hatfield, H. S. (1924). Dielectric separation: a new method for treatment of ores. *Trans. Instn. Min. Metall.* **33**, 336 (Disc. 350).

Haultain, H. E. T. (1937). Splitting the minus-200 with the superpanner and infrasizer. *Trans. Can. Inst. Min. Metall.* **40**, 229.

Hess, H. H. (1956). Notes on operation of Frantz isodynamic magnetic separator. Instrument instruction booklet of S. G. Frantz Co. Inc., U.S.A.

Hickling, N., Cuttitta, F. and Meyrowitz, R. (1961). *NN*-Dimethylformamide, a new diluent for bromoform used as a heavy liquid. *Am. Miner.* **46**, 1502.

Hilderbrand, F. A. (1952). Use of aerosol in grain sorting. *Am. Miner.* **37**, 129.

Holman, B. W. and Shepherd, St. J. R. C. (1924). Dielectric mineral separation: Notes on laboratory work. *Trans. Instn. Min. Metall.* **33**, 343 (Disc. 350).

Holmes, A. (1930). "Petrographic Methods and Calculations." Thomas Murby and Co., London.

International Critical Tables (1929). Vol. VI, 1st Ed., pp. 82–106. McGraw-Hill, New York.

Jahns, R. H. (1939). Clerici solution for the specific gravity determination of small mineral grains. *Am. Miner.* **24**, 116.

Jones, J. M. (1961). Method of establishing a liquid column of graded density. *J. scient. Instrum.* **38**, 367.

Jones, M. P. (1965). A continuous laboratory-size density separator for granular materials. *Mineralog. Mag.* **35**, 536.

Jones, M. P. and Fullard, R. J. (1966). Mineral liberation by thermal decomposition of a carbonate rock. *Trans Instn Min. Metall.* **75**, Sect. C, C127.

Krumbein, W. C. and Pettijohn, F. J. (1938). "Manual of Sedimentary Petrography." Appleton-Century-Crofts, New York.

Loughnan, F. C. (1957). A technique for the isolation of montmorillonite and halloysite. *Am. Miner.* **42**, 393.

McAndrew, J. (1957). Calibration of a Frantz isodynamic separator and its application to mineral separation. *Proc. Australas. Inst. Min. Metall.* **181**, 59.

Meyrowitz, R., Cuttitta, F. and Levin, B. (1960). *NN*-Dimethylformamide, a new diluent for methylene iodide heavy liquid. *Am. Miner.* **45**, 1278.

Milner, H. B. (1962). "Sedimentary Petrography." Vol. I, 4th Revised Ed. George Allen and Unwin, London.

Muller, L. D. (1958). An apparatus for the gravity concentration of small quantities of materials. *Trans. Instn. Min. Metall.* **68**, Part 1., 1.

Muller, L. D. (1960). Some laboratory techniques developed for ore dressing mineralogy. Proceedings of the International Mineral Processing Congress. London, paper No. 52. Institution of Mining and Metallurgy, London.

Muller, L. D. and Burton, C. J. (1965). "The heavy liquid density gradient and its applications in ore dressing mineralogy." Paper No. 49 presented at the VIII Commonwealth Mining and Metallurgy Congress, Australia and New Zealand,

Murthy, M. V. N. (1957). An apparatus for hand-picking mineral grains. *Am. Miner.* **42**, 694.

Nickel, E. H. (1955). A new centrifuge tube for mineral separation. *Am. Miner.* **40**, 697.

O'Connell, W. L. (1963). Properties of heavy liquids. *Trans. Am. Inst. Min. metall. Petrol. Engrs.* **226**, 126.

Officer, V. C. (1947). A new laboratory separator for mineral sands. *N.Z. Jl. Sci. Technol.* **29(B)**, 133.

O'Meara, R. G. and Clemmers, J. B. (1928). Methods of preparing and cleaning some common heavy liquids used in ore testing. U.S. Bureau of Mines, R.I., 2897.

Ottley, D. J. (1966). Gy's sampling slide rule. *Min. and Miner. Eng.* **2**, 390.

Partridge, F. C. (1934). Methods of handling and determination of detrital grains and crushed rock fragments. *Amer. Min.* **19**, 982.

Patty, F. A. (1962). "Industrial Hygiene and Toxicology." Vol. II, 2nd Revised Ed. Interscience, New York.

Pohl, H. A. and Plymale, C. E. (1960). Continuous separations of suspensions by nonuniform electric fields in liquid dielectrics. *J. electrochem. Soc.* **107**, 390.

Pollack, J. M. (1962). Removal of heavy liquid separates from glass centrifuge tubes – additional suggestions. *J. sedim. Petrol.* **32**, 607.

Pryor, E. J. (1965). "Mineral Processing," 3rd Ed. Elsevier, London.

Pryor, E. J. and Liou, K. B. (1948). A simple flotation cell. *Trans. Instn. Min. Metall.* **58**, 85.

Pryor, E. J., Blyth, H. N. and Eldridge, A. (1953). Purpose in fine sizing and comparison of methods. *In* "Recent Developments in Mineral Processing," pp. 11–30. Institution of Mining and Metallurgy, London.

Raeburn, C. and Milner, H. B. (1927). "Alluvial Prospecting." Murby, London.

Ralston, O. C. (1961). "Electrostatic Separation of Mixed Granular Solids." Elsevier, New York.

Rankama, K. (1936). Purifying methods for Clerici solution and for acetylene tetrabromide. *Bull. Comm. géol. Finl.* **115**, 65.

Revnivtzev, V. I. and Dmitriev, Yu. G. (1964). Ultrasonic cleaning of minerals. Proceedings of the VII International Mineral Processing Congress. New York.

Rodda, J. L. (1951). Microseparation of minerals in heavy liquids. *Am. Miner.* **36**, 625.

Rodda, J. L. (1952). Anomalous behaviour of montmorillonite clays in Clerici solution. *Am. Miner.* **37**, 117.

Rosenholtz, J. L. and Smith, D. T. (1936). The dielectric constant of mineral powders. *Am. Miner.* **21**, 115.

Sarkar, G. G. and Manchanda, S. (1962). An improved device for float and sink tests of coals below ⅛ in. *J. Mines Metals Fuels.* **10**, 17.

Savolanti, A. O. M. and Tyni, M. H. (1960). A new mineral-picking apparatus. *Am. Miner.* **45**, 901.

Sax, N. I. (1963). "Dangerous Properties of Industrial Materials," 2nd Ed. Rheinhold, New York.

Sclar, C. B. and Weissberg, A. (1961). Density chart for the preparation of heavy liquids for mineralogical analysis. *Trans. Am. Inst. Min. metall. Petrol. Engrs.* **220**, 349.

Scull, B. J. (1960). Removal of heavy liquid separates from glass centrifuge tubes – alternative method. *J. sedim. Petrol.* **30**, 626.

Senftle, F. E. (1951). Apparatus for the separation of mineral grains. *Am. Miner.* **36**, 910.

Shcherbak, O. V. (1957). *In* "Sovremennye Metody Mineralogicheskovo Issledovaniya Gornykh Porod, Rud i Mineralov," (E. V. Rozhkovoi, ed.), pp. 103–114. Vsesoy, Nauch-Issled. Instn. Mineral Syr'ya, Min. Geol. Okhranny Nedr. SSSR (unpublished translation).

Sollas, J. W. (1886). On the physical characters of calcareous and siliceous sponge-spicules and other structures. *J. Roy. Geol. Soc. Ireland* **7**, 30.

Sollas, J. W. (1891). A method of determining specific gravity. *Nature, Lond.* **43**, 404.

Sutherland, K. L. and Wark, I. W. (1955). "Principles of Flotation." Australian Institute of Mining and Metallurgy, Melbourne.

Taggart, A. F. (1945). "Handbook of Mineral Processing." Chapman and Hall, London.

Vassar, H. E. (1925). Clerici solution for mineral separation by gravity. *Am. Miner.* **10**, 123.

Verschure, R. H. and Ijlst, L. (1966). Apparatus for continuous dielectric-medium separation of mineral grains. *Nature, Lond.* **211**, 619.

Vincent, H. C. G. (1951). Mineral separation by an electro-chemical magnetic method. *Nature, Lond.* **167**, 1074.

Weavind, R. G. and Linari-Linholm, A. A. (1958). The recovery of diamonds from prospection samples. *Jl. S. Afr. Inst. Min. Metall.* 635.

Woo, C. C. (1964). Heavy media column separation: a new technique for petrographic analysis. *Am. Miner.* **49**, 116.

Woodley, D. J. A. and Duffell, C. H. (1964). A high-tension disc separator. *Min. Mag.* **111**, 313.

# CHAPTER 2

# Microscopy: Transmitted Light

## I. D. MUIR

*Department of Mineralogy and Petrology, University of Cambridge, England*

## I. Introduction

### A. SCOPE AND VALUE OF OPTICAL METHODS

Optical methods of examining transparent and translucent substances with the aid of polarized light have been developed to such an extent over the past century that they have now been introduced successfully into many fields of natural science and technology: nowhere has their contribution been of greater importance than in the development of mineralogy and petrology. Brewster, only a few years after the discovery of the polarization of light first pointed out their application to the problem of mineral identification, and the first compound polarizing microscope was produced by Talbot in 1834. The utility of the instrument was extended greatly in 1845, when Sorby showed how thin sections of rocks and minerals could be prepared and their fine structures studied. Even today, despite the increasing application of powerful new sophisticated physical methods of investigation, the polarizing microscope still retains its unique position as the most generally useful, versatile, and widely used instrument in the modern mineralogical laboratory. It can be used for rapid preliminary reconnaissance of minerals, rock or synthetic specimens for the purposes of identification or for selecting representative material for further study. It can also be used as a precision instrument for the quantitative determination of the fabric of a rock, for measuring size and thickness, and for the quantitative determination of the optical properties of minerals.

To the petrologist the polarizing microscope is an essential tool that enables him to examine rocks to reveal their textures and grain size, to identify the minerals, estimate their chemical compositions and determine their relative amounts and mode of occurrence. After some practice most of the common minerals can be recognized by their appearance using such obvious features as colour, relative relief, alteration products, typical inclusions, order of birefringence, and straight or inclined extinction. In other cases simple qualitative tests that may take only a few seconds, such as determining the relationship of pleochroism to length, the sign of elongation, the optic sign, the size of the optic axial angle, or the orientation of the optic axial plane, are sufficient to enable rapid and unambiguous identifications to be made. If a complete determination of the optical properties of a mineral is required, a composite picture can usually be built up by studying grains in different orientations in a thin section.

The quantitative determination of the optical properties can be carried out with much greater certainty and control, and usually with greater precision if the orientation of the selected crystal can be changed, for then all the optical properties can be determined on the same grain. If the mineral sample to be examined is in the form of grains much can be accomplished by mounting a fragment, suitably oriented, on a glass fibre attached to a simple stage goniometer, but for thin sections of rocks the Universal Stage is to be preferred; this can also be used to study grains in liquid or solid mounts. The uses of this instrument are set out in Section IV.

Optical studies can usually enable a mineral to be assigned to its correct crystal system and in many cases allow a rapid and accurate estimate to be made of its composition. They may also be useful in revealing the extent of order–disorder in the crystal structure, the determination of 2V in the alkali feldspars being the best example of this. Dispersion studies may reveal the true symmetry when all other methods fail, as in staurolite, or suggest that a structure is metastable, as in brookite, and adularia. Twin laws can be determined. If a spectrophotometer is fitted in place of the ocular, quantitative measurements of absorption spectra can be made both in the range of the visible spectrum and outside it.

In this chapter it is assumed that the reader has already acquired a good working knowledge of the basic theory of crystal optics, of the functioning of the compound microscope in its petrographic form, and is competent also in the use of the stereographic projection. For the general theory of transmitted light with the polarizing microscope the reader can refer to the accounts given by Hartshorne and Stuart (1960, 1964), Bloss (1961), Wahlstrom (1962), or Gay (1967). An excellent account of the stereographic projection is given by Phillips (1963). Pertinent aspects of these topics are developed here in the appropriate sections, but the treatment is directed mainly towards the practical application of optical methods, the manipulation of crystals on the microscope stage, and to providing a fuller account of the procedures required to operate a 4-axis Universal Stage than is readily available currently.

## B. SAMPLE PREPARATION

Rock and mineral specimens can be prepared in one of two forms: either by making a thin transparent section about 0·03 mm thick, or by crushing to enable the constituent minerals to be examined as grains. In order to prevent preparations from appearing practically opaque owing to scattering of light at their surfaces, it is necessary to embed them in an inert isotropic medium, which is usually Canada balsam or a synthetic resin in the cases of thin sections or permanent mounts of grains, or in a selected liquid for non-permanent grain mounts. The thin section reveals the texture, mode of occurrence, grain size, and relative proportions of the different minerals in the rock, whereas the grain mount has advantages if the mineral has to be identified by means of its most diagnostic optical property, the refractive index.

The specimen is usually carried on a glass slide, which in the British Commonwealth and the United States usually measures 3 × 1 in (76 × 25 mm) and is from 1 to 1½ mm thick; it is protected by a cover-slip which should have a thickness of 0·17 mm for optimum resolution. In Europe, however, the standard slide is 2 × 1 in (50 × 25 mm) and has a thickness of 1 mm; this is the size and the thickness that must be used with the Universal Stage. Slides and cover-slips must be thoroughly cleaned and dried before use to remove specks of dust or the remains of former liquids. New slides should be washed in a detergent, rinsed, and dried with a linen cloth;

cover-slips may be cleaned by breathing on them and polishing with a fine linen cloth. Once clean, both slides and cover-slips should be handled by the edges only, or forceps should be used. Oily slides can be cleaned by soaking for a few hours in benzene and then briefly in alcohol, after which they should be washed and dried as described above: cover-slips are not worth retrieving.

### 1. Thin sections

As the preparation of thin sections is usually carried out by a trained technician, details of the procedure need not be given here but notes are made on certain aspects. Detailed accounts may be found in a number of text books, e.g. Hartshorne and Stuart (1964, p. 270).

It should be noted that many modern thin sections are prepared by using a synthetic resin such as "Lakeside 70" for securing the section to the slide and Canada balsam, which has the same refractive index ($n \sim 1\cdot54$) to retain the cover-slip. If it is desired to cut out a suitably oriented grain from a section for refractive-index determination it is essential to ensure that all the adhesive is removed.

First the grain to be removed should be marked; to do this, centre the grain in the field viewed through a 1-inch objective and with a mapping pen and indian ink draw a ring around the field of view. Allow the ink to dry, turn the slide over and then with a diamond marker inscribe a circle on the lower surface opposite the ink ring. Then remove the cover-slip by inserting the edge of a razor blade gently beneath one corner; the cover-slip should then spring off. Next, using the diamond marker make a ring around the area to be cut out and place the section on the hotplate. When the mounting medium has melted gently push a needle below the little circle, which can then be lifted out. If the whole of the mounting process has been carried out using Canada balsam, the adhesive can easily be removed by immersing the flake in xylol; if Lakeside 70 has been used, however, the most effective solvent is alcohol, which is unfortunately rather slow in action.

When the flake is clean it can be mounted on a slide, the transfer being carried out by lifting it on the moistened tip of a camel hair brush. The mineral grain required will usually have other grains adhering to it. To isolate the grain and prepare a clean fracture for a Becke test, further dissection of the flake with a half safety razor blade is necessary. This is best accomplished on the microscope stage by using a 1-in objective, by immersing the flake in any liquid of suitable refractive index, which will allow the different minerals to be distinguished, and prevent fragments from flying. If the razor blade is held vertically with both hands, the isolation of the crystal with a clean edge is a simple matter; excess fragments can then be removed.

As normally prepared, the upper surface of the section is left rough, but if a slide is being prepared for use with the Universal Stage and it is desired to study minerals whose refractive indices are substantially greater than balsam, it is an advantage to polish the upper surface to avoid irregular refraction of the emergent rays.

## 2. Grain Mounts

Here, either liquid or solid mounts may be required: ideally all the grains in the sample should be of the same size and their mean diameters should lie between 0·10 and 0·15 mm. This can be achieved if they are passed by a 100-mesh sieve and retained by one of 150 mesh. The useful limits for optical study range from 0·01 to 0·2 mm; grains with diameters in excess of this upper limit prove difficult to cover with immersion oil and fail to produce distinctive birefringence colours when observed between crossed polars. A human hair (diameter around 0·05 mm) is useful when viewed with the sample to give an indication of size.

*Liquid mounts*. These are used if refractive indices are to be determined, or if a mineral fraction is to be examined during a mineral separation, but they have disadvantages if the slide needs to be tilted. Details of the grains are more easily observed if the mounting liquid differs in refractive index from the mean index of the crystals by at least 0·02.

To mount the grains, place a very small quantity in the centre of the slide with a small spatula and cover with a cover-slip, spread the grains by pressing down gently on the cover-slip with the tip of the little finger which must be dry, using a slight circular motion. Next, place a drop of the selected oil at the edge of the cover-slip with a dropping rod, further drops being added until the grains are covered. Excess liquid can be drawn off with a small piece of blotting paper; liquid must on no account be allowed to remain on top of the cover-slip as it may then be transferred to, and attack, the high-powered objective.

*Solid mounts*. These are useful if permanent mounts of grains are required, or if the mineral has one good cleavage and different orientations are desired. Grain mounts are often superior to thin sections for determining the optical properties of minerals that have low birefringence as a greater thickness can be obtained. In general it is advisable to mount the grains in a medium of similar refractive index to avoid total-reflection effects at the borders. Thus Lakeside 70 or Canada balsam are suitable mounts for minerals, such as feldspars, Santolite ($n = 1\cdot559$) for micas, and Arrochlor 4465 ($n = 1\cdot65$) for amphiboles, pyroxenes and olivines. To mount the grains, the slide and cement should be heated as before, a few grains sprinkled on, and stirred with a glass rod to scatter them. After cooling, the upper surface of the mount should be ground flat and polished before the cover-slip is applied.

Non-permanent mounts, in which the grains are embedded in gelatine, have been described by Olcott (1960). This method can prove very useful in refractive-index determination when the section needs to be tilted, or the refractive-index liquid changed without disturbing the grains.

## II. The Polarizing Microscope

As descriptions of the basic instrument are available elsewhere, comments here will be restricted to those necessary to enable the user to make an appropriate selection of a microscope and its accessories, to show how it should be operated to produce the best results, and to ensure that adequate precautions can be taken to protect it from damage.

The modern polarizing microscope may either take the form of the conventional *hinged stand*, which enables the optical axis to be adjusted to the most convenient angle for comfortable viewing or for any other purpose, or the more recently introduced form with a *fixed stand* and large stage, usually equipped with built-in illumination; here the optical axis of the instrument remains vertical (Fig. 1). Although the success of the hinged-stand arrangement has been attested by its long period of service, the fixed stand offers many

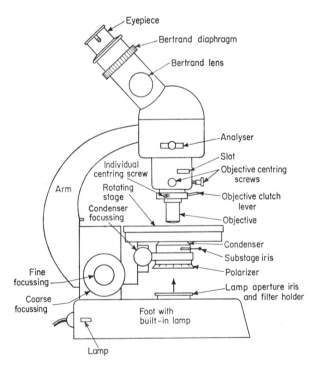

Fig. 1. Fixed-stand polarizing microscope. (After Hartshorne and Stuart, 1964.)

advantages for a research instrument. The principal features of the design are: (1) the focusing is carried out by moving the stage rather than the arm, allowing heavy equipment to be mounted without strain to the rack and pinion; (2) the optical axis remains vertical and the large stage horizontal, which is very desirable when studying specimens mounted in liquids or when using the Universal Stage; (3) the focusing controls are coaxial and are set low, enabling the hands to rest on the bench; (4) a prism in the top of the body tube bends the light beam permitting the use of an inclined eye tube; this gives greater comfort in viewing and free access to the whole of the stage area; (5) a highly efficient 6V, 15 watt lamp with its iris and filter holder is built into the foot; once it has been adjusted it will give a correctly centred illuminating beam that is sufficient for most purposes and is not liable to be disturbed by cleaners. If, however, more powerful lighting is required, a 60 watt lamp can be fitted, or an external high intensity lamp can be used and a mirror fitted to the foot.

### A. OBJECTIVES AND OBJECTIVE CHANGERS

For most petrographic purposes, since very high magnifications are rarely necessary, a choice involving three dry objectives will usually suffice. A low power of 1 in focal length ($\times$ 5, N.A. 0·15) for general inspection, an intermediate power of 1/3 in focal length ($\times$ 20, N.A. 0·50) for studying finer detail, and a high power of 1/6 in focal length ($\times$ 40, N.A. 0·85) for revealing fine detail and obtaining interference figures, make the best choice. It is rarely necessary to resort to oil-immersion objectives ($\times$ 100, N.A. 1·25): if these are used, the cover-slip must first be removed.

To change the objective the most convenient arrangement is the *revolving nosepiece*, in which a different objective can be selected merely by rotating the mounting to the next position, which is located by means of a spring catch. In polarizing microscopes this method suffers from the disadvantage that the centring becomes less precise as the mechanism wears and is even liable to be upset slightly by vibration caused in the ordinary handling of the instrument. Moreover, unless each objective is fitted with its own centring collar it is often quite impossible to centre all the objectives accurately; in such cases it is the high-powered objective that must be centred.

In the other type, in which an *objective clutch* is used, each objective is mounted in its own centring collar which fits accurately into another, fixed to the lower end of the microscope tube, the objectives being retained in position by a strong spring. This arrangement does not make objective changing so rapid, but it is greatly preferred by the writer, since the centring is rarely upset and can in any case be corrected easily; it is the only arrangement that can be used with the Universal Stage.

## B. CONDENSERS AND THE SUBSTAGE

It is the function of the condenser and of the substage aperture diaphragm together to illuminate the object in a suitable manner for the desired type of observation. In a polarizing microscope this may range from a condition where the incident beam is narrowed down to a nearly parallel pencil for refractive-index determination or accurate setting of extinction, to the other extreme where conoscopic methods are to be used and the full aperture of the high-powered objective must be filled with light.

In advanced work it is desirable to have centring as well as focusing controls fitted to the substage to enable both the condenser and the substage iris to be correctly aligned. There is advantage also in selecting an achromatic condenser to reduce glare when the full objective aperture is being used; the author prefers a split condenser system that enables a small supplementary condenser to be inserted into the light path for making observations with high powered objectives. This supplementary condenser, moreover, can easily be replaced by special Universal Stage condensers when required.

Provision should also be made in the substage assembly for a filter holder and a glass diffuser disc and it is sometimes useful to provide a lower iris to collimate the beam as is done in many Leitz microscopes. Slots for auxiliary plates, and provision for mounting a variable-wavelength monochromator should also be made. Since powerful light sources emit a substantial amount of heat radiation which can heat the specimen and immersion liquids and damage the polaroid of the polarizer, it is essential to interpose a filter of heat-absorbent glass into the light path; this can either be placed in the filter holder of the lamp or on the light aperture of the microscope foot if the light source is built-in.

## C. MICROSCOPE TUBE

In addition to the usual slots in the body tube located just above the objective holder, which permit the insertion of auxiliary testing plates, such as the quarter wave, sensitive tint, quartz wedge and Berek Compensator (Emmons, 1943, p. 174), it is an advantage to have others that enable a graduated wedge to be used in conjunction with a Ramsden positive type ocular and a cap analyser for the direct reading of retardations, or very accurate setting of extinction by means of a "twinned" plate such as the Nakamura or Biquartz (Hallimond, 1953, pp. 74–5). For research purposes it is helpful also to have a Bertrand lens fitted that is capable of being centred and focused; this should be fitted with a correctly located diaphragm. The Bertrand lens is used not only for conoscopic examination of crystals but also for checking the adjustment of the illumination.

### D. ILLUMINATION

If the illumination is built-in, instructions for adjustment and for changing to monochromatic light are provided by the manufacturer.

#### 1. The Lamp

If built-in illumination is not provided, an ordinary domestic opal 60- or 100-watt lamp fitted inside a well ventilated housing will provide a large featureless source and even illumination; the opening should be equipped with an iris diaphragm and a filter holder. Greater control and much more light is provided by a low-voltage high-intensity lamp of 48 or 60 watts. In the discussion that follows it is assumed that the main aims are to obtain maximum resolution and to study interference figures.

If a large source, such as an opal bulb or high-intensity lamp fitted with a diffuser, is employed it is best to employ *critical illumination* which permits easy control of the degree of convergence of the incident beam and of the area of the object to be illuminated. To achieve this, the lamp is first placed with its opening about 6–8 in from the mirror and adjusted to illuminate the field evenly. Next, an object showing plenty of detail, such as a fine-grained rock section, is sharply focused by using a low- or medium-power objective ($\times$ 5 or $\times$ 10), and then the lamp iris is partly closed. By moving the condenser, the image of the lamp iris is focused in the object plane and the sharpest possible image obtained. After this, the lamp iris is opened to illuminate the whole field. Next, the back of the objective is inspected, either by removing the ocular or by inserting the Bertrand lens. If the condenser iris is now closed, a sharp image of it should appear and if the objective has been correctly centred, this image should also appear to be centrally located. The condenser iris is now opened until the whole of the back of the objective is just filled with light: this should eliminate glare. If the field is still too bright after the Bertrand lens has been removed the condenser iris may be closed slightly to fill about two-thirds or three-quarters of the objective aperture; this will increase contrast but sacrifice resolution. Once "critical illumination" has been established it can quickly be re-adjusted to suit higher-power objectives by making the necessary changes to the openings of the lamp iris and condenser iris, and by inserting and focusing the supplementary condenser if this is provided.

If a high-intensity lamp is provided it should be of the type with a group of closely spaced filaments; its use enables much stronger illumination of the object to be obtained without glare, and this is needed for photomicrographs, work in monochromatic light, detection of very low birefringence, dispersion studies with convergent light and for refractive-index work. This light source is best controlled by using "Köhler illumination" (Fig. 2), in which the image of the lamp filament is focused on the plane of the condenser iris with the lamp condenser, and the image of the lamp iris on the object plane with the microscope condenser: the microscope condenser therefore acts as the effective light source. With this method it is preferable to have a "corrected" microscope condenser. To avoid the filament being imaged at the rear focal plane of the objective, which would obscure interference figures, a hazening screen of very lightly frosted glass located just in front of the lamp bulb should be employed; this can be swung in when viewing interference figures. An

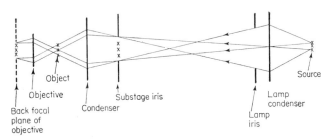

FIG. 2. Köhler illumination. (After Hartshorne and Stuart, 1964.)

alternative is to haze a small area of the bulb with fine carborundum powder and place a spot of ink on the bulb opposite the filament to enable the source to be imaged. The insertion of a glass diffuser disc in the filter holder far from the lamp will destroy the Köhler condition. The procedure with a separate lamp is as follows:

(1) Centre the lamp filament with respect to the lamp condenser. To do this, set the lamp body horizontal, close the iris and focus by means of the lamp condenser the image of the filament on the sheet of card placed vertically some 2–3 ft distant. Check that the filament image is in line with the axis of the lamp, and make any necessary adjustments by means of the centring screws fitted to the bulb holder. Next, find and measure the shortest distance at which the narrowest dimension of the filament will just fill the lower aperture of the microscope condenser. (2) Place the lamp in its position in front of the microscope so that the distance, condenser iris–mirror–lamp iris, is equal to this distance. Close down the lamp iris and focus its image on the object plane with the microscope condenser. Next, open the lamp iris until the whole of the field is just filled. (3) Inspect the rear focal plane of the objective by inserting the Bertrand lens and focus the filament on this plane with the lamp condenser. Apart from minor corrections to microscope and lamp condenser, Köhler illumination is now established and the condenser iris can be adjusted to suit whatever observations are necessary.

*Parallel light.* For determining refractive indices and retardations, a nearly parallel beam of light is required; this is produced by reducing the condenser iris to the smallest size that produces adequate illumination. It is sometimes helpful to lower the condenser, but better results can be obtained if a second iris is provided at some distance from the first, either below the polarizer or in the rear focal plane of the objective. Its use will cause a considerable reduction in resolution.

*Dark-field illumination.* In normal use, as we have seen, the incident illumination is centrally directed to illuminate much or all of the objective opening; this arrangement is often called "bright-field illumination". But another method, dark-field illumination, may be employed with a white-light source to produce a hollow cone of light. Dark-field conditions are produced by inserting an opaque central stop below a wide-angle condenser, so that the minimum angle of the cone of light produced will exceed slightly the angular opening of the objective. Under these conditions, the field will be dark, but if grains are mounted and focused they will refract and reflect light into the objective opening and, hence, will appear luminous against a dark background. The principal uses of dark-field conditions in mineralogy are in refractive-index determination and in checking sample purity.

*Monochromatic light.* This may be provided either by means of a sodium vapour lamp ($\lambda = 589$ nm) or by passing white light from a powerful source through interference filters calibrated for different wavelengths. A very useful form is the *graduated interference* filter, about 10 in long and mounted in a metal frame calibrated in peak wavelengths from 385 nm (violet) to 710 nm, the limit of visible red (Harrison and Day, 1963). It may be inserted in the opening in the microscope foot if the illumination is built-in, or else held in position as near to the substage mirror as possible with a heat-absorbing filter fitted to the light source. Band widths and percentage transmissions for different slit openings are given in Table 1; they apply approximately

TABLE 1

*Characteristics of a typical graduated interference filter*

| Slit width (mm) | Maximum transmission % | Band width at half maximum (nm) | Band width at 1/10 maximum (nm) |
|---|---|---|---|
| 2 | 40 | 25 | 45·5 |
| 4 | 39·5 | 25·5 | 47 |
| 6 | 38·5 | 26 | 48·5 |

throughout the whole wavelength range, as can be seen from the Table. Opening the slit to 6 mm causes an insignificant increase in the half width of the transmission band.

c*

## III. Determination of the Optical Properties

### A. UNPOLARIZED LIGHT

#### 1. Colour

Colour is an important property, but it varies greatly with thickness and in some crystals with vibration direction; crystals which may be deeply coloured in grains a few millimetres in diameter, are almost colourless in the smallest grains.

#### 2. Cleavage and Crystal Shape

In grains, cleavage is revealed by the presence of straight edges that determine the shapes of crushed fragments, or by a preferred orientation. Good examples are the triangular shapes of fluorite grains cleaved on the octahedron or the rhomb-shaped fragments of calcite cleaved on the rhombohedron. Where a single perfect cleavage is found, as in mica, the surface may have a step-like character; in feldspars grains tend to lie on one cleavage plane and reveal the presence of the other as a straight edge. In thin sections cleavages are revealed as fine parallel lines, and a study of the crystal shape can be useful in deciding the crystal system. In crushed grains, however, shape merely indicates the presence or absence of cleavages.

#### 3. Size and Thickness

Measurements of size are required to determine the grain size of rocks, the magnification of drawings or photomicrographs, or to solve problems of size analysis where other methods fail. Linear measurements may be made with a graduated mechanical stage or by means of a graduated scale located at the external focus of a Ramsden-type ocular. By placing a known scale upon the microscope stage the ocular scale may be calibrated for different ocular–objective combinations. In drawings for publication or other purposes, the magnification, the ratio between the actual size of the object and its final representation, should always be stated.

The thickness may be determined within an accuracy of 0·001 mm by using the scale on the fine-focusing adjustment and a $\times 40$ objective which has a very small depth of focus. First, the full objective aperture is just filled with light to obtain a still shallower focus, and then a small scratch on the upper surface of the slide, or a small black speck on the lower surface of the specimen is sharply focused. Next, the upper surface of the specimen is focused, the screw on the micrometer always being rotated in the same direction to avoid any possible lost motion. This procedure may be repeated with different parts of the thread by altering the coarse adjustment, and the results averaged. The difference between the two readings gives the apparent thickness $t'$; with a dry objective:

$$t = nt'$$

where $n$ is the refractive index of the crystal for the particular wave normal (if this is not known it can often be estimated to $\pm 0.01$) and $t$ is the true thickness. Better results can be obtained by using an oil-immersion objective, for the use of oil increases somewhat the distance $t'$ which is usually too small: the relation then is:

$$t = \frac{nt'}{n_0}$$

Where $n_0$ is the refractive index of the oil. If oil is selected that matches $n$ within $\pm 0.01$, then $t = t'$. Even with these precautions the method is insufficiently accurate for use with thin rock sections, but for thick sections of about 0.05 mm and for grains, the results are comparable with those obtained by using path-difference measurements (p. 57).

(p. 57)

## B. PLANE POLARIZED LIGHT – ONE POLAR

First, the vibration plane of the polarizer must be identified; in most British microscopes it is parallel to the E–W crosswire of the ocular and in Continental types it is N–S. The vibration plane can be located easily by examining a thin rock section containing a crystal of biotite that shows the basal cleavage. Biotite shows a deeper absorption colour for the vibration direction parallel to the basal cleavage, so if the crystal is darkest when it is oriented with the cleavage parallel to the E–W crosswire, the vibration plane of the polarizer is E–W, if not it is N–S.

### 1. Colour, Pleochroism and Absorption

Colour and pleochroism are two distinctive properties of minerals in thin section. Colour is the response of the eye to changes of radiation in the visible region (400–700 nm) of the electromagnetic spectrum; however similar changes may occur with radiation outside the visible region and these may not be apparent to the eye. Apart from internal scattering and reflection phenomena, the most common source of colour in minerals is by absorption of radiation through electronic processes. These include: internal electron transitions *within* ions of the transition-metal and rare-earth elements as in titanaugite, chrome diopside, green ferromagnesian silicate minerals, epidote and piemontite; inter-element electron transitions, or "charge transfer", as in biotite, augite, hornblende, staurolite, and hypersthene; and electron transitions induced by imperfections in the crystal structure as in natural and deformed halite, fluorite, and calcite.

In anisotropic minerals polarized light may be absorbed differentially in different directions, an effect known as *pleochroism*. This arises from the presence of ions in anisotropic electrostatic fields. Such fields exist in distorted co-ordination sites and whenever there is strong next-neighbour interaction between ions in a crystal structure. It is important to realize that pleochroism may occur in all regions of the spectrum and that the effect is apparent as colour change only when the differential absorption affects radiation in the visible region. In minerals such as tourmaline and biotite, the effect can be very striking, since the specimen, illuminated by plane polarized light, changes colour as the microscope stage is rotated. Colour in uniaxial minerals may range between two extremes, in which case the term *dichroism* is used. One colour applies to the ordinary vibration direction and the other to the extraordinary; thus iron tourmaline is deep brown and nearly opaque for the ordinary vibration direction, and pale yellow for the extraordinary. This is described as $o$ = dark brown, $e$ = pale yellow; absorption $o > e$. Since uniaxial crystals cut parallel to the basal plane exhibit only the $o$ vibration direction, they can show no colour change on rotation; the maximum colour change is shown by crystals cut parallel to the optic axis.

Biaxial crystals have three principal absorption axes, each with a different intensity and colour; the absorption axes in the orthorhombic system coincide with those of the indicatrix; in the monoclinic system one absorption axis, together with an indicatrix axis coincides with the morphological diad, and the other two frequently, but not necessarily, coincide with the other two indicatrix axes in the $xz$ crystallographic plane.

In the triclinic system the absorption axes need not coincide with any of the indicatrix axes but they usually lie close to them. For this reason it is usual to express the pleochroic scheme of biaxial crystals in terms of the colours shown for each of the principal vibration directions, or their associated refractive indices. Thus for green hornblende $X$ (or $\alpha$) = pale yellow, $Y$ (or $\beta$) = olive green, $Z$ (or $\gamma$) = grass green; absorption $Y = Z > X$ (or $\beta = \gamma > \alpha$).

If the pleochroic effect is weak, as it is in common pyroxenes, it can be more easily observed by leaving the stage in a fixed position and rotating the polarizer than by the usual method of rotating the stage. If the polarizer is rotated in this way it must be returned to its zero position, usually located by a click stop, before passing on to the next stage in the investigation.

## 2. Absorption Spectra

Colour, pleochroism, and degree of absorption may be measured quantitatively by means of a spectrophotometer. If continuous radiation from a tungsten lamp is passed through the specimen, at each wavelength the percentage of radiation absorbed by the specimen may be measured and recorded as an absorption spectrum. One technique developed by Burns (1965, 1966a) utilizes a polarizing microscope equipped with a Universal Stage and mounted horizontally in a spectrophotometer, with a second identically equipped microscope to serve as a control. This method is particularly suitable for measuring polarized absorption spectra of oriented crystals in a thin rock section. If spectra are to be examined beyond the visible range it is essential to ensure that the microscope is equipped with calcite polarizing prisms, for the polarizing action of polaroid discs is restricted to the visible wavelengths.

The orthopyroxene series provides an interesting example of spectrophotometric measurements leading to an interpretation of colour and pleochroism. Polarized absorption spectra show that orthopyroxenes, and indeed all ferromagnesian silicates, are distinctly pleochroic in the shortwave infrared (700–1400 nm), but the effects are frequently not apparent optically (Burns, 1965). The green colours observed in ferromagnesian silicates containing $Fe^{2+}$ in sixfold co-ordination result from absorption of light in the red region by shoulders of absorption bands with maxima in the infrared. Absorption of this radiation induces electron transitions within the $Fe^{2+}$ ions. However, many orthopyroxenes in the hypersthene composition range, particularly those occurring in granulite facies rocks, are green in polarized light vibrating parallel to $Z$, but red in that vibrating parallel to $X$. The red colour associated with the $X$ vibration direction arises from absorption of blue and green radiation by shoulders of charge transfer absorption bands located in the ultraviolet. Charge transfer, or photochemical oxidation–reduction, is facilitated in polarized light vibrating parallel to $X$ by unbalance of charge along the $X$ vibration direction ($y$ crystallographic axis) brought about by the presence of $Al^{3+}$ ions in the orthopyroxene structure (Burns, 1967).

## 3. Refractive Indices

The refractive indices are the most diagnostic of all the optical properties of a substance; approximate determinations are invaluable in enabling identifications to be made with the help of tables, such as those of Tröger (1959), or Larsen and Berman (1934); they are useful also for checking on sample purity in the course of a mineral separation. In suitable cases their accurate determination can permit a reliable estimate to be made of a mineral's composition; olivines, pyroxenes and feldspars provide good examples of this. In binary solid solution series, such as the low-temperature plagioclases, abrupt changes of slope in the curves showing the variation of the principal refractive indices with composition may be used to indicate regions where structural changes in the series may be expected, and which are absent in the high-temperature series.

The determination of the principal refractive indices with a polarizing microscope is usually carried out by using the "immersion method", whereby crushed grains are examined in liquid mounts and the optical reactions tested either with the Becke line or by utilizing the oblique illumination shadows of the Schroeder van der Kalk method. In solid mounts, such as thin sections or permanently mounted grains, only a qualitative estimate of the refractive index can be obtained by assessing the relative relief or by testing the Becke effect against the mounting medium or an adjacent grain of a known mineral, such as quartz or feldspar. Recently, however, the difficult problem of accurate index determination in solid mounts has been solved (provided that the crystal is initially in the correct orientation and its thickness is known) by the application of interference contrast methods involving the use of a beam splitter, a ray combiner, and the determination of the path difference between the unknown test substance and an isotropic test material whose refractive index is known. If the operation is repeated with a second known standard the thickness term in the equations can be eliminated (Pillar, 1960). This method, which is very rapid and is claimed to be at least as accurate as the Becke method, unfortunately requires special equipment, and to apply it a thin section must first be cut into strips whose width ranges between the limiting values of 0·546 mm and 0·054 mm depending on the magnification.

The application of dark-field methods to refractive-index determination has been discussed by Dodge (1948). He has shown that if solid and liquid differ greatly in refractive index, the grains will appear white; however when the refractive indices of the two media differ by only a few units in the second decimal place, or less, the grains will appear to be illuminated by yellow light if their refractive indices are greater than that of the liquid, and by blue light if they are lower. If the refractive indices for $Na_D$ light of the two media match within $\pm 0·001$, the grains will appear to be illuminated by a purplish-blue light with irregular deep-red patches near the margins. The dark-field method is not as accurate as the Becke method under carefully controlled conditions, and has a maximum accuracy of $\pm 0·001$, but it has obvious applications in detecting impurities in powdered materials, particularly if they are very fine grained.

New developments in index determination with liquid mounts come from the application of the "focal screening" methods of Cherkasov (1957) which confer many of the advantages of wavelength variation without the need to employ a variable wavelength monochromator: it is claimed that these methods permit better visibility of the grains at the match point and less disturbance from inclusions. Focal screening involves the fitting of masks made from blackened photographic film in the rear focal plane of the objective. These methods are claimed to be rather more sensitive than the Becke test if high-dispersion liquids are employed (Wright, 1965). For most purposes, however, the Becke method is sufficiently accurate.

*Becke method.* When a crystal immersed in a transparent liquid of different refractive index is illuminated by a parallel beam of light, it acts as a rough lens, a convergent one if its refractive index is greater than that of the liquid, a divergent one if it is lower. If the focus of the microscope objective is raised and the crystal has a higher refractive index than the liquid, a bright line of light, the Becke line, will be seen to move inwards from the edge of the crystal (Fig. 3a and b); conversely, when the crystal has a lower refractive index than

(a)                        (b)

FIG. 3. Becke test for a crystal with refractive index greater than the surrounding medium: (a) where focus is raised; (b) where focus is lowered.

the surrounding medium the Becke line moves outwards. This test may also be used to determine which of two adjacent minerals in a thin rock section has the greater refractive index. When the refractive indices of crystal and medium approach, the contrast between them is reduced, and in monochromatic light a liquid may be found where the refractive indices of the two match exactly; the boundary between them then becomes invisible, and on raising or lowering the focus no Becke effect is seen.

Since in anisotropic crystals the refractive index varies with the vibration direction, a crystal placed so that one of its vibration directions is parallel to the polarizer will show the relief appropriate to the refractive index of that direction, and when rotated through 90° will show the relief appropriate to the other refractive index. Usually this difference in relief is too small to be detected but even with a mineral like quartz, where $n_e - n_o = 0.009$, it can

C**

be observed if the mounting medium is within 0·002 of either refractive index. The effect is striking if cleavage rhombs of calcite are immersed in a liquid of refractive index 1·56.

The Becke test is most sensitive if a medium-power objective ($\times 10$) is used, with the aperture diaphragm stopped down as far as possible, and if a double diaphragm is employed; the best results are obtained if the crystal–oil boundary is nearly vertical and if its azimuth is close to that of the vibration direction of the polarizer (Saylor, 1935). The width of the diaphragm opening at which the bright line can just be seen is proportional to the square root of the difference in refractive index between crystal and medium. The Becke line is never very obvious when the object is sharply in focus; however, if observations are made while the focus is being raised, the line becomes increasingly apparent. In very small crystals where the line cannot be resolved its inward movement is indicated by a brightening and its outward movement by a darkening of the crystal. The Becke line can be observed more easily if a high-power ($\times 40$) objective is used, but according to Saylor the accuracy is reduced (Table 2). In white light, since liquids nearly always have much

TABLE 2

*Errors of refractive-index methods in measuring the refractive index of glass wool (from Saylor 1935)*

| Objective | Becke | van der Kalk | Saylor double diaphragm |
|---|---|---|---|
| $\times$ 10, N.A. 0·25 | $\pm$ 0·0005 | $\pm$ 0·0005 | $\pm$ 0·0005 |
| $\times$ 20, N.A. 0·40 | $\pm$ 0·0009 | $\pm$ 0·0010 | $\pm$ 0·0005 |
| $\times$ 45, N.A. 0·85 | $\pm$ 0·0012 | $\pm$ 0·0020 | $\pm$ 0·0006 |

greater dispersion than solids, a refractive-index match for one colour will not result in a match for another, so the crystal cannot entirely disappear. If a match is made for yellow light the grain may appear to be fringed with red and blue bands that will move in opposite directions, the red inwards and the blue outwards as the microscope focus is raised. Sometimes, usually in large grains, two white Becke lines may be observed that move simultaneously in opposite directions; this effect may occur if the edges have surfaces sloping in different directions. The false line may often be eliminated by closing further the aperture diaphragm, or be identified by using the Schroeder van der Kalk test.

In monochromatic light, an accuracy of $\pm 0·001$ is attainable and if the refractive index of the immersion medium is determined at stage temperature the accuracy can be increased to $\pm 0·0005$.

Emmons and Gates (1948) have pointed out the use that can be made of the dispersion fringes if the Becke test is carried out in white light; it allows an accuracy of $\pm 0.002$ to be attained without recourse to monochromatic illumination. As the mean centre for white light is 550 nm, the reddish Becke line (the brighter one) is composed of wavelengths greater than 550 nm, and the blue one (darker) of the shorter wavelengths. Hence, if the focus is raised the brighter line will move towards the medium of greater refractive index. If the refractive index is matched at a wavelength of 550 nm then the two Becke lines should have equal intensity. However, as refractive indices are usually quoted for the wavelength of the $Na_D$ line (589 nm) the brighter Becke line entering the crystal should be somewhat more intense and be a yellowish orange in colour, while that entering the oil should be greenish blue. Bloss (1961, p. 58) has shown that for binary mixtures of petroleum oil ($n = 1.470$), α-monochloronaphthalene ($n = 1.633$), and methylene iodide ($n = 1.739$), the colours observed give an important indication of the refractive-index difference for sodium light between a crystal of average dispersion ($n_F - n_C \sim 0.015$) and the immersion medium; Fig. 4 summarizes these

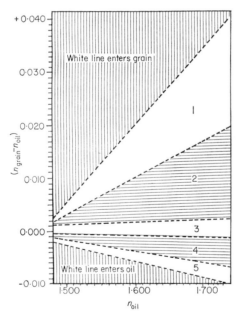

FIG. 4. Approximate relationship of the Becke line colours and movements to the refractive index difference for sodium light between crystal and immersion oil (microscope focus raised). Area 1: whitish yellow line enters crystal; dark blue-violet enters oil. Area 2: Lemon yellow enters grain; blue violet enters oil. Area 3: orange yellow enters grain; sky blue enters oil. Area 4: reddish orange enters grain; whitish blue enters oil. Area 5: dark reddish brown enters grain; blue white enters oil (Bloss, 1961).

relations if a white tungsten bulb and blue "daylight" filter are employed. Its use enables a rapid choice to be made of a liquid that will closely match the crystal, the final determination being carried out in monochromatic light.

*The Schroeder van der Kalk test.* This method utilizes oblique illumination, which may be obtained by opening fully the condenser aperture diaphragm and inserting a piece of card part of the way across the optical path just above the fixed element of the condenser, or by inserting the analyser frame part of the way: for best results a low-power objective should be used. The effect is to cause one side of the crystal to darken and the other to become bright (Fig. 5).

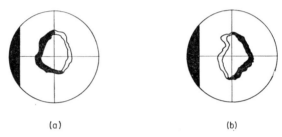

(a)                                        (b)

FIG. 5.   Schroeder van der Kolk test with low-power objective; field shaded by partial insertion of analyser frame (shaded sector on left). (a) Crystal higher in index than immersion liquid. (b) Crystal lower in index than immersion liquid.

If the crystal has a higher refractive index than the medium it should brighten on the side from which the light appears to come, the side opposite the shadow in the field, and the appearance is like a landscape viewed from above in the light of the setting sun. If the crystal is lower in refractive index than the medium, the effect is reversed. The focusing of the condenser is critical for this test, since the reaction can be reversed by faulty positioning; for this reason it is advisable to try the method first with crystals of known refractive index and to ensure that subsequently the condenser and stop occupy the same position. The optical reaction can be intensified and the sensitivity of the test increased if a semicircular diaphragm made from blackened photographic film is inserted into the barrel at the back of the objective on the same side as that from which the analyser frame is introduced.

In white light if the refractive indices of crystal and liquid are nearly matched, dispersion effects may be observed, the crystal being bordered on one side by red and on the other by blue. This does not indicate an exact match for yellow. If yellow is not exactly matched, however, it may appear with the red and produce a bright orange colour, or with the blue and produce a bright steel-blue border.

This test has advantages over the Becke test if there is a very large refractive index difference between the crystal and the mounting medium, it is also very useful as a rapid check for revealing impurities in grain mounts when making mineral separations, since a rapid and almost simultaneous testing of all the grains within the field of view may be accomplished. Its disadvantage lies in the need to apply the test under carefully controlled conditions to avoid reversal of the shadows, whereas the Becke test is nearly always unambiguous. The van der Kalk method, moreover, appears to be less sensitive than the Becke test unless Saylor's double-diaphragm modification of the method is applied (Table 2).

*Procedure.* The refractive-index determination is carried out either on grains or on a crystal, known to be suitably oriented, which has been cut out of a thin section and from which all traces of adhesive have been removed (p. 34). The crystals are immersed in a suitable liquid and a cover-slip may or may not be placed on top. If the liquid is left uncovered, evaporation may change its refractive index after only a few minutes. If the mineral has low birefringence this can be an advantage and avoid the need to change the liquid to match each principal refractive index in turn, but the front lens of a high-power objective can easily become covered with the liquid. The writer always prefers to use a cover-slip, and if a liquid has to be changed, most of it can be removed by inserting the torn end of a small piece of blotting paper just below the edge of the cover-slip. This operation, if carried out carefully, will not disturb the grains. Since the amount of liquid remaining is small, and the new liquid will usually only differ in refractive index by 0·01 or less from the old one, any contamination effect, after the liquids have had time to mix (about 5 min) can be ignored.

The orientation of crystal fragments so mounted can easily be changed by a gentle movement of the edge of the cover-slip with a needle or pencil point; this method is particularly effective if there is a fairly thick layer of liquid present. Even platy crystals can often be caused to stand on edge sufficiently long to enable a Becke test to be made. If the interval between successive liquids is 0·01, the refractive index can, with practice, be estimated to within 0·003 by observing the relief by which the crystal differs from its surroundings. Since many liquids are usually required to determine all three refractive indices of a biaxial crystal, it is sometimes useful to embed the lower parts of the grains in softened gelatine (Olcott, 1960); the liquids can then easily be washed off without altering the orientation of the crystals.

*Orientations required for determining principal refractive indices.* The ordinary refractive index of a uniaxial crystal may be determined on any grain but the extraordinary can only be obtained from a crystal viewed perpendicular to the optic axis. The choice of a suitably oriented crystal in such cases, unless it is of prismatic habit, is made by obtaining interference figures on grains that exhibit the highest interference colours in relation to their thickness. Since the maximum accuracy of the immersion method is $\pm 0\cdot 0005$, the need to ensure that the crystal selected is viewed exactly perpendicular to the optic axis will only arise if the birefringence of the crystal is very high. If $\mid n_o - n_e \mid < 0\cdot 025$, an error of up to $15°$ in the orientation will only affect the determined value of $n_e$ by a maximum error of $0\cdot 001$.

In biaxial crystals a section cut parallel to the optic axial plane, i.e. normal to $\beta$, can be used to determine the values of $n_\alpha$ and $n_\gamma$; sections cut normal to the bisectrices or to the optic axes are suitable for determining $n_\beta$. No section can have both indices above or both below $n_\beta$, and this may prove to be an easy way of determining $n_\beta$. For example, if one grain has $1\cdot 536$ as the value of its greater refractive index, and another has this value as its lower one, the value of $n_\beta$ is $1\cdot 536$. The remarks made above on uniaxial crystals concerning the methods of establishing the orientation, and the accuracy with which it need be determined, apply equally well to the biaxial case.

After a principal refractive index of the crystal and the liquid have been matched in monochromatic light, a small drop of liquid is removed with a micropipette and transferred to a refractometer, which should be either a Leitz-Jelley type or an Abbé type instrument. The refractometer should always be taken out of its case before the refractive-index determination is commenced, and allowed to come to room temperature (Fisher, 1958). For the Abbé type this may take half an hour or so. A serious error can arise if this is not done owing to temperature change in the liquid when it is transferred to the refractometer, especially if the liquid has a high index and a heat filter has been omitted.

This difficulty of establishing the exact refractive index of the liquid at the time the Becke test is made, can be avoided if the whole operation is carried out with a Zeiss Microscope Refractometer (Fig. 6) which is secured to the microscope stage. The crystals are placed in the small lenticular cavity of the instrument along with the selected liquid, and covered with a cover-slip. After crystal and liquid have been matched, the refractive index of the liquid can be determined directly by focusing the objective on the refractive-index scale on the lower side of the measuring body. Next, the end of the scale is aligned with the intersection of the cross-wires of the ocular by means of two set-screws (Fig. 6a). Finally, the measuring body is displaced until it comes to rest against a magnetically secured stop, and the refractive index is read off at the intersection of the scale with the ocular crosswires (Fig. 6b). This apparatus has two cells, which are interchangeable, one for the range $1\cdot 3 – 1\cdot 7$ and the other for $1\cdot 6 – 2\cdot 0$. As the scales are divided into intervals of $0\cdot 005$, values correct to $\pm 0\cdot 001$ can be either estimated, or obtained more certainly by using an ocular screw micrometer.

*Double variation of immersion liquids.* One of the most tedious parts of the refractive-index determination is the use of successive liquids to approach, and then match, first one of the principal refractive indices and then the others; for a biaxial crystal of moderate birefringence up to ten changes of liquid may be required. Since the refractive indices of liquids decrease with rising temperature to a very much greater extent than do those of solids, the

FIG. 6 (a). Zeiss microscope refractometer.

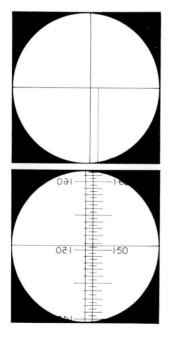

FIG. 6 (b). Scale as viewed through microscope. $n = 1\cdot507$.

provision of temperature control to the cell on the stage confers many advantages. Even greater flexibility can be obtained if a variable wavelength monochromator can be used, since the dispersion of liquids is generally greater than that of solids, the more so in the high index range.

Three methods have been proposed that utilize these properties, the *dispersion* method of Merwin and Larsen (1912), in which the matching is accomplished by means of a variable wavelength monochromator; the *single variation* method of Emmons (1928), in which hot water circulating through a cell on the stage enables the temperature to be altered; and the *double variation* method of Emmons (1929), in which both temperature and wavelength controls are used. Of these, the double variation method is by far the most convenient and flexible; its advantages are even greater if the orientation of the crystal can be altered, hence it is usually carried out with the Universal Stage (p. 86).

*Refractive-index liquids.* A set of immersion liquids of known refractive index covering the range from 1·44 to 1·78 at 0·01 intervals forms an essential part of the optical equipment, each liquid being miscible with those above and below. The liquids ideally should be colourless, stable, and inert towards the materials with which they are used, but since many of them decompose in light they should be kept in dark bottles of about 5 ml capacity fitted with ground-glass stoppers with dropping rods attached; they must be stored away from light and heat. Kept in this way the lower index liquids may not change over periods of years but those containing methylene iodide may drift over much shorter periods.

Although it is possible to obtain sets of immersion liquids commercially, it is just as satisfactory to mix the liquids and make up a set in the laboratory, because much smaller intervals may then be selected over particular ranges, such as 1·52–1·58 for work on feldspars, where an interval of 0·002 is useful, and 1·66–1·74 for pyroxenes and olivines, where an interval of 0·005 is recommended. In any case, liquids should be checked on the refractometer before use and sets should be checked at three-monthly intervals. To mix two liquids of known refractive index to obtain a desired intermediate one, two burettes are required: the respective volumes may be obtained from the following formula, which applies strictly to ideal mixtures where $v_2$ and $v_2$ are the volumes of the two known liquids respectively, $n_1$ and $n_2$ their refractive indices, and $n$ the desired refractive index:

$$v_1 n_1 + v_2 n_2 = (v_1 + v_2)n.$$

A list of immersion liquids of similar volatilities that form a miscible series is given in Table 3.

To cover the range above 1·74, Meyrowitz and Larsen (1951) have suggested a series of liquids consisting of mixtures of methylene iodide (1·74), a solution

TABLE 3

*Liquids for the preparation of a set of immersion media*

|  | Refractive index | Temperature coefficient |
|---|---|---|
| Isoamyl valerate | 1·41 | − 0·00044/°C |
| Petroleum oil | 1·47 | − 0·0004/°C |
| α-Monochloronaphthalene | 1·63 | − 0·00044/°C |
| α-Monobromonaphthalene | 1·66 | − 0·00048/°C |
| Methylene iodide | 1·74 | − 0·00070/°C |
| Methylene iodide saturated with sulphur | 1·78 |  |
| Phenodi-iodoarsine | 1·84 |  |

of 10% sulphur in arsenic tribromide (1·814), and a solution of 20% sulphur and 20% arsenic disulphide in arsenic tribromide (2·00); these have high temperature coefficients ($-0.0006$ to $-0.0007/°C$) and very high dispersions that increase rapidly with rising index. Other liquids, using yellow phosphorus, sulphur and methylene iodide, can be used to cover the range between 1·74 and 2·06 (West, 1936), but they are unpleasant to use. For refractive indices above 2·06 solids must be used, and Larsen and Berman (1934) list a non-crystalline series made by melting together mixtures of sulphur ($n = 2.0$) and selenium ($n = 2.7$), but the end members are difficult to use, for those high in sulphur crystallize easily, and those high in selenium are deep red in colour and nearly opaque.

### C. CROSSED POLARS – PLANE POLARIZED LIGHT

If the analyser is inserted with its vibration plane normal to that of the polarizer, a distinction can be made between isotropic and anisotropic substances. The former remain dark whatever their orientation, whereas the latter are illuminated in most positions and only become dark or "extinguished" in four positions 90° apart during a complete rotation of the microscope stage. The maximum intensity of illumination occurs at 45° to the extinction positions, and the colours observed are called *polarization* or *interference colours*. Crystals pass into extinction when their vibration directions become parallel to the polars, for in this position light from the polarizer is not resolved by the crystal and hence there is no component that can be transmitted by the analyser.

### 1. Polarization Colours

If an anisotropic crystal is placed on the microscope stage in a general position with respect to the polars the plane polarized light transmitted by

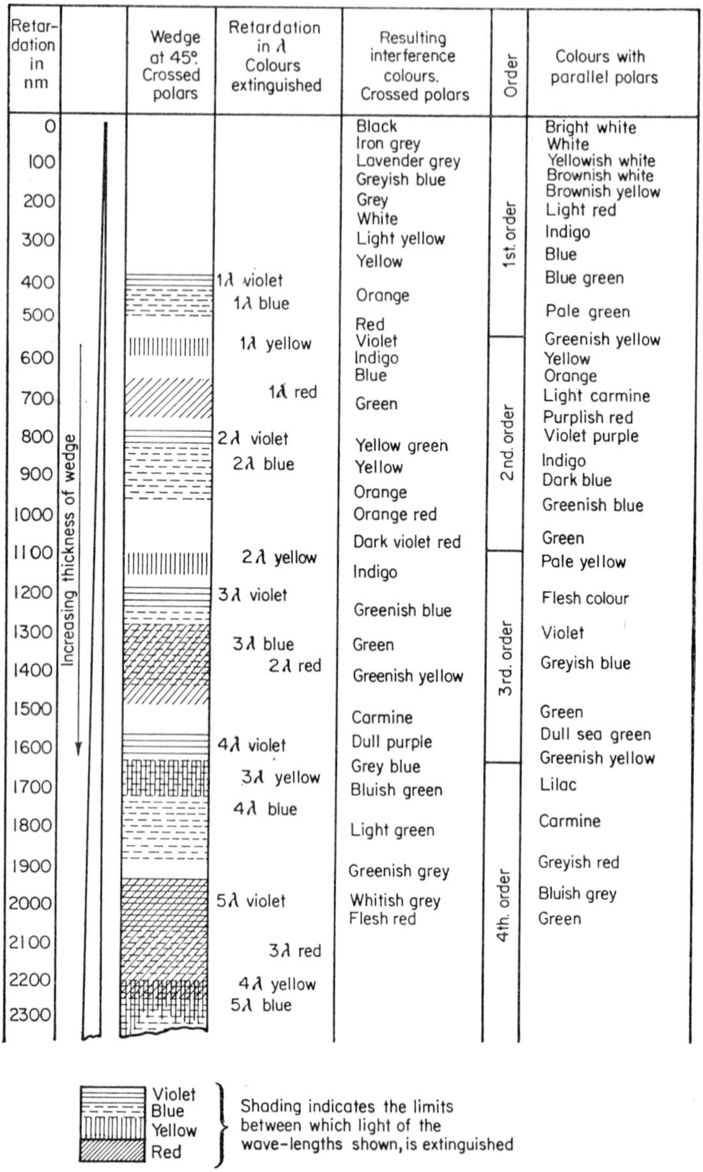

I. D. MUIR

| Retardation in nm | Wedge at 45°. Crossed polars | Retardation in λ Colours extinguished | Resulting interference colours. Crossed polars | Order | Colours with parallel polars |
|---|---|---|---|---|---|
| 0 | | | Black | | Bright white |
| | | | Iron grey | | White |
| 100 | | | Lavender grey | | Yellowish white |
| | | | Greyish blue | | Brownish white |
| 200 | | | Grey | | Brownish yellow |
| | | | White | 1st. order | Light red |
| 300 | | | Light yellow | | Indigo |
| | | | Yellow | | Blue |
| 400 | | 1λ violet | | | Blue green |
| | | 1λ blue | Orange | | |
| 500 | | | Red | | Pale green |
| 600 | | 1λ yellow | Violet | | Greenish yellow |
| | | | Indigo | | Yellow |
| 700 | | 1λ red | Blue | | Orange |
| | | | Green | | Light carmine |
| 800 | | 2λ violet | | 2nd order | Purplish red |
| | | 2λ blue | Yellow green | | Violet purple |
| 900 | | | Yellow | | Indigo |
| | | | Orange | | Dark blue |
| 1000 | | | Orange red | | Greenish blue |
| 1100 | | 2λ yellow | Dark violet red | | Green |
| | | | Indigo | | Pale yellow |
| 1200 | | 3λ violet | Greenish blue | | Flesh colour |
| 1300 | | 3λ blue | Green | | Violet |
| 1400 | | 2λ red | Greenish yellow | 3rd. order | Greyish blue |
| 1500 | | | Carmine | | Green |
| 1600 | | 4λ violet | Dull purple | | Dull sea green |
| | | 3λ yellow | Grey blue | | Greenish yellow |
| 1700 | | 4λ blue | Bluish green | | Lilac |
| 1800 | | | Light green | | Carmine |
| 1900 | | | Greenish grey | | Greyish red |
| 2000 | | 5λ violet | Whitish grey | 4th. order | Bluish grey |
| 2100 | | 3λ red | Flesh red | | Green |
| 2200 | | 4λ yellow | | | |
| 2300 | | 5λ blue | | | |

Increasing thickness of wedge

Violet
Blue
Yellow
Red
} Shading indicates the limits between which light of the wave-lengths shown, is extinguished

FIG. 7. Newton's scale of colours observed in a wedge between crossed polars, and the complementary set of colours observed between parallel polars: as the thickness of the wedge increases the colours in the third, fourth and succeeding orders become fainter and more complex. Above the fourth order only pale pinks and greens are seen. (After Hartshorne and Stuart, 1964.)

the polarizer will be resolved into two mutually perpendicular plane polarized disturbances having the same wave normal; these will travel through the crystal with different velocities, and on emergence the components that pass through the analyser will have acquired a phase difference. The retardation produced is the product of the *partial birefringence* (refractive-index difference of the two vibrations) of the substance and of the thickness. Owing to the great range of light wavelengths in the visible spectrum, different colours may suffer partial or complete destructive interference, and this may cause the crystal to appear coloured. The complete sequence of colours produced with increasing retardation in white light, if dispersion is negligible, is known as *Newton's scale of colours* (Fig. 7).

The colours are divided into orders by the recurring violet tint to which the eye is particularly sensitive, the first-order violet is equivalent to one wavelength retardation for yellow light.

Occasionally it is useful to study the colours between parallel polars which enables colours complementary to the normal interference tints to be observed. The method can be useful in distinguishing between low and high order whites, and in confirming the colour observed between crossed polars as shown by Fig. 7. The true complementary colours will only be observed if the wedge or crystal is set accurately in the 45° position, since in other positions the amplitudes of the components passing the analyser are unequal.

*Michel-Lévy chart.* This chart, whose outline is given in Fig. 8, gives a graphical solution to the relationship between the polarization colour, the partial birefringence of the crystal and the thickness. The original chart is unfortunately not obtainable, but can be found in Iddings' "Rock Minerals" (1906) as a folder; smaller versions can be found in several text books of optical crystallography.

The partial birefringence found in this way must not be confused with the *birefringence* of the substance, often quoted as one of the optical constants, for this refers to the maximum birefringence $| n_e - n_o |$ for uniaxial crystals and $n_\gamma - n_\alpha$ for biaxials, observed for wave normals lying perpendicular to the optic axis in the former case, and parallel to the optic axial plane in the latter. Methods of recognizing such sections using convergent light are described in Section IIID.

*Abnormal polarization colours.* Anomalous colours that do not match the Newton scale are sometimes observed; they are due to one or more of the following causes: strong colour due to selective absorption of part of the spectrum; dispersion of the birefringence; or dispersion of the optic axes or extinction directions. If the birefringence is moderate or high, the polarization colours differ little from those observed in ordinary light, except for very strongly absorbent crystals, and the abnormal effects are most commonly observed in minerals of low birefringence or in wave normals that have low

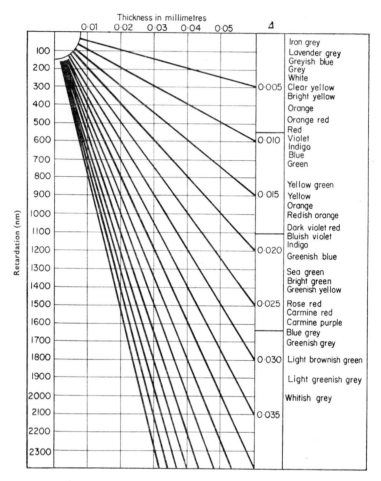

FIG. 8. Outline diagram of Michel-Lévy's chart giving a graphical solution to the relationship between the retardation (in nanometers), thickness (in millimetres), and birefringence (Δ). (After Hartshorne and Stuart, 1964.)

partial birefringence. Low-order grey colours are replaced by anomalous deep blues or by dark-reddish tints. In extreme cases of crystals belonging to the less symmetrical systems there may be a failure to pass into extinction altogether.

## 2. Determination of Fast and Slow Vibration Directions

A distinction between the two vibration directions associated with a common wave normal can be made by superimposing on the crystal a birefringent crystal plate or wedge called a *compensator*, in which the fast- or slow-vibration directions are known. This can be placed above or below the crystal in the auxiliary slots inclined at 45° to the vibration planes of the polars. If the crystal is now placed with its vibration directions inclined at 45° to those of the polars, and the plate or wedge is inserted, the polarization colours will appear either to rise or to fall depending on whether or not the fast and slow vibration directions in the crystal and plate are respectively parallel to each other. Superimposing fast on fast appears to make the crystal thicker; fast on slow would make it appear thinner. Superimposing fast on slow is called *compensation*, and fast on fast is called *addition*. If the retardation in the crystal, and in the compensating plate are equal, compensation will produce darkness in place of the former polarization colour. In practice, complete darkness will only be produced in white light if the dispersion of the bire-fringence in the crystal and the compensating plate are equal; this is unusual.

Three simple compensators are needed; in each, the two vibration directions are respectively parallel to the long and short edges of the slide, the slow direction being indicated by an inscribed arrow. The retardations mentioned refer to yellow light.

*The λ/4 mica plate.* The retardation of the λ/4 plate is 150 nm giving a pale grey polarization colour. It is used for determining fast- and slow-vibration directions in crystals of very low birefringence that exhibit very low first-order grey; compensation produces near darkness, addition first-order white. It is also useful with second- or third-order colours.

*The 1λ sensitive tint plate.* The retardation is 575 nm, giving sensitive violet polarization colours. This produces a striking change with first-order greys and whites giving yellow or orange with compensation and blue with addition, so it is useful also with interference figures, but it can be confusing with third- or fourth-order colours, since it moves them by a whole order and for these the mica plate is preferred.

*The quartz wedge.* Retardation varies over a distance of about 2 inches from about 150 nm at the thin end to about 4λ, 2300 nm, at the other. This is useful for crystals showing fourth-order colours or higher, interference figures on crystals with very high birefringence, such as calcite and sphene, and on grains. Since grains vary in thickness they may exhibit polarization colours as contours; in "addition" the colours move outwards and more contours appear, whereas in "compensation" the colours move inwards.

*Procedure.* The crystal to be examined is placed in extinction and then rotated through 45° bringing its vibration directions parallel to those in the testing instrument, which is selected as described above according to the nature of the polarization colour displayed. The compensating plate is then inserted and the rise or fall in the polarization colour noted.

## 3. Sign of Elongation

In prismatic crystals that have the vibration directions respectively parallel to and normal to their length, i.e. orthorhombic and uniaxial crystals, those that have the slow vibration direction parallel to the length are said to have a *positive sign of elongation*, those elongated fast are said to have a *negative sign of elongation*. If uniaxial crystals are elongated parallel to their unique crystallographic axis, their sign of elongation is the same as the optic sign.

## 4. Determination of the Birefringence

*Determination of the retardation or phase difference.* The determination of the retardation can be carried out very simply by superimposing on the crystal a compensating instrument, such as a quartz wedge, into the auxiliary slots above the objective, or if a Huyghenian ocular is available, in conjunction with a cap analyser, in the ocular slots.

First the crystal is set with its vibration directions at 45° to those of the polars, and then the testing instrument is inserted. As its vibration directions are parallel to those of the crystal, the interference colours will rise if the slow direction of the testing instrument (usually marked on it) and the crystal coincide. If now the crystal is set so that its fast vibration direction is parallel to the slow one of the wedge, the interference colours will fall as the instrument is inserted and there will appear a position of compensation where the retardations of the crystal and the wedge are exactly balanced and the crystal will appear dark. The crystal should be removed and the interference colour noted; the wedge should next be removed slowly and the number of red bands observed. For example, if the crystal exhibits a blue colour, with the wedge or a $1\lambda$ plate, the fast and slow directions are identified, and the position of complete compensation is found, the crystal is then removed and the original blue colour is observed. Now, as the wedge is gradually withdrawn two red bands are observed; the interference colour is, therefore, third-order blue. If a graduated wedge is used inserted in the ocular slots the retardation can be read directly in the position of compensation.

*Measurement of thickness.* Although the retardation can be determined very accurately it is much more difficult to ascertain the thickness. For qualitative purposes it can be obtained from the Michel-Lévy chart if the birefringence is known, but if the birefringence has to be determined from the retardation, the thickness must first be determined. The simplest way is to use the scale on the fine-focusing adjustment (p. 42), but this is not sufficiently accurate for thin-section work. More accurate results are obtained if the upper surface of the section is polished and the thickness determination made using retardation measurements on several adjacent grains of a mineral, such as quartz, whose birefringence is known; this operation is better done with the Universal Stage.

*Relation between birefringence and partial birefringence.* From the geometry of the indicatrix it can be shown to a close approximation that for uniaxial crystals:

$$\frac{\Delta'}{\Delta} = \sin^2 \theta$$

and for biaxials:

$$\frac{\Delta'}{\Delta} = \sin \theta \sin \theta'$$

where $\Delta'$ represents the partial birefringence for the particular wave normal, $\Delta$ the total birefringence, and $\theta$ the angle between the wave normal and the optic axis for uniaxial crystals. In the biaxial case $\theta$ and $\theta'$ represent the angles between the wave normal and one and the other of the two optic axes, respectively. A graphical solution to this relationship is given in Fig. 9.

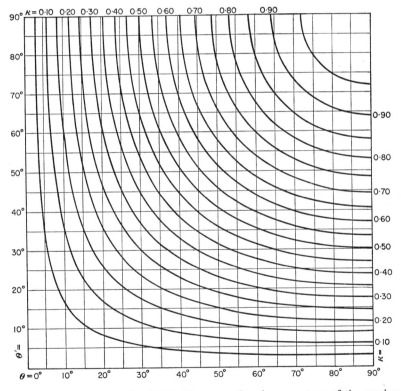

FIG. 9. Graphical solution after Wright for computing the percentage of the maximum birefringence ($K$) in any section if the angular distances ($\theta$ and $\theta'$) between the wave normal and each of the optic axes is known. (After Johannsen, 1918.)

## 5. Extinction Angles

As we have seen, the vibration directions of a birefringent crystal section may be located by rotating it to extinction. The extinction directions for any wave normal are defined by the orientation and shape of the indicatrix and may be determined if this is known by means of the Biot-Fresnel construction.

If one of the vibration directions is parallel to the length of the crystal, or a single cleavage, extinction is said to be *straight*. If it bisects the angle between the traces of two prominent cleavages such as {110} in pyroxenes or amphiboles when viewed down the *z* crystallographic axis it is said to be *symmetrical*. If it is neither straight nor symmetrical, it is said to be *inclined* or *oblique*.

*Straight extinction*, with respect to the trace of the *z* crystallographic axis (usually the length), is exhibited by all non-basal (isotropic) sections of uniaxial crystals. It is also shown by orthorhombic crystals with wave normals lying in the zones [100], [010], and [001], since for all of these one of the vibration directions corresponds with a principal refractive index. The same relationship applies to wave normals in the [010] zone of the monoclinic system, directions normal to the *y* crystallographic axis.

*Symmetrical extinction* usually refers to prismatic cleavages and to wave normals coincident with the *z* crystallographic axis, in which orientation, easily checked by the cleavage traces being vertical, a rapid distinction can be made between orthorhombic and monoclinic crystals on the one hand (*symmetrical*), and triclinic on the other (*inclined*).

*Inclined extinction* is shown by all triclinic crystals regardless of orientation unless by chance their vibration directions happen to coincide with a prominent crystallographic direction, by all monoclinic crystals for wave normals lying outside the [010] zone, and by all orthorhombic crystals for wave normals other than those in the zones [100], [010] and [001].

*The Biot-Fresnel construction.* This may be used to determine the vibration directions associated with any wave normal, provided that its angular position is known with respect to the crystallographic axes, and that the positions of the optic axes are specified. The relations may be illustrated best by means of a stereographic projection of a monoclinic crystal, of augite (Fig. 10). In this $X$, $Y$ and $Z$ represent the poles of the principal vibration directions and A and B those of the optic axes; N is the wave normal (direction of the microscope axis) and PQR the plane of advance of the wave front (plane of the microscope stage). The traces of the two prismatic cleavages are indicated by dashed lines. The vibration directions in the plane PQR, $V_1$ and $V_2$ may be located at the intersections of PQR with the planes that internally and externally bisect the angle between the two planes that contain the wave normal and the two optic axes respectively (NAQ and NBP). $V_1$ and $V_2$ also externally and internally bisect the angle made by the circular sections

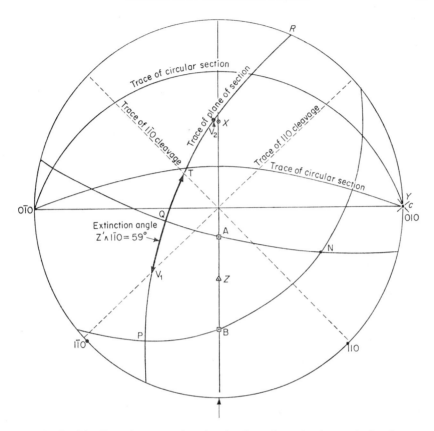

FIG. 10. The Biot-Fresnel construction showing how the extinction angle for the wave normal N, Z′ to the trace of the (1$\bar{1}$0) cleavage, the angle V₁T, may be determined for a clinopyroxene whose optical orientation is known. In the diagram X, Y, and Z represent the poles of the principal vibration directions, and A and B those of the optic axes. The extinction directions V₁ and V₂ in the plane RTP normal to N, are located by bisecting the arc PQ subtended by the great circles containing the wave normal and each of the optic axes. V₁ and V₂ also bisect the arc subtended between the circular sections in the plane of the section.

with PQR. Since the trace of the (1$\bar{1}$0) cleavage passes close to N, this cleavage, although somewhat inclined should be observed, whereas the (110) cleavage will not be visible. The extinction angle measured on this section therefore will be the angle V₁T. From inspection, since V₁ lies very close to the great circle containing Y and Z it must be the slow vibration direction; this extinction angle therefore is quoted as Z′ ∧ (1$\bar{1}$0) cleavage = 59°.

*Angles to be measured.* In the *orthorhombic system* small extinction angles may be shown by generally oriented sections; non-uniform extinction in such cases may indicate a variation in optic axial angle across the crystal. In the *monoclinic system* the characteristic extinction angle is that on (010) between the trace of the z crystallographic axis and the fast- or slow-vibration direction which must be specified; it should also be indicated, if possible, whether the vibration direction specified lies in the acute or obtuse crystallographic angle β. If the optic plane is parallel to (010), such sections will display the highest birefringence, but the orientation must be checked with an interference figure (a centred flash figure); if the optic plane is normal to (010), sections parallel to (010) will yield a centred bisectrix figure. In the *triclinic system* extinction angles are specified for certain directions lying in the (010) plane, usually identified by twinning, or by the existence of a (010) cleavage, as in the plagioclases; sometimes they are specified on cleavage flakes, such as (010) or (001) in the feldspars, where the angle measured is from the trace of the faster vibration direction $\alpha'$ to the trace of the (001) cleavage on a (010) flake, or to the (010) cleavage on a (001) flake.

*Measurement of extinction angles.* As extinction angles should be measured in parallel light, the condenser diaphragm should be well closed; white light will usually suffice, but if dispersion is evident monochromatic light should be used. If the mineral has very low birefringence and exists as discrete grains in an isotropic mount, extinction can be set more accurately by using the sensitive tint plate. At extinction there should be no contrast between the crystal and the background, whereas a slight rotation from this position will produce yellow or blue colours. Other accurate methods of setting extinction make use of "twinned plates", such as the Biquartz or Nakamura plate (Hallimond, 1953, p. 74).

To measure an extinction angle remove the analyser and bring the cleavage, crystal edge or twin composition plane parallel to the N–S crosswire; next, read the microscope-stage scale against its vernier, insert the analyser, rotate clockwise to a position of extinction and take the stage reading again. The difference between the two readings gives the extinction angle. Extinction should be checked up to half a dozen times and the average taken. Next, the stage should be rotated anticlockwise 45°, a compensating plate inserted and the fast- and slow-vibration directions identified so that it can be stated from which vibration direction the extinction angle was measured, e.g. for plagioclase $\alpha' \wedge$ trace of (010) cleavage on grain lying on (001).

## 6. Twinned Crystals

If the crystals are anisotropic, twin composition planes are easily recognized because adjacent sub-individuals exhibit differences in extinction position or in degree of illumination. If the twin composition plane is vertical (sharp and does not move laterally on focusing down) it can provide a very useful morphological reference direction. In *normal twins* if the twin composition plane is vertical both sub-individuals are equally illuminated when the trace of the composition plane is made parallel to either of the polars; moreover, since the extinction angles in the two sub-individuals are equal and opposite in sign, this provides a useful method of increasing the accuracy with which extinction angles may be determined. Twinning may be studied in a more adequate way if a Universal Stage is employed (p. 84).

### D. CROSSED POLARS – CONVERGENT LIGHT

In convergent light it is possible to study the optical reactions in a wide range of directions at the same time by means of the *interference figure*, a real image formed at the rear focal plane of the objective. Interference figures are best studied with a condenser that illuminates the whole of the back of the objective, and an objective of wide aperture; hence a high power, about ×40, objective and strong illumination are necessary; under favourable conditions a magnified image of the central part of the figure may be obtained by using a lower-powered objective. The interference figure may be inspected either by removing the ocular and looking down the tube, when it will be small and inverted with respect to the normal image of the object, or by retaining the ocular and inserting the Bertrand lens, which magnifies the figure and brings it into the focal plane of the ocular. The Bertrand figure, although larger, is much less distinct than that obtained by direct inspection and is not inverted; its resolution can be improved if a focusing control is fitted. When the microscope is used in this way it is often referred to as a *conoscope*; observations made on interference figures are referred to as *conoscopic observations*. In parallel light, observations are described as *orthoscopic*.

If the crystals have a diameter of about $\frac{1}{4}$ mm and have moderate ($> 0\cdot010$) or high birefringence, excellent figures are produced by the Bertrand lens; if the birefringence is lower direct inspection is more satisfactory. With small crystals, particularly if they occur in a rock section, it is necessary to isolate the rays passing through the crystal from those formed by adjacent grains. This can be done either by inserting a pinhole stop in place of the ocular or by stopping down the diaphragm of the Bertrand lens. When grains are very small, having diameters down to less than 10 $\mu$m, the only satisfactory device is the Wright ocular (Wright, 1910; Hallimond and Taylor, 1948), which is inserted in place of the normal eyepiece. It has a diaphragm that can restrict the field to the grain under observation. If the upper lens is tilted out, a small bright interference figure is observed, but the isogyres are broad and rather diffuse.

Interference figures enable a distinction to be drawn between isotropic crystals and isotropic sections of anisotropic crystals. Anisotropic crystals may be divided into uniaxial and biaxial, and the optic signs of each may be determined. In biaxial crystals the size of the optic axial angle may be estimated, and studies may be made of dispersion. Perhaps of even greater importance is the use of the interference figure to establish the orientation of a crystal, for measurements of the principal refractive indices, of the birefringence, or to determine the orientation of the indicatrix.

### 1. Uniaxial Crystals

Here the most useful orientations to select are those normal to or nearly normal to the optic axis; they may be recognized since they are either isotropic, or display low polarization colours, such as grey or white, when examined in parallel white light between crossed polars.

Sections cut *normal to an optic axis* show a centred optic axis figure consisting of a black cross with concentric circles of polarization colours rising in order from the centre to the margin of the field. In this figure the traces of the ordinary vibration directions are tangential and those of the extraordinary, radial. If the microscope stage is rotated no change can be detected in the appearance of the figure. As the thickness increases or the birefringence rises, the colour rings become more numerous and more closely spaced and the black cross more distinct. The optic sign may be determined by inserting a compensator, such as a sensitive tint plate or a quartz wedge, when the reactions will be as shown in Table 4.

Sections inclined to the optic axis at angles of up to about 30° produce an un-centred optic axis figure that is the same as the centred figure except that the centre of the cross is displaced from the centre of the field and may even lie outside it. Upon rotating the microscope stage, the whole figure describes a small circle about the centre of the field, but the arms of the cross remain parallel to the crosswires. Once the position of the optic axis has been established and the quadrants identified, the optic sign may be determined as for a centred optic axis figure.

Sections cut parallel to an optic axis when viewed in parallel white light between crossed polars exhibit the highest polarization colours of the mineral for a given thickness. Such sections will produce a *flash figure*. When the crystal is in the orthoscopic extinction position, the figure appears as a large and exceedingly diffuse black cross with four sets of very vague hyperbolic isochromatic bands. Upon a very small rotation of the microscope stage, the cross breaks up into hyperbolae whose arms flash out into opposite quadrants and disappear. The optic axis lies in these quadrants and hence the optic sign may be determined. The birefringence $\mid n_o - n_e \mid$, and the $n_o$ and $n_e$ refractive indices may be determined on a crystal that gives a flash figure.

### 2. Biaxial Crystals

Here again the most useful and distinct interference figures may be obtained from sections that in parallel white light exhibit dark grey or other low-order polarization colours. These will either be normal or nearly normal to one or other of the optic axes or to the acute bisectrix; from such sections, the optic sign may be determined, the size of the optic axial angle estimated, and the orientation of the optic axial plane established with respect to morphological reference directions, such as cleavage traces or twin composition planes. Observations can also be made on the birefringence and dispersion. Obtuse bisectrix figures are generally of less use, but if obtained must be recognized as such. Flash figures, obtained from sections cut parallel to the optic axial plane are useful for establishing the orientation of a suitable crystal for measurement of the $n_\alpha$ and $n_y$ refractive indices and for determining the birefringence, extinction angles, etc.

The optic axis figure consists of a single isogyre and a series of isochromatic curves. The isogyre straightens or curves as the stage is rotated, the amount of curvature depending on the size of the optic axial angle (Fig. 11). The trace of the optic axial plane is normal to the tangent of the isogyre where it passes through the melatope (point of emergence of the optic axis).

The acute bisectrix figure is given by a crystal displaying a slightly higher polarization colour. In the extinction position the isogyres form a distinct cross and the isochromatic bands are centred about the two melatopes that mark the points of emergence of the optic axes; the narrower isogyre marks the trace of the optic axial plane. Upon rotating the stage, the cross breaks up and the isogyres become hyperbolae centred on the melatopes. In the 45° position the convex sides of the isogyres where they pass through the melatope point in the direction of the acute bisectrix; hence the orientation of the optic plane can be established. The angular distance between the melatopes depends on the optic axial angle, the $n_\beta$ refractive index, and the numerical

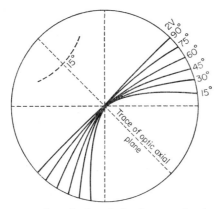

FIG. 11. Curves showing approximately the relation between the shape of the isogyre and the optic axial angle (2V) in a section normal to one optic axis and set into the 45° position.

TABLE 4

*Interpretation of interference figures*

| Type of figure | Orientation | Polarization colour | Orientation of figure for sign determination | Reactions with accessory plates for positive mineral | | | Vibration directions observed | |
|---|---|---|---|---|---|---|---|---|
| | | | | Sensitive tint | Quartz wedge | On crystal | Positive crystal | Negative crystal |
| | | | | Movement of polarization colours | | | | |

**Uniaxial Crystals**

These are distinguished from biaxial crystals by the failure of the centre of the isogyre cross to break as the stage is rotated, or if the centre of the cross is not visible, by the isogyres remaining parallel to the crosswires. This distinction does not apply to flash figures.

*Centred optic axis*
Isogyres form a black cross which does not alter during rotation of the stage — Orientation: Optic axis coincides with microscope axis — Polarization colour: Zero — Quartz wedge: None — Vibration: Only o vibration directions observed

*Uncentred optic axis*
Centre of cross displaced from centre of field; vertical and horizontal isogyres alternately sweep across the field — Orientation: Optic axis is inclined to microscope axis — Polarization colour: Low or moderate $|e'-o|$ — On crystal: Rise — Vibration: Positive e', o; Negative e, o

*Flash figure*
Large diffuse black cross which breaks into hyperbolae that disappear from the field after 5°–10° of rotation of stage into quadrants occupied by optic axis — Orientation: Optic axis parallel to microscope stage — Polarization colour: Maximum $|e-o|$ — On crystal: Rise — Note: Not useful but sign may be determined by orthoscopic test with accessory plate to determine whether e is fast or slow — Vibration: slow/fast, Positive e, o; Negative e, o

**Biaxial Crystals**

These differ from uniaxial crystals in that the isogyre cross seen at the orthoscopic position of extinction breaks into two hyperbolae as the stage is rotated; if only a single isogyre is visible it will show a flexure and will not remain parallel to the crosswires as the stage is rotated. This distinction does not apply to flash figures.

*Acute bisectrix*
Well defined cross at extinction which breaks into two hyperbolae centred on melatopes as stage is rotated. Isogyres remain in the field if $2V \leqslant 50°$; if it is higher they will leave the field after a — Orientation: Acute bisectrix, γ if positive sign, α if negative sign, coincides with microscope axis — Polarization colour: Low $\beta-\alpha$ if positive $\gamma-\beta$ if negative — On crystal: Fall — Vibration: Positive γ, α, β; Negative α, γ, β

| Type of figure | Orientation | Polarization colour | Orientation of figure for sign determination | Reactions with accessory plates for positive mineral | | | Vibration directions observed | |
|---|---|---|---|---|---|---|---|---|
| | | | | Sensitive tint | Quartz wedge | On crystal | Positive crystal | Negative crystal |
| *Obtuse bisectrix* Diffuse black cross at extinction breaks into hyperbolae that disappear from field after stage is rotated by 10° to 35°. Figure indistinguishable, except by apparent sign from acute bisectrix figure if $2V \sim 90°$, and from uniaxial flash figure if $2V \sim 0°$. | Obtuse bisectrix $\alpha$ if positive sign, $\gamma$ if negative sign, coincides with microscope axis | Moderate. $\gamma-\beta$ if positive $\beta-\alpha$ if negative | Trace of optic axial plane | Fall / Rise / Rise / Fall | | Rise | | |
| *Optic axis* A single isogyre, straight, and parallel to the N–S or E–W crosswires at extinction. It curves during rotation of the stage and 2V may be estimated from the curvature in the 45° position. | Optic axis coincides with microscope axis | Very low | Trace of optic axial plane / To acute bisectrix | Blue / Yellow | | None | Only $\beta$ vibration direction | |
| *Flash figure* Large diffuse black cross at extinction which breaks into hyperbolae that leave the field after 5° to 10° of rotation of microscope stage occupied by acute bisectrix. Uniaxial and biaxial flash figures cannot be distinguished | Third mean line, $\beta$, vibration direction coincides with microscope axis | Maximum $\gamma - \alpha$ | Quadrants occupied by acute bisectrix have lower colours / Higher colours / Acute bisectrix / Trace of optic axial plane / Colours fall | Not useful but sign may be determined by orthoscopic test with accessory plate to determine whether acute bisectrix is $\gamma$ or $\alpha$. | | Rise | | |
| *Generally oriented section* A single isogyre which crosses the centre of field at extinction. As the stage rotates isogyre will curve. It narrows in direction of melatope and in 45° position is convex towards acute bisectrix | Optix axis or acute bisectrix inclined at small or moderate angle to microscope axis | Low | | Orange / Blue-green | | Rise | | |

D

aperture of the objective. With a $\frac{1}{6}$-in objective N.A. 0·85, and a mineral such as barytes $n_\beta = 1\cdot637$, 2V = 38°, the melatopes lie just within the field of view. If the optic angle is larger the isogyres may disappear altogether as the stage is rotated.

The obtuse bisectrix figure is obtained from a section that shows higher polarization colours; it resembles the acute bisectrix figure, but the cross at extinction is much less distinct and the melatopes lie well outside the field of view. Upon rotating the stage the isogyres move rapidly, their speed increasing as the optic axial angle decreases. If 2V = 90° a rotation of some 35° from the extinction position is necessary to cause the isogyres to clear the field; if 2V is very low, however, the figure resembles a flash figure, and the isogyres will disappear after a rotation of about 10°.

The flash figure is very similar to the uniaxial one, the principal feature being the large diffuse black cross. As the stage is rotated from the extinction position, the isogyres move out very rapidly (usually requiring about 5° of rotation) into the quadrants occupied by the acute bisectrix. If the movement of the isogyres can be discerned, the optic sign may be determined from this figure.

*Generally oriented sections.* These produce a single isogyre, usually curved, that obliquely crosses the field. To utilize such a figure for the determination of optic sign, the stage should first be rotated to make the isogyre parallel to one of the crosswires, provided it still remains within the field of view. If the isogyre in this position passes through the centre of the field, the section is normal to an optical symmetry plane. Next, the stage should be rotated 45° and the isogyre watched to note any change of shape. If one remembers that the isogyre narrows in the direction of the melatope and is convex towards the direction of the acute bisectrix then the more complete figure can be imagined and the optic sign determined. Care must be exercised in estimating 2V from such a figure, since the curvature of the isogyre as it passes through the melatope may not be visible. The optical reactions of interference figures are summarized in Table 4.

### 3. Dispersion

The effects of dispersion on interference figures are described in many text books (Wahlstrom, 1962, pp. 287–300; Hartshorne and Stuart, 1964, pp. 221–228) and will not be discussed here; a practical example of the use of dispersion studies is given on p. 91.

## IV. Rotation Apparatus – The Universal Stage

If rotation apparatus is available the optical determinations can be carried out in a more controlled, systematic and quantitative manner and with a much greater degree of certainty. Two main types of apparatus are currently in use, the single axis *spindle stage*, in which the crystal, immersed in a cell filled with liquid, is secured with adhesive to the end of a thin horizontal needle which is capable of being rotated through at least 180°, and the much larger and more complex *Universal* or *Fedorov Stage* which can be used both with thin rock sections, and with grains.

Although many variants of the spindle stage have been proposed, one of the most useful is that of Roy (1966), which permits a fine adjustment of the setting of the crystal after mounting, together with a very short working distance which allows the use of standard high-power objectives for conoscopic observation; this enables direct measurement of the optic axial angle to be carried out. In fact all the optical properties may be determined on a single grain.

A spindle attachment for use with the Universal Stage has been described by Wilcox (1959). If it is used with single grains, it overcomes most of the restrictions on rotation that are inherent in the design of the instrument. Further, since it becomes part of the Universal Stage, most of the troublesome centring adjustments encountered with spindle stages can be avoided.

In ordinary petrographic work many grains of a mineral must be examined to derive its optical properties, and there is an essential dependence on random orientation. Moreover, in determining optical properties, such as extinction angles, optic axial angle and dispersion, the orientation and shape of the indicatrix are merely inferred. In Universal Stage procedures, however, one suitably oriented crystal will often suffice for the determination of all the optical properties; moreover it is with the direct determination of the indicatrix itself that one is mainly concerned. The orientation of the indicatrix and its optical character are then related to morphological reference directions such as crystal faces, cleavage planes, twin composition planes and twin axes. Mechanically all the Universal Stage axes, together with the axis of the microscope stage must intersect in a point that must lie in the plane of the object, and also be centred with respect to the field of view. The object, usually a thin rock section, is mounted on a $2 \times 1$ in ($50 \times 25$ mm) glass slide retained in position between segments that allow normal incidence at glass–air interfaces. The simplest arrangement that permits a full study of biaxial crystals is, therefore, a 3-axis Universal Stage having one vertical axis of rotation supported inside two mutually perpendicular horizontal ones, to which must be added the microscope axis; more complex instruments having 4-, 5-, and 6-axes are also manufactured. The standard instrument (Fig. 12), the use of which will be described here, is the 4-axis stage in which an additional vertical auxiliary axis is provided between the two horizontal tilting axes: in practice the additional axis is little used.

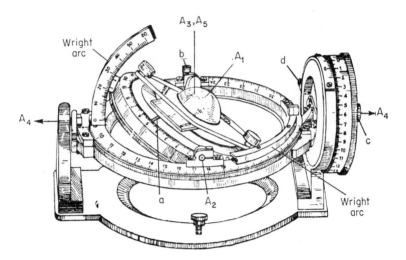

FIG. 12. Leitz 4-axis Universal Stage showing the dispositions of the rotation axes $A_1$, $A_2$, $A_3$, $A_4$, and of the microscope axis $A_5$. (After Naidu, 1958.)

## A. EQUIPMENT

Most modern microscopes fitted with objective clutch changing systems and with large stages can be adapted to take a 4-axis Universal Stage. The central ring of the stage should first be removed and the upper lens of the

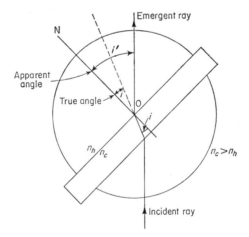

FIG. 13. Refraction by an inclined crystal plate enclosed between two glass segments. $n_c$ = refractive index of plate; $n_h$ that of segments. $n_c > n_h$. From Snell's Law, $n_h \sin i' = n_c \sin i$.

condenser system replaced by the special long-focus Universal Stage condenser: a strong light source is needed. Relatively low power special objectives fitted with diaphragms must be used because the normal segments have a radius of 12·5 mm. A series, such as those of Leitz, giving magnifications of ×5, ×10, ×20 and ×30 make the best choice. Because of the relatively low resolving powers of the higher power objectives, a ×20 ocular is useful in setting cleavages or composition planes. A useful accessory for determining birefringence and optic sign is a Berek Compensator; this can also be used to detect strain birefringence in objectives, segments, or glass central plates. A glass-stoppered bottle fitted with a dropping rod for nujol or glycerine, and a bottle of alcohol for cleaning glass surfaces must also be provided. As the data are recorded by stereographic projection, a 20-cm diameter stereographic net with a central pin that allows the tracing paper to rotate is essential.

## 1. The Segments

If a crystal plate is inclined (Fig. 13), incident light would first be refracted toward the normal of the plate and then upon emergence be refracted away from it. Under these conditions rotation of the crystal through a given angle ($i$) in air results in a very much smaller change in the wave normal in the crystal ($i'$). The relationship for isotropic refraction is given by Snell's Law where $n$ is the refractive index: $n \sin i' = \sin i$.

The use of glass segments reduces, and in many cases eliminates, these shortcomings, and also increases the magnification, for then:

Total magnification = Objective × ocular × refractive index of segments.

The magnification quoted for Universal Stage objectives assumes the use of segments with $n = 1·554$. The requirement that the refractive index of crystal and segments should be identical could only obtain with isotropic substances, but if the birefringence is less than 0·02 and the refractive index of the segment lies between the maximum and minimum refractive index of the mineral, no correction is required. To reduce the size of these corrections and reduce other errors due to refraction, three standard sets of segments having $n = 1·516$ for minerals of low refractive index, $n = 1·554$ the middle of the plagioclase range, and $n = 1·649$ for ferromagnesian minerals, are supplied. Segments should be selected whose refractive index lies closest to the mean refractive index of the mineral being examined. The method of applying the refractive-index corrections is described later (p. 79).

## 2. Nomenclature and Uses of Axes

The positions of the four axes, $A_1$, $A_2$, $A_3$, and $A_4$, in Berek's notation (1924) are shown in Fig. 12; the microscope stage ($A_5$) gives a third vertical axis. $A_1$, the innermost axis (inner vertical), is used for setting cleavage and composition plane traces parallel to the N–S crosswire and for setting crystals to extinction: it may be locked by means of the lug "a". $A_2$, the N–S axis is used to set cleavage, composition planes and optic-symmetry planes vertical. Inclinations about this axis are read as left and right tilts against either of two hinged arcs; the Wright arcs, which lie flat on the $A_3$ circle when not in use. The $A_2$ axis is clamped by means of the screw "b". $A_3$ is an auxiliary axis used mainly in the indirect determination of 2V; it is not required for the simple operations and in its zero-position the reading is 90° against its divided scale. It is clamped by means of the screw "c". $A_4$ is the E–W horizontal axis and carries at its right-hand extremity a large graduated control drum. This axis is used for testing that optic symmetry planes have been set vertical, to check changes in polarization colours, and to set optic axes vertical. Readings on $A_4$ are made against its divided circle as tilts N (stage tilted towards the observer), where the reading is subtracted from 360°, and tilts S (stage tilted away from the observer). This axis may be clamped by means of the screw "d". $A_5$ is used to set the stage into the *Normal position*, when $A_1$ is vertical, $A_2$ is N–S, and $A_4$ E–W, and into the *Diagonal position* at 45° to this where 2V is measured and the optic symmetry axes made coincident with $A_4$ determined. It is also used for the direct measurement of extinction angles.

### 3. Preliminary Adjustment and Mounting

(a) First centre the objectives and then remove them, raise the tube, and place the Universal Stage in position. Bring the two holes in the base plate into coincidence with the threaded holes in the microscope stage and screw the stage down with the screws provided.

(b) Rotate about $A_5$ so that the main control drum of $A_4$ is set on the right-hand side; clamp $A_5$ and set the stage into the Normal position, with $A_3$ reading at 90°.

(c) Fit a low-power objective, insert the central glass plate, place a thin section on the stage and focus on a grain at the centre of the field of view.

(d) To centre the stage rotate about $A_1$ to determine whether this axis coincides with that of the microscope; if the grain is uncentred, adjust by means of the two diagonal screws in the base plate until a given point remains centred during rotation about $A_1$. As there is inherently more play in the $A_1$ axis than in the others, time should not be wasted in securing perfect centring.

(e) To mount the section, place a few drops of nujol ($n = 1·48$) on the glass plate and press down the thin section firmly with the cover-slip uppermost to spread the liquid and remove air bubbles. Next, place a large drop of liquid on the cover-slip.

(f) To mount the upper segment, place the selected upper segment in position and screw it down sufficiently far to distribute the liquid over the whole of the cover-slip. The screw must not be overtightened; both segments must have the same refractive index.

(g) To mount the lower segment, rotate the stage far over about $A_4$ to bring the lower side of the glass plate uppermost and place a few drops of liquid on it. Press the lower segment firmly into place and rotate the stage back until the reading on the $A_4$ drum is zero; it will be retained in position by suction.

(h) To adjust the height of the section; since the thickness of microscope slides varies, the point of intersection of the $A_2$ and $A_4$ axes must be adjusted to coincide with the plane of the section. To do this bring a definite object such as a tiny inclusion to the intersection of the ocular crosshairs.

(j) Rotate the stage about the $A_4$ axis. If the height is correctly adjusted the image will remain stationary. If it moves in the opposite sense to that of the rotation about $A_4$ the section is too high and its height must be reduced. This is done by screwing down the central plate mount by means of the lugs on the lower side of the $A_1$ circle. If the object moves in the same sense as the motion of $A_4$ it is too low and must be raised. Before this is done the tension on the retaining screws of the upper segments should be slackened a little.

(k) To set the $A_4$ axis E–W and determine the zero position of $A_5$; raise the microscope tube and focus on a dust particle on the surface of the upper segment. Bring this into coincidence with the N–S crosshair by rotating about $A_2$. Rotate about $A_4$ and restore the point to coincide with the N–S crosshair by adjusting $A_5$. Repeat until the point remains coincident with the crosshair during rotation about $A_4$. Take the reading on $A_5$ and clamp this circle. This is the permanent setting for the Universal Stage in the Normal position. Its position, however, should always be checked before measurements are begun or if the Universal Stage has been dismounted.

## B. ORIENTATION PROCEDURES

In practice as it is usually known whether the crystal to be studied is isotropic, uniaxial or biaxial, the Universal Stage procedures for effecting these distinctions will not be given here. A beginner will find it advisable at first to work with a relatively thick section of some mineral rather than a thick rock slice, for this will avoid the puzzling effects caused by overlap when the section is tilted. In thick sections extinction can be set more easily and the co-ordinates of cleavage and twin planes can be more easily located.

Correct illumination is essential for accurate results; for setting extinction the aperture diaphragm should be well stopped down and the objective diaphragm closed as well. Because of the long working distance imposed on the objectives, conoscopic methods of determining the indicatrix (for procedures here, see Hallimond, 1953, p. 144), are not generally applied, unless special small upper segments, a special condenser, and special objectives are used. The standard methods of determining the indicatrix involve ortho-scopic procedures in which a definite sequence of manipulations is followed to obtain extinction.

## 1. Uniaxial Crystals

The principal reason for studying uniaxial crystals with the Universal Stage is for petrofabric studies, in which the position of the optic axis of every grain must be established with respect to the orientation of the rock section. Procedures are therefore required that will locate the optic axis either by bringing it vertical, or if that is not possible by making it horizontal and coincident with the $A_4$ axis.

*To set an optic axis vertical and determine the optic sign.* (a) With the Stage in the Normal position rotate the crystal to extinction about the $A_1$ axis. (b) Rotate about $A_4$ and the crystal will either remain in extinction or become illuminated. If it remains in extinction the optic axis lies in the N–S vertical plane. If the crystal becomes illuminated, rotate it about $A_1$ to the alternative extinction position; it will then remain in extinction during rotation about $A_4$. (c) Rotate 45° anticlockwise about $A_5$ into the Diagonal position; the crystal should now be illuminated again. (d) Rotate now about the $A_4$ axis; the polarization colours should fall as the optic axis approaches the vertical. When the crystal is in extinction the optic axis is parallel to the microscope axis. The rotation to extinction about $A_4$ should be repeated several times and the results averaged. At this stage the readings on the $A_1$, and $A_4$ circles should be recorded. (e) *To determine the optic sign.* Rotate appro-ximately 15° about $A_4$ in the direction that reduces the inclination of the inner stage; in this position the crystal should exhibit a grey polarization colour; insert the sensitive tint plate and determine whether $e'$ (parallel to the trace of the optic axis) is fast or slow; if $e'$ is fast the mineral is optically negative, if it is slow it is positive.

*To set the optic axis horizontal and parallel to $A_4$.* This method makes use of the maximum birefringence and may be used when the optic axis is inclined at less than 40° to the plane of the section; it is a more satisfactory procedure when the position of the optic axis, as indicated by extinction on rotation about $A_4$, is ill defined, or if the birefringence is very low. (a) With the Stage in the Normal position rotate the crystal to extinction about the $A_1$ axis with the trace of the optic axis in the E–W vertical plane; check that this is the case by rotating about $A_4$, when the crystal should become illuminated again. (b) Restore the crystal to extinction by rotating about $A_2$, which will bring the optic axis either coincident with $A_4$, or vertical. If the optic axis is hori-zontal the crystal will remain in extinction during rotation about $A_4$. Having established which is the orientation, record the readings on $A_1$ and $A_2$.

(c) Rotate the Stage anti-clockwise about $A_5$ into the Diagonal position, insert a compensator and determine the optic sign. If the optic axis is vertical incline slightly about $A_4$ before inserting the compensator.

## 2. Biaxial Crystals

In biaxial crystals two of the three mutually perpendicular optical symmetry planes must be located as their poles and intersections fix the positions of the principal vibration directions; optic axes are located by rotating the appropriate optical symmetry plane about $A_4$ with the Stage in the Diagonal position. Finally, a bisectrix is set parallel to the microscope axis and the optic sign is determined in the usual way with a compensator. As the optical symmetry planes are located they should be plotted on the stereographic projection where they can be related to observed crystallographic reference directions.

*To determine the orientation of the indicatrix.* (a) With the Stage in the Normal position set the crystal to extinction by rotating about $A_1$. (b) Rotate alternately about $20°$ in either direction about $A_4$; in general the crystal will become illuminated. The Stage should be left inclined to that side on which the greater illumination was observed. (c) Restore extinction by rotating about $A_2$ to the nearer extinction position and rotate about $20°$ to the other side about $A_4$. Usually this will produce feeble illumination. (d) Restore extinction by making a slight adjustment to $A_1$. (e) Make further slight adjustments to the settings of $A_1$ and $A_2$, with intervening rotation about $A_4$ through an arc of about $40°$ until the crystal remains in extinction. When this is obtained, one of the optical symmetry planes has been set vertical and perpendicular to $A_4$ and one of the principal vibration directions is coincident with $A_4$. At this stage note the settings of $A_1$ and $A_2$. (f) *To identify the vibration direction set parallel to $A_4$.* Rotate the Stage about $A_5$ into the Diagonal position, rotate alternately in either direction about $A_4$, and observe the polarization colours. Two cases are possible. (g) If the principal vibration direction $Y$ coincides with $A_4$ and the plane made vertical is the optic axial plane. Upon rotation, one or both of the optic axes may become vertical; their approach can be recognized by the fall in the polarization colours to extinction. If an optic axis is located, the reading on $A_4$ should be noted and a search made for the second one also. If an optic axis cannot be made vertical its approach may often be detected by the fall in the polarization colours upon great rotation about $A_4$. (h) If no optic axis is detected but a strong rise in polarization colour accompanies rotation about $A_4$, then either $X$ or $Z$ is coincident with $A_4$. To distinguish between these, first set $A_4$ at zero. Insert a compensator (length fast); if its length is NE–SW, parallel to the $A_2$ axis, a fall in the polarization colour indicates that $Z$ is coincident with $A_4$, a rise that $X$ is coincident with it. (j) Restore the Stage to the Normal position and rotate about $A_1$ to the other extinction position and proceed as before to locate the

D*

second optical symmetry plane and the character of the principal vibration direction normal to it.

It is advisable to locate first the symmetry plane that is most nearly normal to the plane of the section; this can usually be identified by setting the crystal to extinction about $A_1$, testing by rotating about $A_4$, setting the other extinction position on $A_1$ and testing this with $A_4$ also. The plane selected is that which produces the least illumination during rotation about $A_4$. If an optic axis is located, the Stage should be rotated 90° clockwise about $A_5$ into the other Diagonal position and the $A_4$ setting checked; a difference of more than $1\frac{1}{2}$° in its position, reveals a faulty setting of the optic axial plane.

### 3. Location of Cleavages and Twin Composition Planes

In setting vertical cleavages and twin composition planes, careful adjustment of the illumination is essential: parallel light is required, so the Universal Stage condenser is removed, a high-power objective ($\times 20$) fitted with its diaphragm fully open, and the aperture diaphragm closed until the plane selected appears sharpest. (a) With the Stage in the Normal position rotate about $A_1$ to bring the trace of the plane parallel to the N–S crosswire. (b) Rotate about $A_2$ until the trace makes the finest and sharpest possible line, check the setting by lowering or raising the focus slightly, and observe the Becke line. If the cleavage has been set correctly two lines of equal intensity should move out uniformly from the cleavage trace. If the twin is a normal one the illumination in both sub-individuals should match exactly if the composition plane is vertical. (c) Record the readings on $A_1$ and $A_2$.

### C. PLOTTING THE RESULTS

The readings made about $A_1$, $A_2$ and $A_4$, whose rotation axes are mutually perpendicular must be transferred to the stereographic projection; in this discussion it will be assumed that no refractive-index corrections to the $A_2$ and $A_4$ readings are necessary. On the trace the original position of a plane or direction is plotted with respect to the pole of the section (centre of the stereogram) and the zero reading of the $A_1$ circle, which is marked by an arrow on the trace. Hence, readings about $A_1$ are set off clockwise around the primitive from the zero point, the south end of the N–S diameter. Since a clockwise rotation about $A_1$ ($\alpha_1$ angles) increases the reading, the zero point of the trace should be rotated clockwise over the underlying net to record the azimuth of a reading made on the $A_1$ circle (the angle at 30° in Fig. 14). Angular rotations made about $A_2$ ($\alpha_2$ angles) are recorded on the trace by using the N–S trending great circles inscribed on the net. A great circle set vertical at a reading of 32° made on the right-hand Wright arc is recorded as 32°R, and the plane is traced over the great circle inclined at 32° to the right

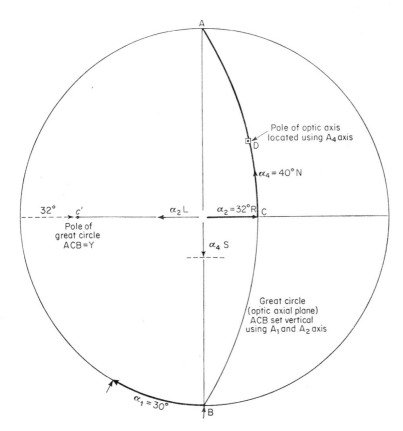

FIG. 14. Stereogram to show method of plotting angular rotations $\alpha_1$, $\alpha_2$ and $\alpha_4$, made on the $A_1$, $A_2$, and $A_4$ circles respectively, using a trace that may be rotated over a fixed stereographic net on which the $\alpha_1$ angles are inscribed clockwise starting from the S end of the N–S diameter of the stereogram.

of the N–S diameter after the $\alpha_1$ rotation has been made. Readings made on $A_4$ ($\alpha_4$ angles) are recorded by using the E–W trending small circles inscribed on the net, and directions made vertical are located by the intersection of the appropriate small circle with the great circle corresponding with the tilt on $A_2$. A reading 40°N on $A_4$ (stage tilted towards the operator) is plotted by marking on the trace the intersection of the small circle inclined at 40° to the E–W diameter of the net on its northern side, with the 32° great circle already recorded on the trace.

### D. REFRACTIVE-INDEX CORRECTIONS

If the crystal and segments differ in refractive index, a radial distortion is introduced; in all cases the angle to be corrected is that between the pole of the section (the normal to the $A_1$ axis), and the microscope axis. If either the $A_2$ or $A_4$ axes are inclined alone then the readings made on these axes may be corrected before the planes are plotted, but if both axes are inclined, as in locating an optic axis, the *uncorrected* direction should first be plotted, its inclination determined as a radial angle and its corrected position located inwards or outwards along the radius through the uncorrected pole after determining the correction from Fig. 15. Since the birefringence of most rock-forming minerals is less than 0·03, it suffices for most purposes to use the mean refractive index or the $n_\beta$ refractive index if this is known. With minerals of higher birefringence, such as acmite or fayalite, it is better to use the actual refractive index for the wave normal and vibration direction concerned, as significant errors may be incurred by using an assumed mean index. In minerals of low birefringence, such as the plagioclase feldspars, where the difference between the mean refractive index $\left( \dfrac{n_\alpha + n_\gamma}{2} \right)$ and $n$ for the

TABLE 5

*Example of refractive-index corrections for a titanuaugite*

The specimen was *Titanuaugite* from alkali olivine basalt 68215, flow of 1750; Haleakala, Maui, Hawaiian Islands

| Reference planes | Symbol of pole | Readings recorded | | | Remarks |
|---|---|---|---|---|---|
| | | $A_1$ | $A_2$ | $A_4$ | |
| Optical-symmetry planes | | | | | |
| Z | △ | | | | Not located directly |
| Y | -¦- | 265 | 26½L | 44N 6S | Uncorrected readings $A_2$ 28L, $A_4$ 46N and 6S, give radial angles of 28°, 53°, and 29° which are corrected to 26½°, 50°, and 27½°, respectively |
| X | ⊙ | 346 | 18R | | Uncorrected $A_2$, 19°R corrected to 18°R |
| Cleavage plane | #¹ #² | 045 136 | 3R 36R | | {110} cleavages No correction required for $\alpha_2$ angle. Uncorrected $\alpha_2$ angle 39½° corrected to 36° |
| Zero on $A_5$ | 15° | $n_\beta$ refractive index 1·715 (by immersion method) Segments $n = 1·649$.    $2V_\gamma = 49°$ Extinction angle $Z \wedge z = 44\frac{1}{2}°$ (from plot) | | | |

segment rarely exceeds 0·015, the need to apply a correction may be ignored except at the very high inclinations sometimes needed for determining the positions of the optic axes.

## 1. Example

Details of the method of plotting and the application of the corrections are illustrated by the example on page 80 (Table 5) in which the uncorrected readings are noted in the Remarks column.

The optic axial plane whose pole is $Y$ was plotted first; to do this the reference mark on the trace was set over the $A_1$ reading of 265°, the uncorrected great circle ACB drawn in lightly and its pole plotted (Fig. 16). The correction

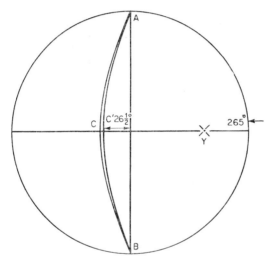

FIG. 16. Method of applying corrections to the stereographic plots. Uncorrected readings $\alpha_1$ 265°, $\alpha_2$ 28°L give the uncorrected position of the optical symmetry plane ACB; after reference to Fig. 15 and Table 5 the corrected $\alpha_2$ angle of 26° is determined. This enables the corrected plane AC′B to be plotted.

to the $A_2$ reading of 28° was obtained using the chart (Fig. 15) for segments $n = 1·649$ and $n_\beta = 1·715$; the corrected angle is $26\frac{1}{2}°$. Next, the corrected great circle AC′B was drawn in and its pole labelled. The uncorrected positions of the optic axes were then plotted on the great circle ACB at the intersections with the 46°N and 6°S small circles respectively: these are the points E and F (Fig. 17). The inclinations of these poles may be read from the net as 53° and 29°, respectively, which give corrected values of 50°, and $27\frac{1}{2}°$. The corrected poles were plotted inwards along the radials and are shown as the poles E′ and F′ on the great circle AC′B. The angle E′F′ = 49° gives the true value of 2V.

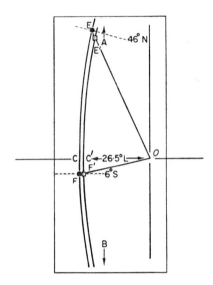

Fig. 17. Enlargement of central portion of Fig. 16 to show application of corrections to the poles of the optic axes: the uncorrected radial angles OE and OF correct to OE' and OF', respectively.

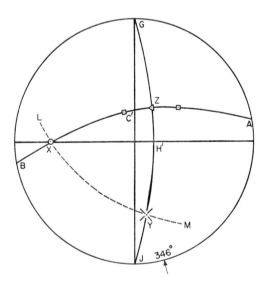

Fig. 18. Plotting the second optic symmetry plane GH'J: here only the corrected plane need be plotted. The third optic symmetry plane LM was constructed to pass through the poles of the other two.

Since the corrected positions only of the other two planes are required, the second optical symmetry plane whose pole is $X$ (marked $\odot$) may be plotted directly after the $\alpha_2$ angle has been corrected (Fig. 18). Note that the pole of $Z$ lies in the first optical symmetry plane AC'B. The third plane whose pole is $X(\odot)$ may be drawn in through $Y$ and $Z$ which have already been located. In an ideal plot each part of the spherical triangle should measure $90°$; if any of the sides amounts to $88°$ or less the plot is unsatisfactory and the co-ordinates of the optical symmetry planes must be redetermined. For the plot of the data in Table 5 the first cleavage plane is drawn in as a dashed great circle and its pole marked $\#^1$, and the corrected positions of the second plane and its pole marked $\#^2$ (Fig. 19). The intersection of the two cleavage planes with the optic axial plane locates the $z$ crystallographic axis, and if the

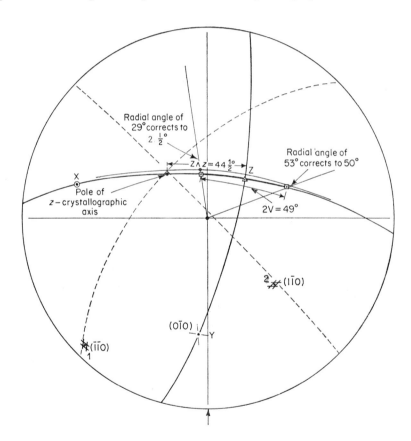

Fig. 19. The completed stereogram showing how the position of the $z$ axis is located by the intersection of the prismatic cleavages with the optic axial plane, and how the extinction angle, $Z \wedge z = 44\frac{1}{2}°$, may be determined.

great circle containing the pole $Z$ is now placed above the E–W diameter inscribed on the net the extinction angle $Z \wedge z = 44\frac{1}{2}°$ can be read off. Note that the great circle of which the $z$ crystallographic axis is the pole contains the poles of the two cleavage planes and the principal vibration direction $Y$, these will be the face poles (110), (010), and ($\bar{1}$10) or their opposites. Since this pyroxene does not display (001) exsolution lamellae, it is not possible from the data here to state whether $Z$ lies in the obtuse or acute crystallographic angle $\beta$. In this case such information could only be established from an X-ray single crystal study of the crystal, which is unnecessary since it is well known that in common clinopyroxenes such as this, $Z$ always lies within the obtuse crystallographic angle $\beta$. With this knowledge the (110) and (11$\bar{0}$) cleavages may be identified. The measured angle (110) $\wedge$ ($\bar{1}$10) between the cleavage poles, 89°, compares well with the angle calculated from X-ray data.

### E. LOCATION OF TWIN AXES – PLOTTING TWIN CRYSTALS

If twinned crystals are encountered, the crystallographic elements of twinning, the twin axis, composition plane, and twin plane can supply useful crystallographic reference directions. Moreover, because of the symmetrical relationship about the twin axis, the study of such crystals affords an easy method of obtaining increased accuracy. If cleavage planes are also located the twin law may be determined. Twin crystals are always used in the study of plagioclase feldspars and should be sought when measurements have to be made on pyroxenes and amphiboles.

First, the crystal should be sketched and the cleavages and composition planes in the two sub-individuals labelled (Fig. 20 and Table 6) and plotted. Next, the co-ordinates of the optical symmetry planes are determined in each sub-individual and the principal vibration directions in each $X_1$, $Y_1$, $Z_1$; $X_2$, $Y_2$, $Z_2$ are plotted (it is convenient to use different colours for each sub-individual). Next, the twin axis is located. (a) Rotate the trace until any pair of corresponding poles $X_1X_2$, $Y_1Y_2$ or $Z_1Z_2$, are superimposed on one of the great circles inscribed on the net and trace this. (b) Rotate the trace to bring the remaining pairs above their respective great circles and trace them also. (c) If the orientation in each sub-individual is perfect, the three great circles should intersect in a point. Usually, however, a triangle of error will be formed whose centre of gravity is taken as the pole of the twin axis. (d) If the twin axis coincides with (or lies close to) the plotted pole of the twin composition plane the twin is a Normal one. If the twin axis lies in the composition plane the twin law is of Parallel or Complex type; these can usually be distinguished and the twin law identified by the known angular relationship to observed cleavage planes.

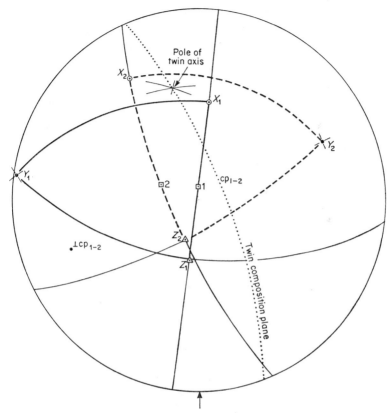

Fig. 20. Plot of a twinned crystal of plagioclase using data in Table 6. $X_1Y_1Z_1$ represent the first sub-individual and $X_2Y_2Z_2$ the second. The twin axis is located at the intersection of the arcs $X_1X_2$, $Y_1Y_2$ and $Z_1Z_2$. Since this lies in the composition plane the twin is of Parallel or Complex type.

If a substantial triangle of error occurs, Berek's (1924) modification of the method leads to more satisfactory results, particularly in the feldspars where all that is required is the mean angular relationship between the indicatrix of each sub-individual and the twin axis. (a) Rotate the trace until a pair of equivalent poles such as $X_1X_2$, which are well separated, lie above a great circle inscribed on the net, and plot the bisecting point $P_1$ between them; plot also the other bisecting point $P_2$ 90° distant from $P_1$. (b) Rotate the trace until $P_1$ lies above the E–W diameter of the net and trace a short arc that passes through the pole $P_2$. (c) Repeat operation (b) with $P_2$ above the E-W diameter and draw in an arc passing through $P_1$. (d) Repeat operations (b) and (c) with $Y_1Y_2$ and $Z_1Z_2$ to locate the poles $P_3$, $P_4$, $P_5$ and $P_6$ and their arcs. (e) Three of the short arcs will intersect in a point, which is the pole of the twin axis.

TABLE 6

*Example of twinned plagioclase*

The specimen was plagioclase from ferrogabbro EG 3655; Layered Series, Skaergaard Intrusion, E. Greenland

| Reference planes | Symbol of pole | Readings | | | Remarks |
|---|---|---|---|---|---|
| | | $A_1$ | $A_2$ | $A_4$ | |
| Twin composition plane | | 201 | 18L | | $CP_{1-2}$ Cleavage parallel to CP in both sub-individuals. No other cleavage visible. |
| Optical-symmetry planes. $Y_1$ $X_1$ $Z_1$ | $-\!\!\mid\!\!-_1$ $\odot_1$ $\triangle_1$ | 172 261 263 | 1L 34R 54L | 85 | Constructed. |
| $Y_2$ $X_2$ $Z_2$ | $-\!\!\mid\!\!-_2$ $\odot_2$ $\triangle_2$ | 203 298 253 | 16R 18R 66L | 165 | Constructed. Twin axis lies in composition plane so twin is of parallel or complex type. |

Angles over pole of twin axis.

$$X_1 \wedge X_2 = 37°$$
$$Y_1 \wedge Y_2 = 145°$$
$$Z_1 \wedge Z_2 = 169°$$

$Z \wedge \perp CP_{1-2} = 59°$ hence CP is (001), or rhombic section (Slemmons, 1962, Table 2 case 1), hence twin law is Acline $A$, or Pericline, or Manebach-Acline $A$ complex.

Twin is Ala $A$ or Manebach-Acline complex and composition $An_{45}$, low or low transitional structural state (Slemmons, 1962, Plate 7).

## F. DOUBLE-VARIATION METHOD

In the double variation method the Universal Stage is used to orient a crystal with $X$, $Y$ or $Z$ parallel to the microscope axis for the determination of two of the principal refractive indices by the immersion method: the third index is determined by extrapolation. Temperature control within a range of about 50°C is effected by means of a water cell that replaces the central plate of the stage; an Abbe refractometer, connected in parallel to the water-circulatory system enables the refractive index of the mounting liquid to be determined at any time. The most convenient type of monochromator is the continuous-running filter mounted in a metal frame that can be inserted in a guide that fits the light aperture in the microscope base. A powerful light source is required, and the special high dispersion liquids listed by Emmons (1943, p. 63) form an essential part of the equipment.

### 1. Procedure

(a) To mount the grains, place a drop of the refractive-index liquid, whose refractive index lies between $n_\alpha$ and $n_\gamma$ of the mineral on the glass plate of the water cell and lay on it a large thin cover-slip to protect the surface. In the centre place a few grains of the mineral and a further drop of liquid, cover with a second cover-slip, and lower the upper segment position with a drop of liquid on its lower surface. Carefully tighten the retaining screws; the largest grains will be held in position by pressure, the rest may drift aside as the stage is inclined. If a suitable grain is not centrally placed it can be brought closer to the centre by aligning it with the E–W crosswire and by inclining about $A_2$, upon loosening the upper segment slightly the grain will appear to move upwards towards the centre of the field. Its movement can be arrested by returning $A_2$ quickly to its zero position and tightening the screws again. A starting temperature of, say, 25°C is selected. (b) Next, the orientation of the indicatrix is determined and plotted (uncorrected) and the principal vibration direction nearest the centre of the stereogram set parallel to the microscope axis. This can be done by rotating the trace over the stereogram until the pole lies above the E–W diameter inscribed on the net below. At this point the $A_1$ and $A_2$ readings are noted and set on the appropriate axes of the Universal Stage. It is not usually necessary to correct the $A_2$ reading. (c) Next, the stage is rotated about $A_3$ to extinction when one principal vibration direction is N–S and the other E–W. (d) The monochromator is now adjusted until a match is obtained; with highly dispersive liquids a change of 20 nm on either side of the correct value should give Becke lines with opposite movements. (e) The circle ($A_3$) or polarizer is then rotated 90° to obtain the other principal vibration direction and the monochromator is adjusted again to obtain a match. If a second monochromator is used for the refractometer, the matching refractive index and wavelength in each case can be recorded. A correction must be applied to the refractometer reading for the wavelength at which the observation was made. Correction tables are supplied by most manufacturers of Abbe refractometers. (f) The temperature of the system is now changed. If the first match was made for long wavelengths the temperature should be raised; if the match was made for short wavelengths the temperature should be reduced. A change of about 5°–10°C in temperature is usually satisfactory; about 10 min should be allowed to enable the system to become adjusted to the new temperature. (g) Next, a second set of readings should be made and recorded. (h) The results are plotted on a Hartmann chart (Fig. 21) which reduces normal dispersion

curves to straight lines. This has wavelength as abscissa in divisions of dimi-
nishing size from blue to red and refractive index as ordinate. The gradua-

tion of the abscissa is made proportional to $\dfrac{1}{200 \text{ nm}}$ (Wülfing, 1924, p. 61).

The intersection of the dispersion curves with the wavelength of the $\text{Na}_D$

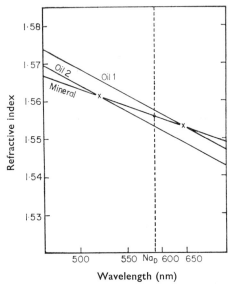

FIG. 21. Dispersion curves for two liquids and a mineral plotted on a Hartmann scale.
Since the Hartmann scale reduces the dispersion curve to a straight line, the refractive
index for the $\text{Na}_D$ line can be deduced.

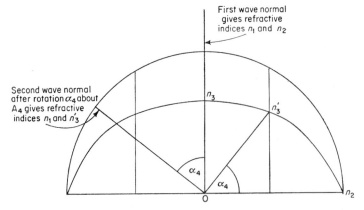

FIG. 22. Construction to extrapolate the third principal refractive index from a principal
and an intermediate index. (After Hallimond, 1953.)

line at 589 nm gives the conventional values of the refractive indices $n_1$ and $n_2$, for two of the principal refractive indices. (j) If the Stage is now rotated about 50° about $A_4$ away from $n_2$ (with the polarizer set N–S) in the direction that reduces the inclination of the inner stage, a third refractive index $n'_3$ may be determined. This is proportional to the intercept made by a line at right angles to the wave normal in an ellipse with axes proportional to $n_2$ and $n_3$ (Fig. 22). If the total birefringence is less than about 0·060, $n_3$ may be obtained from the following formula (Hallimond, 1953, p. 159):

$$n_3 = n_2 - (n_2 - n'_3) \sin^2 \alpha_4.$$

A graphical solution to this is given by Emmons (1943, plate 10), but the calculation, with four-figure tables is not very tedious.

An example, for which olivine was used, is given in Fig. 23.

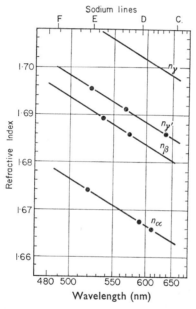

Fig. 23. Dispersion curves of an olivine crystal ($Fa_{16}$). Plotted points indicate wavelengths where liquid and crystal were matched; $n_\gamma$ determined by extrapolation from $n_\gamma'$.

## V. Examples

### A. BIREFRINGENCE DETERMINATION ON SCAPOLITE

Scapolite (tetragonal) exhibits good {110} cleavages and has a theoretical range of composition from $Na_4[Al_3Si_9O_{24}]Cl$ for marialite to $Ca_4[Al_6Si_6O_{24}]CO_3$ for meionite. There is a considerable difference in the birefringence of the two end members and an approximately linear variation between them (Deer *et al.*, 1962, Vol. 4, p. 329): a determination of the birefringence is sufficient to estimate the composition.

Scapolite occurs in a slide as prismatic crystals up to 2 mm in length and $\frac{1}{2}$ mm in width along with quartz which shows polarization colours up to first-order yellow. The quartz grains displaying the highest colours are first examined in turn in convergent light between crossed polars until one showing bright yellow is found that gives a centred or nearly centred flash figure; this grain has the maximum birefringence of the mineral, 0·009. Next, the retardation is determined with a compensator, by using either a graduated quartz wedge or a Berek compensator; the retardation is 350 nm. From Fig. 10, by interpolation, the thickness is given as 0·042 mm. It is best to check the thickness by measuring retardations on about three quartz grains in the same part of the slide and as near as possible to the scapolite crystal selected. The scapolite crystal selected also shows the highest polarization colour, in this case a second-order blue, and the orientation is checked with a nearly centred flash figure. The centre of the cross appears to be displaced by an amount estimated to be about half the radius of the field. As the objective has a numerical aperture of 0·85, the visible cone in the crystal is about 60° and the displacement of the centre of the cross must be about 15°. Next the retardation is determined as 625 nm which gives $\Delta'$, the partial birefringence, as 0·015. Correcting for the 15° displacement, $\dfrac{\Delta'}{\Delta} = \sin^2 75°$, we get $\Delta = 0·016$. From this we deduce that the composition is 35% meionite $\pm$ 5%. If a Universal Stage had been used the exact inclinations of the optic axes could have been determined, and moreover the thickness could have been estimated on grains of quartz adjacent to the scapolite, irrespective of their orientation.

### B. OPTICAL DETERMINATION OF THE COMPOSITION OF AN ORTHOPYROXENE

Orthopyroxenes vary in composition chiefly between $MgSiO_3$ (enstatite) and 88% $FeSiO_3$, there being linear variations in the refractive indices and birefringence, and considerable variations in 2V, including two changes of optic sign (Deer *et al.*, 1962, Vol. 2, Fig. 10). In the absence of a Universal Stage, the composition can best be estimated by determination of one or more of the principal refractive indices. As the minerals are ortho-rhombic, have good {110} cleavages, and the optic orientation is $X = y$, $Y = x$, $Z = z$, it is clear that crushed fragments that lie on {110} will yield the $n_\gamma$ refractive index (slow vibration direction parallel to the cleavage trace) and $n_\alpha'$, an intermediate refractive index; $n_\gamma$ can be used to determine the composition. Orthopyroxenes however, are frequently zoned and this can be checked only by examining a thin section, and noting whether any crystals exhibit non-uniform extinction (see Biot-Fresnel construction, p. 62) between the core and the margins. The effect is more easily ascertained if a Universal Stage is used and then 2V can be determined on core and margins also. Care must be exercised however, if the composition is estimated using 2V determinations alone, since the 2V curves for volcanic and plutonic pyroxenes differ slightly and complications may also arise if the $Fe_2O_3$ or $TiO_2$ contents are high. Because of this both the $n_\gamma$ refractive index and 2V determinations should be carried out; if they agree the result can be accepted with confidence. An example of this anomalous behaviour is described by Muir and Tilley (1957, p. 249) for the hypersthene from the Uwekahuna gabbro porphyry. Chemical analysis revealed a composition of $Of_{22\cdot3}$ and the refractive indices $n_\alpha$ 1·685, $n_\beta$ 1·692, $n_\gamma$ 1·696 correspond with $Of_{26}$; $2V_\alpha =$ 67°–63°, mean value 65° and corresponds with $Of_{28}$. The cell dimensions $a = 18\cdot31$, $b = 8\cdot867$, $c = 5\cdot201$ Å, agree fairly well with the graph given by Hess (1952) for volcanic orthopyroxenes $Of_{22}$ high in alumina (in this case $Al_2O_3$ 2·14, $TiO_2$ 1·12, $Fe_2O_3$ 1·10%).

### C. DISPERSION STUDIES TO RESOLVE THE TRUE SYMMETRY OF STAUROLITE

Staurolite has in the past been considered to be orthorhombic and holo-symmetric with $X(Bx_a) = y$, $Y = x$, $Z = z$, but an analysis of morphological and X-ray single-crystal data, together with a negative pyroelectric test, indicates either of two space groups *Cmmm* orthorhombic, or *C 2/m* mono-clinic with $\beta = 90° \pm 3'$ (Hurst *et al.* 1956). Optical observations were made in monochromatic light on plates 0·4 mm thick, cut parallel to (001) and to (010). The (001) section was shown to extinguish sharply in white light in a position that exactly bisects the angle subtended by the traces of the {110} prism faces; this excludes the possibility of triclinic symmetry. A section cut parallel to (010) showed a small but unmistakable variation of 1° in the extinction position for red ($\lambda = 690\cdot7$ nm) and for blue ($\lambda = 466\cdot7$ nm) light. On the Universal Stage $2V_\alpha$ was found to be 88° for red light and 84° for blue. These results indicate the presence of both optic axial and crossed dispersion, the latter of which cannot occur in an orthorhombic crystal, and the true symmetry is therefore monoclinic.

### D. OPTICAL DETERMINATION OF A CLINOPYROXENE BY USING A TWINNED CRYSTAL

Clinopyroxenes, depending on their compositions, may have their optic axial planes oriented either parallel or normal to (010). The acute bisectrix $\gamma$ in either case, lies in the (010) plane normal to the $y$ axis. For pyroxenes, the characteristic crystallographic reference directions are the {110} cleavages which intersect at nearly 90°, and a {100} parting, which is often seen. Other useful features which are commonly observed are a normal twin on {100}, and fine exsolution lamellae of a second pyroxene phase, which can be parallel either to (100) or to (001), or to both if two sets are present.

In the monoclinic system the extinction angle quoted is the "characteristic extinction angle" on (010), between the nearer of the two vibration directions and the trace of the $z$ crystallographic axis. In all cases it should be specified whether the vibration direction concerned lies in the acute or obtuse crystallographic angle $\beta$.

To determine the orientation of the clinopyroxene indicatrix, a crystal was first sought that was cut normal to the $z$ axis. [Such a section displays grey interference colours and shows the cleavages as two parallel sets of sharp lines that do not move laterally as the focus of a high power objective is lowered (Fig. 24a). The intersection of the cleavage planes defines the $z$ axis. When the polars are crossed the extinction directions bisect the cleavage angles.] Fast and slow vibration directions were next identified and marked on the diagram. Next, an interference figure was obtained and this showed the emergence of an optic axis about half way between the centre and the margin of the field (Fig. 24b), estimated as about 15° from $z$; as the stage was rotated the isogyre rotated and curved as shown, and by comparing the curvature near the melatope when the isogyre was in the 45° position with the chart (Fig. 11), it was estimated that 2V was about 50°. In this same position the sensitive tint plate (length slow) was inserted parallel to the trace of the optic plane. The optical reaction indicated addition (second-order blue) on the concave side of the isogyre adjacent to the melatope, and compensation (first-order orange) on the convex side, indicating that the optic sign was positive (Fig. 24c). Inspection of the orthoscopic image in this orientation showed that the optic axial plane bisected the cleavage angle, i.e. passed through the $z$ axis. Hence, since the mineral is monoclinic, and shows inclined extinction on (010), $\beta$ must coincide with the $y$ axis and the optic axial plane must be oriented parallel to (010).

Next, the extinction angle was determined, and for this a (010) section was found. [This should show the highest interference colour, in this case second-order yellow. Ideally, the presence of a centred flash figure should establish that the crystal is oriented parallel to (010). However, flash figures obtained from crystals in thin section, even if the birefringence is high (0·030 in this case), are usually so broad and diffuse that departures of up to 10° or more of their centres from the centre of the field may not be noticed.]

The only certain way to identify a correctly oriented section was to find a crystal twinned on {100}, with the composition plane vertical and which had, in addition, well developed exsolution lamellae parallel to (001) with sharp boundaries that could be tested to check that the lamellae were vertical. A "herring-bone" twin that shows these features is sketched in Fig. 24(d). As the

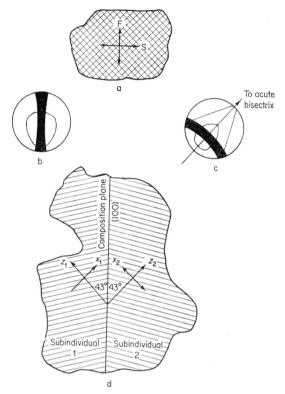

FIG. 24. Optical determinations on a clinopyroxene. (a) Crystal cut normal to the $z$ crystallographic axis, showing vibration directions bisecting the angles subtended by the {110} prismatic cleavages. (b) Interference figure obtained while crystal is in the extinction position; this shows a melatope emerging about one-third of the way between the centre of the field and the S border. This straight isogyre bisecting the field indicates that an optic symmetry plane, in this case the optic axial plane, is N–S and vertical. (c) Interference figure after microscope stage had been rotated clockwise through 45°. (d) Twin crystal of clinopyroxene cut normal to (010), displaying fine (001) exsolution lamellae.

twin is a normal one, the birefringences in the two parts of the twin should match exactly if the composition plane is vertical. Fast and slow vibration directions were sought in each component and the extinction angles measured in each. If the section is correctly oriented the two values should be equal; in this case $\gamma \wedge z = 43°$ and from the dispositions of the (001) planes in the two parts of the twin it can be seen that this was measured in the obtuse angle $\beta$. The orientation of the indicatrix was therefore established and the optical properties of this clinopyroxene could be quoted as: optic plane parallel to (010), positive sign, 2V approximately 50°, $\gamma \wedge z = 43°$ in obtuse $\beta$, birefringence approximately 0·030.

The (010) section could be cut out to measure the refractive indices $n_\alpha$ and $n_\gamma$ and the section normal to [001] would give $n_\beta$ for the vibration normal to the optic plane.

If a Universal Stage had been available it would have been possible to determine 2V and the extinction angle graphically on the crystal cut normal to the $z$ axis and to check the accuracy of the plot with the extinction angles measured on the (010) twin section, secure in the knowledge that if the latter section was not cut fortuitously exactly normal to (010), the orientation could easily be adjusted before the extinction angles were measured. Moreover, the determination of the optical orientation could have been carried out with more certainty and in approximately the same time.

If the pyroxene examined had been pigeonite, which is characterized by a low optic axial angle, the distinction between augite and pigeonite could have been effected by inspection of the interference figures obtained on the section cut normal to the $z$ axis. For augite, with 2V $\sim 50°$, the optic axis would be displaced by about $15°$ (in the crystal) from the centre of the field, whereas for pigeonite (with 2V $< 20°$) it would appear to lie near or just beyond the margin. To determine the orientation of the optic plane in pigeonite a section in the (010) plane cut normal to the acute bisectrix is required. This would show straight extinction to the length of the crystal and very low partial birefringence. An interference figure would establish the approximate size of 2V and whether the optic plane was oriented parallel to the length of the crystal (and to the vague traces of the cleavage planes), or perpendicular to these features.

### E. DETERMINATION OF THE OPTICAL INDICATRIX OF WOLLASTONITE

The determination of the optical indicatrix of the triclinic mineral wollastonite, $CaSiO_3$, affords an excellent example of the difficulties encountered in the study of a triclinic mineral and of the limitations imposed on the information that can be obtained without having recourse to rotation apparatus.

In triclinic minerals the unit cell is now defined by using X-ray data; these show that wollastonite has the following interaxial angles: $\alpha$ 90°02′, $\beta$ 95°22′, $\gamma$ 103°26′. It has a perfect (100) cleavage, good cleavages parallel to (001) and ($\bar{1}$02), is usually elongated parallel to the $y$-axis, and often develops simple twins on $\{100\}$. As the twin axis is [010], these must be Parallel twins.

### 1. Determination with a Standard Petrographic Microscope

In a thin section 0·003 mm thick, a crystal cut normal to the $y$ axis can easily be identified as it shows the intersection of the three cleavages, each appearing as sharp traces that do not appear to move laterally as the focus is lowered. The perfect cleavage intersects the others at angles of $+84\frac{1}{2}°$ and

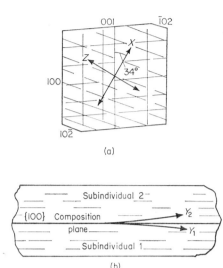

(a)

(b)

FIG. 25. Crystals of wollastonite, CaSiO₃, selected for optical study using standard petrological microscope. (a) Crystal cut normal to [010] showing well developed (100), and (001) cleavages, with a third but poorer cleavage parallel to (Ī02). (b) Parallel twin on (100), with twin axis [010] cut normal to [001].

−70° (Fig. 25a). As the α angle hardly departs from 90°, the trace of the (100) cleavage defines the direction of the z axis but as the γ angle is 103°, the {001} cleavage merely marks the trace of the x axis. Sections cut normal to the y axis display the highest interference colours for the mineral, in this case a first-order orange, and an interference figure obtained from this crystal was a rather indistinct flash figure that appeared to be centred. This confirmed that β was parallel or nearly parallel to [010], that the refractive indices associated with the fast and slow vibration directions were $n_\alpha$ and $n_\gamma$ to a very close approximation, and that the birefringence ($n_\gamma - n_\alpha$) was close to 0·015.

Next, the fast and slow vibration directions were sought by using a sensitive tint plate, and their directions were recorded as shown in Fig. 25(a). From the known value of the crystallographic angle β, the (001) and (Ī02) cleavages were identified and the optic orientation established as shown. The extinction angle determined, α′ ∧ (001) was measured in the acute angle β. Later this crystal was cut out of the thin section and the following refractive indices determined: $n_\alpha = 1\cdot619$; $n_\gamma = 1\cdot634$.

Thus far it had not been possible to determine whether β coincides with the y axis or is inclined to it at a small angle. This could only be established by examining crystals cut in another orientation, preferably in the zone perpendicular to [010]. This zone is not easy to locate, since (010) or prismatic cleavages are absent. Since the optic axial plane must be in or close to this zone, suitably oriented crystals would be expected to display grey interference colours and, depending on their orientation, might show the traces of a cleavage running parallel to the length of the crystal. Examination of a number of such sections revealed small extinction angles varying from zero up to 5°, which suggested that $\beta$ was inclined at a small angle to the $y$ axis, thus establishing the triclinicity. Further than this it is not possible to proceed, except to determine the refractive index $n_\beta$ from a section containing the $y$ axis as 1·631, to identify $\alpha$ as the acute bisectrix, to establish that 2V was moderate and probably about 40°, and to confirm that the optic axial plane was oriented nearly perpendicular to the $y$ axis.

In the very unlikely event of a $\{100\}$ twin being cut fortuitously normal to the $z$ axis it would be possible to confirm that $\beta$ was inclined at about 5° to the $y$ axis in the trace of the acute angle $\gamma$. This section, whose appearance is sketched in Fig. 25(b), could be identified because both components of the twin would exhibit the same interference colour and, moreover, equal but opposite-in-sign extinction angles could be measured.

## 2. Determination with the Universal Stage

The operations here follow the same sequence as with the normal microscope. It would only be possible to carry out all the measurements on the same crystal if a well developed single crystal was used and examined either by using a Waldmann Sphere inserted in place of the central part of the Universal Stage, or if the grain was first mounted on a spindle stage in one orientation, demounted, and remounted in another orientation. If the crystals are homogeneous the same result can be obtained much more easily, as in the clinopyroxene example, by making observations on two or more crystals in different initial orientations on the Universal Stage and assembling the results on a composite stereogram.

First the specimen (in this case a thin section) should be examined with an ordinary petrological microscope in the usual way and suitably oriented crystals selected for further study. In this case two crystals were selected, one a twin crystal (Fig. 26a) cut perpendicular to the $y$ axis and the second an untwinned crystal that gave an off-centred optic axis figure and showed one cleavage which was thought to be $\{100\}$ (Fig. 26b).

Next the thin section was mounted on the Universal Stage, the twin crystal was brought to the centre and the cleavages in each of the two sub-individuals and the composition plane were labelled and sketched (Fig. 26a). The cleavages and the twin composition plane were each set vertical in turn and plotted, as in Fig. 27(a). Their intersection defined the $y$ axis, which appeared to be common to the two sub-individuals within the limits of accuracy of setting the inclinations of the cleavage planes ($\pm 1\frac{1}{2}°$). However, when the $y$ axis was set

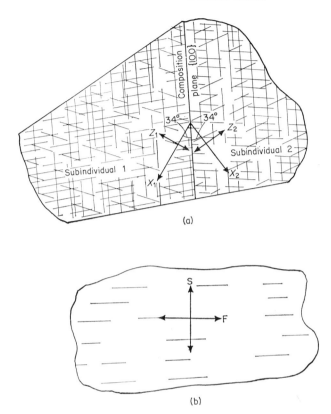

(a)

(b)

FIG. 26. Crystals of wollastonite selected for optical study with the Universal Stage. (a) Twin crystal cut nearly normal to the y-axis; a stereogram of the twin is given in Fig. 27(a). (b) Crystal showing one cleavage. A stereogram of this crystal is given in Fig. 27(b).

vertical, extinction angles measured about $A_5$, and the fast and slow vibration directions had been identified, it was noted that $\alpha'_1 \wedge z$ and $\alpha'_2 \wedge z$, the respective extinction angles in the two parts of the twin, were equal and opposite in sign. From this it was concluded that although (100) was the composition plane, [010] (as reported in text books) was *not* the twin axis. Moreover, when the composition plane, which had already been set vertical was made parallel to the N–S crosswire, the analyser inserted, and the stage rotated about $A_4$, the partial birefringences in the two parts of the twin varied slightly with respect to each other. This suggested that the twin was a parallel one and that the twin axis was [001], which in wollastonite is perpendicular to [010].

After this, two optical symmetry planes were located in each component of the twin and the results plotted as shown in Fig. 27(a). This plot confirms that

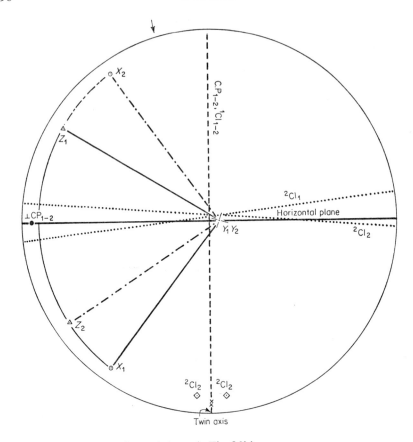

FIG. 27. (a) Stereogram of crystal shown in Fig. 26(a).
$Y_1Y_2$:$A_1$ 172°, $A_2$ 6°R.    $CP_{1-2}$:$A_1$ 168°, $A_2$ 2°R.
$X_1$ :$A_1$ 225°, $A_2$ 86°L.    $2CL_1$ :$A_1$ 263°, $A_2$ 5°R.
$Z_1$ :$A_1$ 133°, $A_2$ 86°L.    $2CL_2$ :$A_1$ 255°, $A_2$ 5°R.
$X_2$ :$A_1$ 112°, $A_2$ 88°L.    Twin Axis: $A_1$ 258°, $A_2$ 89°L.
$Z_2$ :$A_1$ 202°, $A_2$ 85°L.

the twin axis is [001], and as $X_1X_2$, $Z_1Z_2$ all lie in the same great circle which
passes through [001], it follows that $Y_2$ and $Y_2$ must coincide as indeed
must all directions that lie in the zone circle [001]. The angle $Y \wedge [010] = 4°$.
As the angle $X_1X_2$ over the pole of the twin axis is 68°, and as $X$ in each case
lies in the trace of the acute angle $\beta$ (Fig. 26a), it follows that $X \wedge [001] =$
$-34°$.

It is still not known whether $Y$ lies in the trace of the obtuse angle $\gamma$
(between [010] and the face pole (010) ) or in the acute angle $\gamma$, nor has 2V
been determined.

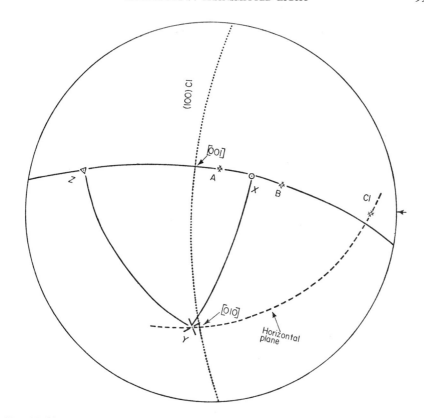

Fig. 27 (b). Stereogram of crystal shown in Fig. 26(b). $Y$:$A_1 = 350°$, $A_2$ $64°L$, $A_4$ $1°N$, $38°S$. $2V_\alpha = 39°$. Cleavage: $A_1$ $270°$, $A_2$ $9°L$.

This can be done quite simply by making the necessary measurements on the second crystal (Fig. 26b). The results are plotted in Fig. 27(b). From this the pole where the optic plane cuts the (100) cleavage can be identified. The great circle with [001] as pole is the horizontal crystallographic plane and both [010] and $Y$ lie on this.

The data plotted on Figs. 27(a) and (b) must now be assembled on a single stereogram (Fig. 28) in the conventional orientation. First the directions of the crystallographic axes and the forms {001}, {100}, and {010} are marked in and then the optic data are transferred.

To do this the stereograms, Figs. 27(a) and (b), are superimposed on the net and rotated with respect to each other until $Y$ of the twinned crystal (Fig. 27a) and $Y$ of Fig. 27(b) are both superimposed on the E–W diameter of the net. The angle between these poles is then measured; it is 68°. The poles of $X$ and $Z$ in the orientation of sub-individual 1, but averaged for the

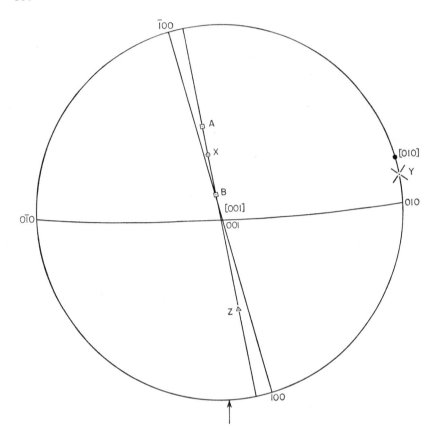

Fig. 28. Composite stereogram incorporating the results plotted in Figs. 27(a) and (b), after the plane of projection had been rotated into the conventional orientation with [001] in the centre and (010) at the E end of the E–W diameter. [001], (100), and [010] located from X-ray data.

two components of the twin, are likewise rotated 68° along their corresponding small circles. In this way the optical and crystallographic data of Fig. 27(a) are rotated into the orientation of Fig. 27(b). This new composite stereogram is then rotated about the centre until the pole [001] comes to lie above the E–W diameter of the net. When it does this it is superimposed on a pole 28° to the left of the centre of the stereogram. All the poles $X$, $Y$, $Z$, A, B, [001] and [010] are now rotated 28° to the right along their corresponding small circles. Finally, the trace is rotated until the pole [010] lies above the point so marked on Fig. 28. All the poles are then transferred to the new plot.

Finally, the directions $X$, $Y$, $Z$, A, and B are defined as polar co-ordinates as in Table 7 (for 010, $\phi = 0°$ and $\rho = 90°$):

TABLE 7

*Polar co-ordinates*

|   | $\rho$ | $\phi$ |
|---|---|---|
| $X$ | 34° | 261° |
| $Y$ | 90° | 271° |
| $Z$ | 56° | 81° |
| A | 50° | 261° |
| B | 11° | 261° |

REFERENCES

Berek, M. (1924). "Mineralbestimmung mit Hilfe der Universal Drehtische-methoden." Gebrüder Borntraeger, Berlin.

Bloss, F. D. (1961). "An Introduction to the Methods of Optical Crystallography." Holt, Reinhard and Winston, New York.

Burns, R. G. (1965). "Electronic Spectra of Silicate Minerals: Application of Crystal-Field Theory to Aspects of Geochemistry." Ph.D. Dissertation, University of California, Berkeley.

Burns, R. G. (1966a). Origin of pleochroism in hypersthene. *Mineralog. Mag.* **35**, 715.

Burns, R. G. (1967). Apparatus for measuring polarized absorption spectra of small crystals. *J. scient. Instrum.* (In press.)

Cherkasov, Yu. A. (1957). "Modern Methods of Mineralogical Investigations of Rocks, Ores and Minerals." Gosgeoltechn., Moscow. (I. Mittin, trans.). *Int. Geol. Rev.* (1960) **2**, 218.

Deer, W. A., Howie, R. A. and Zussman, J. (1962–63). "Rock Forming Minerals." (5 vols.), Longmans Green, London.

Dodge, N. B. (1948). The dark field illumination method. *Am. Miner.* **33**, 541.

Emmons, R. C. (1928). The double dispersion method of mineral determination. *Am. Miner.* **13**, 504.

Emmons, R. C. (1929). The double variation method of refractive index determination. *Am. Miner.* **14**, 415.

Emmons, R. C. (1943). "The Universal Stage." *Mem. geol. Soc. Am.* No. 8.

Emmons, R. C. and Gates, R. M. (1948). The use of Becke line colours in refractive index determination. *Am. Miner.* **33**, 612.

Fisher, D. J. (1958). Refractometer perils. *Am. Miner.* **43**, 777.

Gay, P. (1967). "Introduction to Crystal Optics." Longmans Green, London.

Hallimond, A. F. (1953). "The Polarizing Microscope." 2nd Ed. Cooke, Troughton & Simms, York.

Hallimond, A. F. and Taylor, E. W. (1948). An improved polarizing microscope. III. The slotted ocular and slotted objective. *Mineralog. Mag.* **28**, 296.

Harrison, R. K. and Day, G. (1963). A continuous monochromatic interference filter. *Mineralog. Mag.* **33**, 517.

Hartshorne, N. H. and Stuart, A. (1960). "Crystals and the Polarizing Microscope." 3rd Ed. Arnold, London.

Hartshorne, N. H. and Stuart, A. (1964). "Practical Optical Crystallography." Arnold, London.

E

Hess, H. H. (1952). Orthopyroxenes of Bushveld type, ion substitutions and changes in unit cell dimensions. *Am. J. Sci.*, Bowen Volume, p. 173.

Hurst, V. J., Donnay, J. D. H. and Donnay, G. (1956). Staurolite twinning. *Mineralog. Mag.* **31**, 145.

Iddings, J. P. (1906). "Rock Minerals." Wiley, New York.

Johannsen, A. (1918). "Manual of Petrographic Methods." McGraw-Hill, New York.

Larsen, E. S. and Berman, H. (1934). Microscopic determination of non-opaque minerals. *Bull. U.S. geol. Surv.* No. 848, 2nd Ed.

Merwin, H. and Larsen, E. S. (1912). Mixtures of amorphous sulphur and selenium as immersion media for the determination of high refractive indices with the microscope. *Am. J. Sci.* Ser. 4, **34**, 42.

Meyrowitz, R. and Larsen, E. S. (1951). Immersion liquids of high refractive index. *Am. Miner.* **36**, 746.

Muir, I. D. and Tilley, C. E. (1957). Contributions to the petrology of Hawaiian basalts. I. The picrite basalts of Kilauea. *Am. J. Sci.* **255**, 241.

Naidu, P. R. J. (1958). "4-Axes Universal Stage." (Published privately.) Madras.

Olcott, G. W. (1960). Preparation and use of a gelatine mounting medium for repeated oil immersion of minerals. *Am. Miner.* **45**, 1099.

Phillips, F. C. (1963). "An Introduction to Crystallography." 3rd Ed. Longmans Green, London.

Pillar, H. (1960). "Interferenzeinrichtung für Durchlichtmikroskopie." Carl Zeiss, Oberkochen.

Roy, N. N. (1966). A modified spindle stage permitting the direct determination of 2V. *Am. Miner.* (In press).

Saylor, C. P. (1935). Accuracy of microscopic methods for determining indices by immersion. *J. Res. nat. Bur. Stand.* **15**, 277.

Slemmons, D. B. (1962). "Determination of volcanic and plutonic plagioclases using a three- or four-axis Universal Stage." *Bull. Geol. Soc. Am.* Spec. Publ. **69**, 1.

Suwa, K. and Nagasawa, K. (1961). A correction diagram for measuring an inclination angle with the Universal Stage. *J. Earth Sci.* **9**, 29.

Tröger, W. E. (1959). "Optische Bestimmung der Gesteinbildenden Minerale." Teil I, 3 Aufl. E. Schweizerbart'sche Verlagsbuchhandlung, Stuttgart.

Wahlstrom, E. E. (1962). "Optical Crystallography." 3rd Ed. Wiley, New York.

West, C. D. (1936). Immersion liquids of high refractive index. *Am. Miner.* **21**, 245.

Wilcox, R. E. (1959). Universal Stage accessory for direct determination of the three principle indices of refraction. *Am. Miner.* **44**, 1064.

Wright, F. E. (1910). A new ocular for use with the petrographic microscope. *Am. J. Sci.* Ser. 4, **29**, 415.

Wright, H. D. (1965). The petrographic use of the spindle stage and focal plane screening in the determination of the optical variation of some zoned minerals. *Mineralog. Mag.* **35**, 656.

Wülfing, E. A. (1924). *In* "Mikroskopische Physiographie." Vol. I, part 1. Undersuchsingsmethoden, 5th Ed. (H. Rosenbusch, ed.). Schweizerbart'sche Verlagsbuchhandlung, Stuttgart.

CHAPTER 3

# Microscopy : Reflected Light

## S. H. U. BOWIE

*Atomic Energy Division, Institute of Geological Sciences, London, England*

E*             103

104

# I. Introduction

The systematic study of opaque or "ore" minerals with a reflecting microscope dates from the work of Campbell (1906) who suggested the possibility of identifying minerals in polished section on the basis of colour and etch reactions. Later, Murdoch (1916) made tests with a needle to determine relative hardness and published his well known system of opaque-mineral identification based primarily on colour and scratch hardness. Reflectivity measurements had been made by Drude (1888) and by Försterling (1908) with a spectrometer fitted with rotatable nicols, but it was not until Orcel (1927) and Schneiderhöhn (1928) introduced photoelectric and visual photometer methods of measuring reflectivity that a major advance was made in the quantitative determination of ore minerals. Meanwhile, Davy and Farnham (1920), van der Veen (1925), Schneiderhöhn and Ramdohr (1931) and Short (1931) had developed a whole series of etch tests covering most of the ore minerals. This method of approach culminated in the second edition of Short's "Microscopic Determination of the Ore Minerals" (1940), which was based on etch reactions and microchemical tests, and was widely used until well after the end of the Second World War.

Research carried out by the author (1954) indicated that a system based primarily on the accurate measurement of reflectivity and micro-indentation hardness might make the determination of opaque minerals not only more precise, but quicker than had been possible previously. The difficulties and ambiguities associated with scratch hardness, etch reactions and microchemical tests were largely avoided, and readily observable ancillary properties, such as colour, anisotropism, polarization colours and bireflection were used to differentiate between minerals of similar reflectivity and hardness. This work (Bowie and Taylor, 1958) paved the way for the introduction of quantitative tables for ore mineral identification based on reflectivity and micro-indentation hardness data.

# II. Mounting and Polishing Specimens

The first and one of the most important steps in the study of opaque minerals in reflected light is the preparation of a flat polished surface. Because of the very marked difference in the hardness of mineral species, this is usually a much more difficult task than is the preparation of metallurgical specimens. Much research has therefore gone into the perfection of methods of producing a surface that is flat, and at the same time, free of pits and scratches. It is also important to ensure that in the process of mounting and polishing the specimen changes do not take place in any of the mineral constituents as a result of heat or pressure. For this reason thermo-setting plastics which require temperatures of up to 120°C and pressures of the order 5–10 tons/sq. in.

for polymerization to take place, should be avoided. Instead, any of the modern cold-setting plastics or epoxy resins that are not too soft can be used. A mount that is too soft will pick up abrasive from the grinding lap and will tend to coat the lap with plastic, thus reducing its effectiveness.

The specimen is normally cut with a diamond saw lubricated by a solution of soluble oil in water to prevent over heating. If the slice obtained is friable or porous, the surface to be polished should be covered with a liquid plastic of similar type to that used for the mould and left in a partial vacuum until hard. This surface should then be ground flat on a 44/60 grade (353–251 $\mu$m) diamond-impregnated lap or on a cast-iron lap with silicon carbide abrasive. The edges of the slice should next be trimmed and bevelled so that it becomes firmly fixed in the mould. Specimens should always occupy as much of the surface area of the mould as possible.

### A. SPECIMEN MOUNTS

Numerous types and shapes of mould are available for mounting single specimens or several at a time. These are sometimes of positive flash type (Fig. 1) or a simple metal or polythene ring that is sealed to a glass plate

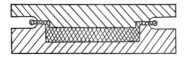

Fig. 1. Positive flash mould designed to produce mounts with front and back surfaces parallel. The depth to which the piston projects into the body of the mould is exaggerated for clarity as is the clearance between the piston and body. Both should be 3–5 thousandths of an inch.

with silicone grease and fixed in position with adhesive tape. It is preferable to use the positive flash type mould, since this produces a mount with the front and back surfaces parallel. With other types of mount the back usually has to be machined; alternatively, the two surfaces can be made parallel by using Plasticine and some form of press. Whatever kind of mould is used, the prepared specimen slice is placed in it ground-side downwards and the mould filled with plastic following the particular manufacturer's instructions. Mounting media that show a tendency to frothing should not be used. However, if frothing does occur, the mould can be placed in a partially evacuated desiccator to remove air bubbles. The specimen is usually labelled at this stage by inserting a paper label with the name or number of the specimen; or, alternatively, the number can be stamped on when the mould is hard. Most "cold-setting" plastics will polymerize at room temperature and it is strongly recommended that the process should not be speeded up by raising the temperature to 70–80°C, as is often practised, unless it is known that the specimen does not contain any temperature-sensitive mineral. Addition of too much accelerator should also be avoided as this causes the plastic to heat up and cracking may result.

### B. GRINDING AND POLISHING

If the mould does not produce a mount with a bevelled edge, the edge should be bevelled on a grinding wheel before grinding and polishing commences. This precaution can prevent serious damage to the lead polishing laps should they become too dry. Also, any flaws or cavities in the surface of the specimen should be filled with plastic at this stage.

Numerous methods of grinding and polishing ore specimens are described in the literature and each method has its advocates. Some, in the opinion of the writer, are not suitable for the preparation of specimens of the quality necessary for quantitative measurements to be made, nor are they adequate for bringing out details in structure and textural relationships (Fig. 2). In the experience of the author, the Graton-Vanderwilt type of machine described by Short (1940) is basically the most satisfactory. A somewhat modified Graton-Vanderwilt machine incorporating some improvements suggested by Phillips (1937) is at present used in the Atomic Energy Division laboratories (Fig. 3). The machine has two rotating laps and rotating specimen heads, mounted over the laps, each capable of carrying three specimens. The laps and rotating heads are driven by separate motors permitting the laps to rotate clockwise at about 80 rev/min, and the heads to rotate anti-clockwise at about 150 rev/min. The heads are fitted to the drive spindles through universal joints which ensure that the specimens always maintain contact with the lap. Two types of head are used: one in which the specimen is held in position by an axial spigot, the other in which the specimens are held in circular holes cut in a steel sheet. The laps are usually made of cast-iron, copper or lead and are interchangeable. The cast-iron and copper laps are grooved radially as well as concentrically, whereas the lead laps are concentrically grooved only. The lead must be of electrolytic quality and the surface kept flat by occasional machining.

Preparation of the specimen for polishing differs somewhat for different minerals, but generally the procedure is to grind the specimen on a 100/200 grade (152/124 $\mu$m) diamond-impregnated lap or to go straight to the cast-iron lap with an abrasive paste of 5 $\mu$m alumina in a lubricant consisting of one part of kerosene to one part of liquid paraffin. The lap is first cleaned with lint soaked in benzene, or other cleaning fluid, and the paste spread smoothly onto the lap with the fingers. The excess is rubbed off until an even film remains. The machine is then run with a load of 15 lb, until the specimens begin to show signs of sticking to the lap. The lap is cleaned and recharged and the process repeated four or five times. Care at this stage will ensure that the specimens have a matt surface free from deep scratches and that no damage is caused to the minerals by shattering. If most of the constituent

FIG. 2. Blades of specular hematite (white) in hisingerite (dark grey) with tracery of pitch-blende (grey) surrounding the hematite crystals. ($\times 1390$)

minerals are hard or very hard, further grinding with 5/20 grade alumina may be necessary, but usually the next stage is to grind as before with 3/50 alumina until a fine matt surface is produced (Fig. 4).

For normal identification and for textural studies, polishing can be completed in a minute or two with a lead lap impregnated with 0–1 $\mu$m diamond dust. The procedure is to moisten the lead lap with a lubricant consisting of

E**

Fig. 3. Modified Graton-Vanderbilt polishing machine used in Atomic Energy Division
laboratories.

kerosene, liquid paraffin and lard oil in the proportion of 20:10:1 and to
work a small amount of diamond dust (one carat is sufficient for about 20
runs) into the surface with the fingers. The lap is then run with the specimens
in position until they begin to show signs of sticking; then the lap is moistened
and thoroughly cleaned with benzene-soaked lint while still running. The
machine is kept running for a further 30 sec, after which time polishing is
complete. In some cases this process may require to be repeated two or three
times (Fig. 5). After a time the lap becomes impregnated with diamond dust
and less is required than for the first few charges. This technique is much more
rapid than that described by Short and produces flat surfaces with a minimum
of Beilby-layer formation.

Soft minerals, such as galena or argentite, do not respond well to this
method of polishing, as they usually finish with a number of fine scratches on
the surface. However, it is relatively easy to finish soft minerals with magnesium
oxide using a lead lap and the same polishing medium as used with the diamond dust. The procedure of polishing is very much the same as with diamond
dust. The lap is first cleaned thoroughly with benzene-soaked lint and a
small quantity of a smooth paste of magnesium oxide and lubricant is spread
on the surface. Three to six runs are sufficient, depending on the hardness of
the constituent minerals, to produce good results with only slight relief
between the hardest and softest of minerals.

The so-called "speedy" methods of grinding and polishing, which involve
heavy pressure and various cloth-covered laps and diamond abrasives for
finishing, are not to be recommended if quantitative measurements are con-

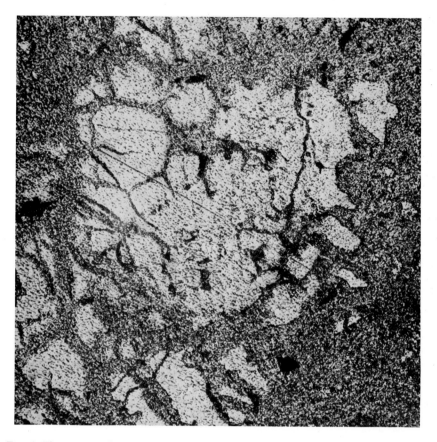

Fig. 4. Fine matt surface produced after grinding with 3/50 grade alumina. Pitchblende (light grey), quartz (dark grey). (×170)

templated, since this type of polishing tends to produce mineral grains with curved surfaces, thick surface flow layers and poor inter-grain definition. The method of Hallimond (1963), which appears to be free from these defects, is an inexpensive method and produces good results in a relatively short time, though the sueded cloth finishing may produce some relief.

Thinned polished sections (Donnay, 1930) can be made by using the technique outlined for normal polished sections by first mounting the specimen and then slicing the mount with a diamond saw. The slice is mounted, with the polished side downwards, on a cover-glass with a liquid thermo-setting plastic of suitable refractive index or an epoxy-resin. This in turn is mounted on a glass slide with slightly "under-cooked" Canada balsam. The section is then treated as a thin section and ground down to normal thickness. Slight heating

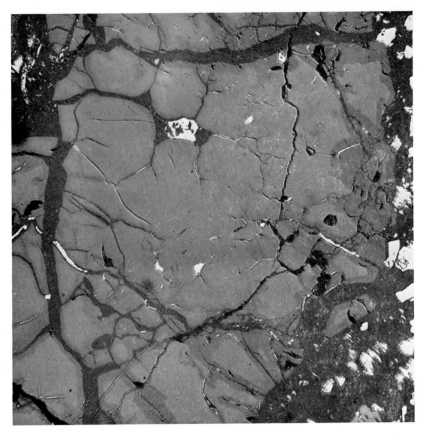

FIG. 5. Same surface as shown in Fig. 4 after being polished for 90 sec on lead lap with 0–1 μm diamond dust. Pitchblende (grey), quartz (dark grey) and pyrite (white). (× 170)

removes the cover-glass from the supporting slide. The cover-glass and attached section are then cleaned and mounted in the normal way. The resulting covered, thinned polished section is best viewed with oil-immersion objectives. It is also possible to prepare polished thin sections (Kennedy, 1945) with the upper surface left uncovered. These are more difficult to prepare than thinned polished sections, and, as they are uncovered, they tarnish fairly readily and are difficult to repolish.

### III. Theory of Reflected Light

As the theory of reflected light has been treated by Hallimond (1953) and more recently by Cameron (1961), it will only be discussed here insofar as it is relevant to the quantitative methods described.

### A. SPECULAR REFLECTIVITY

Most of the light reflected from the surface of a well polished mineral obeys the laws of specular reflection. However, as no real surface is perfectly specular, some light is diffused. Hence specular reflectivity cannot be quoted in absolute terms, but is defined as the ratio of the intensity of light reflected within a small solid angle $\delta r$ about the reflection angle, $r$, to the intensity of the incident beam. This means that the specular reflectivity value is dependent on the angle of acceptance of the photometer (Fig. 6). The direct measure-

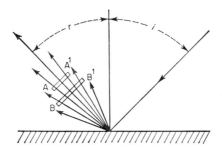

FIG. 6. Diagrammatic illustration of reflection of light from a polished surface. Because of diffusion of the reflected light, a photo-cell at $AA^1$ will register a smaller reflectivity value than one of larger area in position $BB^1$. Reproduced with permission from Bowie and Henry (1964).

ment of specular reflectivity at near-normal incidence is usually made with an integrating-sphere photometer (Fig. 7). In this apparatus light from a stabilized source passes through the lenses, L, illuminating the apertures of adjustable diaphragms, D. With the specimen out of the way, the beam passes into the integrating sphere I in position A, via the prism R and is diffused by the screen E. The light intensity is measured by the photocell, P.

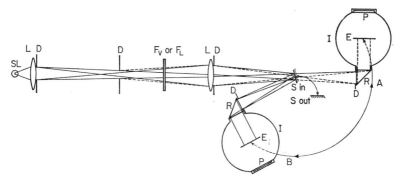

FIG. 7. Integrating-sphere photometer used for the direct measurement of reflectivity at near-normal incidence. Reproduced with permission from Bowie and Henry (1964).

The sphere assembly is then turned to position B and the specimen moved to receive the incident light so that the reflected beam enters the sphere in the same manner as before and the light intensity is again measured.

### B. REFLECTION OF LIGHT FROM MINERALS

When light passes from one transparent medium to another, some of the light is reflected. For an isotropic mineral at normal incidence the reflectivity, $R$, is related to the refractive index of the mineral, $n$, and the refractive index of the immersion medium, $n_0$, by the Fresnel equation:

$$R = \frac{(n - n_0)^2}{(n + n_0)^2}$$

If the mineral is not transparent, the reflectivity is also dependent on the absorption coefficient, $k$,[†] the Fresnel equation then becoming modified to:

$$R = \frac{(n - n_0)^2 + k^2}{(n + n_0)^2 + k^2}$$

In the case of uniaxial minerals, similar relationships hold for the reflection of plane polarized light vibrating respectively perpendicular to and parallel to the $c$ axis.

In the case of bireflecting minerals, similar relationships hold for light vibrating along the principal directions:

$$R_1 = \frac{(n_1 - n_0) + k_1^2}{(n_1 + n_0) + k_1^2} \text{ and } R_2 = \frac{(n_2 - n_0) + k_2^2}{(n_2 + n_0) + k_2^2}$$

Maximum or minimum reflection is obtained when the vibrating direction of the incident light coincides with one or the other principal directions, but a given section may not exhibit the maximum bireflection possible for the mineral because it is not cut in the most suitable direction.

The maximum bireflection of a uniaxial mineral is shown by sections cut parallel to the $c$ axis. For minerals of lower symmetry there is only a single section exhibiting the maximum possible bireflection. In orthorhombic minerals, this is parallel to one pair of crystallographic axes. In monoclinic minerals, it is either perpendicular or parallel to the $b$ axis. In triclinic minerals it is independent of all three crystallographic axes.

The principle refractive indices and absorption coefficients of an absorbing mineral can be determined by measuring the reflectivities of suitably orientated sections in both air and oil. For an isotropic mineral the expressions derived from the Fresnel equation are:

[†] Sometimes the absorption index $\kappa$ is used instead of $k$, the relationship being $k = n\kappa$.

$$n = \frac{\frac{1}{2}(n_0{}^2 - 1)(1 - R)(1 - R_0)}{n_0(1 - R)(1 + R_0) - (1 + R)(1 - R_0)}$$

$$k = \sqrt{\frac{R(n + 1)^2 - (n - 1)^2}{1 - R}}$$

where $n_0$ = refractive index of oil, $R$ = reflectivity of the mineral in air, $R_0$ = reflectivity of the mineral in oil. The errors in $n$ and $k$ occasioned by small errors in the reflectivity measurements are very considerable in some circumstances (Piller and von Gehlen, 1964). For identification purposes, there is little to be gained by deriving $n$ and $k$, though it is useful to measure $R_0$ as well as $R$, as described in a subsequent Section.

## C. ROTATION ANGLE AND PHASE DIFFERENCE

For an opaque anisotropic mineral in the 45° position midway between two positions of extinction with crossed polars, the incident plane-polarized ray, OE, is resolved into two components, $OE_1$ and $OE_2$, which correspond to the principal directions of the mineral (Fig. 8). The components of the reflected ray, $OR_1$ and $OR_2$, differ in phase and combine to give an ellipse of vibration

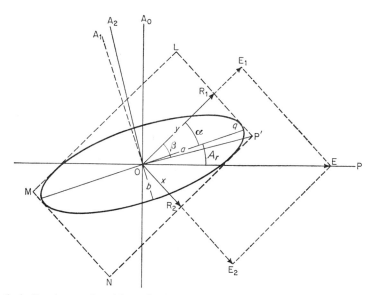

FIG. 8. Elliptical vibration produced by reflection of incident linear vibration OE by an anisotropic mineral at a 45° position. $OE_1$ and $OE_2$ are components of OE parallel to the vibration directions of the mineral; $OR_1$ and $OR_2$ are the corresponding reflected components. OP and $OA_0$ are planes of vibration of the polarizer and analyser (crossed position). $A_1O$ is the vibration plane of the analyser set $\perp$ to $Oq$. $A_2O$ is $\perp$ to OP'. Reproduced with permission from Cameron (1961); also from Cameron (1957).

with semi-axes $a$, along $Oq$, and $b$ at right angles to it. The angle $POq$ is the rotation angle ($A_r$). The components of the reflected ray define the rectangle LMNP′, and the half-diagonal of this rectangle, OP′, gives the direction and amplitude of the linear vibration that would be the resultant of $OR_1$ and $OR_2$ if these components were in phase. Then if the analyser were rotated from position $A_0$ to $A_2 \perp$ OP′, the linear vibration would be extinguished. In general, however, the components of the reflected ray are not in phase, so that the light is elliptically polarized. Then rotation of the analyser from $A_0$ to $A_2$ produces minimum illumination rather than complete extinction.

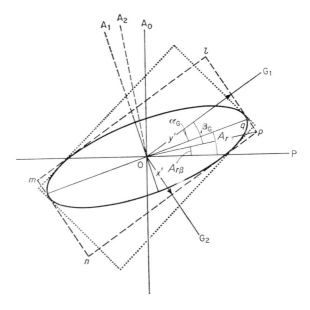

FIG. 9. Relationships at compensation for a right ellipse of vibration. The dotted rectangle is rectangle LMNP′ of Fig. 8. $OG_1$ and $OG_2$ are the slow and fast directions of the mica compensator. Reproduced with permission from Cameron (1961); also from Cameron (1957).

The difference in phase ($\Delta_{x,y}$) between the components $OR_1$ and $OR_2$, the ellipticity ($\delta = \tan^{-1} b/a$), and the angles $\alpha$ and $\beta$ between $OR_1$ and, respectively, the major axis of the ellipse and the diagonal of the rectangle defined by $OR_1$ and $OR_2$, are related by the following equations:

$$\tan 2\alpha = \tan 2\beta \cos \Delta_{x,y}$$
$$\cos 2\beta = \cos 2\delta \cos 2\alpha$$
$$\tan 2\delta = \tan \Delta_{x,y} \sin 2\alpha$$
$$\sin 2\delta = \sin 2\beta \sin \Delta_{x,y}$$

It can be shown that twice the ellipticity, $2\delta$, is equal to the phase difference between the two components of the ellipse of vibration when they are at 45° to the major axis of the ellipse. In this position, $\alpha = \beta = 45°$. The phase difference $2\delta$ is determined by inserting a mica compensator (Fig. 9) and rotating both compensator and analyser together till maximum darkness is produced, or until the polarization cross seen conoscopically is as black as possible.

The maximum value of the rotation angle, $A_r$, through which the analyser must be rotated for extinction or minimum illumination varies with the mineral species, but as it seldom amounts to more than a few degrees its use in mineral identification is limited. The related property of bireflection can usually be determined more precisely. Again, the maximum value of the phase difference, $2\delta$, does not in general vary sufficiently from species to species, in relation to the precision with which it can be determined, for it to be of more than occasional use in identification.

### D. ROTATION SENSE

When a mineral is in the 45° position, the direction in which the analyser has to be rotated to give minimum illumination is known as the rotation sense. Rotation sense is only meaningful, however, when referred to a recognizable direction in the mineral. Suppose a mineral with a basal cleavage exhibits straight extinction between crossed polars and that the grain under observation has the cleavage parallel to the vibration direction of the polarizer. When the stage is turned anti-clockwise through 45° the grain becomes illuminated. If the illumination is then minimized by rotating the analyser anti-clockwise, the rotation sense is said to be positive with respect to the cleavage. The rotation with respect to the $c$ axis (normal to the cleavage) would be of opposite sense, that is, negative. As the rotation sense may change with wavelength, it should be determined in monochromatic light and the wavelength specified.

### E. SIGN OF PHASE DIFFERENCE

For uniaxial minerals, the sign of the phase difference between the components $OR_1$ and $OR_2$ (Fig. 8) can be determined in one of the 45° positions in which the mineral gives an anti-clockwise rotation. The phase difference is said to be positive if $x$ is the fast component and $y$ the slow component. Thus if a "fast-length" gypsum plate is inserted in a NW–SE direction, the phase difference due to the mineral will be added to that of the plate if the phase difference is positive; and if negative it will be subtracted.

The sign of the phase difference is the same for all orientations in the case of uniaxial minerals, but for minerals of lower symmetry it may differ for different orientations.

## IV. Measurement of Reflectivity

It is more than 50 years now since the relative brightness of two adjacent mineral grains was used as an aid to the identification of opaque minerals. However, quantitative techniques were not available until Orcel (1927) used a potassium–silver photo-emissive cell in much the same way as the modern selenium barrier-layer cell is now employed. A visual photometer designed by Berek was put on the market by E. Leitz in 1930. Later Hallimond (1953) described an improved visual photometer with three instead of two polariz-ing discs. The latter instrument, when used carefully, is capable of detecting differences of about 2 % in the reflectivity of two adjacent areas. The author (1957) described a simple selenium barrier-layer photometer from which

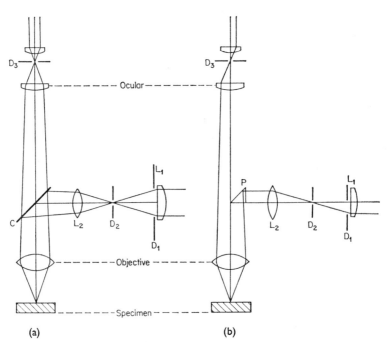

(a)                            (b)

Fig. 10. Two types of reflector used in ore microscopes: (a) whole-field cover-glass (C) illuminator; (b) half-field prism (P) reflector. $L_1$ is the lamp lens and $L_2$ the illuminator lens. $D_1$ is the aperture diaphragm (sometimes termed the lamp diaphragm or condenser aperture diaphragm), $D_2$ is the field diaphragm and $D_3$ the ocular diaphragm. Reproduced with permission from Bowie and Henry (1964).

more sophisticated instruments have been developed (Bowie and Henry, 1964). Photomultiplier-tube photometers are now in use in several laboratories for measuring the reflectivity of small mineral grains as well as for studying coal constituents (Murchison, 1964).

## A. REFLECTING MICROSCOPES

Two main types of reflecting unit are used in modern ore microscopes. In one, the beam entering the vertical illuminator is reflected down onto the specimen by a coated cover-glass; in the other, a total-reflection prism is used (Fig. 10). With the former type of illuminator the incident light is normal, but with the latter it falls somewhat obliquely on the object. An improved reflecting unit of the cover-glass type has recently been described by Smith (1964). In this illuminator light falls on a metallic surface mirror at the back of the unit and is reflected upwards onto a coated cover-glass which directs the beam down onto the specimen (Fig. 11). An important advantage of this design is that it gives almost homogeneous extinction over the whole of the field.

For quantitative reflectivity work, the microscope should have coated strain-free objectives which should be kept as clean as possible. The microscope should be fitted with an ocular diaphragm, or series of stops, that can be made to coincide with the area delineated by the field iris. The lamp should be external and fitted with an iris diaphragm. The light source should be of high intensity, preferably a halogen-vapour-filled tungsten-filament lamp which has an output of 100 watts at 12 volts. This permits adequate light for spectral reflectivity measurements in the wavelength range 400–700 nm.†
Whether microscopes are fitted with cover-glass or prism illuminators, care should be taken to ensure that the vibration direction in the polarizer coincides with the symmetry plane of the illuminator, otherwise there will be appreciable loss of light.

### B. MICROSCOPE CONDITIONS FOR REFLECTIVITY MEASUREMENTS

As all microscope objectives are convergent, it is only the axial ray illuminating the centre of the specimen that is normally incident. For all other rays the proportion of light reflected is dependent on the angle of incidence. The departure from normal incidence increases with the convergence of the

TABLE 1

*Variation in measured reflectivity of pyrite with numerical aperture; field and aperture diaphragm settings constant*

| Numerical aperture | Pyrite reflectivity, %* |
|---|---|
| 0·15 | 54·3 |
| 0·28 | 53·3 |
| 0·45 | 53·8 |
| 0·85 | 54·9 |

* "White" plane-polarized light (2850°K) compared with carborundum standard at 20·2%.
† nm = nanometre; $10^{-9}$ metre.

F

objective; thus, as might be expected, there will be differences in the reflectivity value depending on the numerical aperture (n.a.) of the objective used.
However, these will be relatively small unless there is a marked difference
between the reflectivity of the mineral and the standard with which it is compared. The apparent increase in reflectivity of pyrite (Table 1) is directly proportional to increasing n.a., 0·28 to 0·85, within the accuracy of the measurements. The objective with n.a. 0·15 is anomalous because it is different in
construction from the other three objectives.

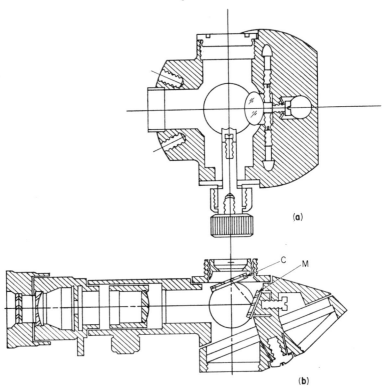

(a)

C  M

(b)

FIG. 11. Smith illuminator: (a) plan showing the two Allen screws that control the E–W
movement of the mirror; (b) side elevation showing reflector tube with revolving objective holder. M is a metallic surface mirror and C a coated cover-glass. Reproduced with
permission from Hallimond and Payne (1964).

Increase in numerical aperture also results in greater divergence in the
angle between the vibration of the polarizer and the symmetry plane of the
vertical illuminator. This resulted in Berek (1937) recommending that measurements of reflectivity should be made with the polarizer in the 45° position.
Recent tests with a cover-glass illuminator do not confirm Berek's findings
for a comparison of the reflectivities of pyrite and carborundum (Table 2).

TABLE 2

*Variation in measured reflectivity of pyrite with numerical aperture; polarizer in 0° and 45° positions, field and aperture diaphragm settings constant*

| Numerical aperture | Pyrite reflectivity, %* | |
| --- | --- | --- |
| | 0° position | 45° position |
| 0·15 | 54·3 | 54·5 |
| 0·28 | 53·3 | 55·4 |
| 0·45 | 53·8 | 54·6 |
| 0·85 | 54·9 | 55·8 |

* "White" plane-polarized light (2850°K) compared with carborundum standard at 20·2 %.

From the above results there is clearly no advantage of using the 45° position. It is therefore recommended that reflectivity measurements are made with the polarizer in the 0° position and that the reflectivity values are determined with the mineral at each of the two extinction positions 90° apart—the principal directions.

The reflectivity obtained when comparing a highly reflecting unknown with a standard of appreciably lower reflectivity will apparently increase with field diameter (Table 3).

TABLE 3

*Variation in measured reflectivity with increasing field diameter; numerical aperture (0·28) and aperture diaphragm setting constant*

| Field diameter, μm | Pyrite reflectivity, %* |
| --- | --- |
| 385 | 51·6 |
| 460 | 53·8 |
| 625 | 54·5 |

* "White" light (2850°K) compared with carborundum at 20·2 %.

In addition to the above discrepancies, minor variations are caused by changes in the setting of the aperture diaphragm (Table 4). In obtaining the results given in Tables 1 to 4 a correction was made for "primary glare" (mainly due to light reflected by the back lens of the objective) by placing a black-box beneath the objective and subtracting the galvanometer reading from the values obtained for both carborundum and pyrite. This correction, however, does not take into account "secondary glare", which is proportional to the reflectivity of the specimen, and is due mainly to

TABLE 4

*Variations in measured reflectivity of pyrite with aperture diaphragm setting;
numerical aperture (0·28) and field diaphragm setting constant*

| Approximate proportion of back lens of objective illuminated | Pyrite reflectivity, %* |
|:---:|:---:|
| 0·75 | 54·1 |
| 0·5 | 54·2 |
| 0·25 | 55·1 |

\* "White" light (2850°K) compared with carborundum at 20·2 per cent.

the reflection and scattering of light at glass surfaces in the optical system.
Secondary glare varies with the type and construction of the illuminating
unit as well as with the wavelength of the light used (Table 5).

TABLE 5

*Variation in measured reflectivity of chromium with cover-glass reflector units
of different construction; aperture, field and ocular diaphragm settings similar*

(a) Reflector unit with 2 collimating lenses, 0·28 n.a. bloomed objective

| Wavelength, nm | Carborundum standard $R\%$ N.P.L. value | Chromium $R\%$ N.P.L. value | Chromium $R\%$ by microscope |
|:---:|:---:|:---:|:---:|
| 670 | 19·9 | 69·0 | 82·0 |
| 460 | 21·0 | 59·8 | 64·3 |

(b) Reflector unit with 1 collimating lens, 0·28 n.a. bloomed objective

| Wavelength, nm | Carborundum standard $R\%$ N.P.L. value | Chromium $R\%$ N.P.L. value | Chromium $R\%$ by microscope |
|:---:|:---:|:---:|:---:|
| 670 | 19·9 | 69·0 | 70·5 |
| 460 | 21·0 | 59·8 | 62·0 |

With carborundum as standard and with a microscope not fitted with an
ocular iris diaphragm the error due to glare in measuring a high-reflectivity
unknown is substantial (curve (a), Fig. 12). However, if an ocular iris is fitted
and made coincident with the field iris, the glare error is greatly reduced
(curve (b), Fig. 12). The error can be further reduced by using Köhler illumination and stopping the lamp diaphragm down until about half of the back lens
of the objective is illuminated. (This normally coincides with the area occupied
by the image of the lamp filament.) Stopping down too far results in a marked

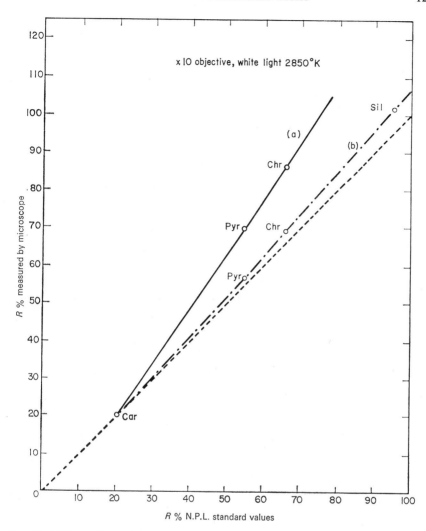

FIG. 12. (a) Curve illustrating the effect of secondary glare for a large field area on values of reflectivity measured with a microscope not fitted with ocular diaphragm; (b) Curve obtained for same field area but with ocular iris fitted and adjusted to coincide with the area delineated by the field diaphragm. Car = carborundum standard, Pyr = pyrite, Chr = chromium and Sil = silver.

reduction in illumination and a relative increase in primary glare. The field iris should also be stopped down as far as possible and the ocular iris at all times kept coincident with it. If these precautions are taken the error due to secondary glare even when comparing silver with a carborundum standard is usually less than 5 %.

## C. PHOTOMETERS

Photoelectric photometers are now widely used for the measurement of reflectivity. Two main types are available, one based on a photomultiplier tube, the other on a selenium barrier-layer cell. Both have been used successfully in measuring the reflectivity of ore minerals and have their advantages and disadvantages. Photomultiplier tubes enable the primary current to be amplified by a factor of $10^7$ or more, but they generally have to be specially selected otherwise they exhibit fairly marked fatigue effects unless kept switched on. They tend to have a greater sensitivity towards the blue end of the spectrum, although tubes with a more uniform spectral response are now available. The selenium-cell photometer, however, is relatively insensitive to a small light flux and has to be amplified by some other means or another. So far the most stable amplifier available (Bowie and Henry, 1964) is one in which the output from the cell is fed into a primary galvanometer. This has an intense light source so arranged that the spot falls on two large selenium cells connected in opposition. The current generated when there is uneven illumination of the two cells is in turn fed into a secondary spot-galvanometer which is read by the observer. The sensitivity achieved with this arrangement is about twenty times that obtainable with an unamplified selenium-cell photometer. In addition, a high degree of stability and linearity is obtained.

Silicon and cadmium sulphide photocells have been tested for reflectivity measurement, but both have considerably greater sensitivity at the red end of the spectrum than at the blue end; and the cadmium sulphide cell has the disadvantages of a slow rate of response and marked fatigue. Use of either of these cells with "white" light will result in values for dispersive minerals that will be at variance with those obtained by visual photometer or selenium barrier-layer cell. Similarly, photomultiplier tubes with high sensitivity at the blue end of the spectrum will not give comparable values. This is illustrated below for galena and gold with the same tungsten-filament lamp operating at a colour temperature of 2850°K and taking pyrite as standard with $R = 54 \cdot 5 \%$. The values for the selenium-cell photometer are taken as standard, since these approximate closely to values obtained by visual photometer.

|        | Selenium Cell | Photomultiplier Tube | Silicon Cell |
|--------|---------------|----------------------|--------------|
| Galena | 43·2          | 45·0                 | 39·5         |
| Gold   | 73·5          | 59·0                 | 89·0         |

The high value for galena and low value for gold with the photomultiplier tube results from the greater sensitivity of the photomultiplier tube at the blue end of the spectrum, whereas the low value for galena and the high value for gold with the silicon cell results from the greater sensitivity of the cell at the red end of the spectrum.

On the basis of several years experience with both types of photometer, the following appear to be the main advantages and disadvantages of instruments so far developed:

| Amplified Selenium Cell | Photomultiplier Tube |
|---|---|
| 1. Relatively insensitive; not suitable for spectral reflectivity measurements on grains less than 30 $\mu$m. | 1. Highly sensitive; can give spectral data on grains down to 1 $\mu$m. |
| 2. Shows little tendency to fatigue at light fluxes used in reflectivity measurement. | 2. Tubes have to be specially selected for minimum fatigue, otherwise tube has to be left switched on. |
| 3. Relatively inexpensive. | 3. Requires specially designed optical system for measurements of small grains and therefore expensive. |
| 4. Suitable for routine use and teaching purposes. | 4. Suitable for research work. |
| 5. Not suitable for spectral reflectivity measurements at half-height widths of less than 10–20 nm. | 5. Can be used with narrow-band interference filters with half-height widths of 10 nm or less. |

### D. CALIBRATION OF APPARATUS

Before the photometer can be calibrated it is essential to ensure that the light source is stable. This can be achieved by using either a transistorized stabilizer or a continuously charged heavy-duty accumulator. The next step is to test whether the combined response of the cell and galvanometer is linear. This can be done in three ways without the use of reflectivity standards. First, a highly reflecting mineral is placed on the stage and the field iris opened until a reading of 100 is obtained on the galvanometer. Discs from which $\frac{1}{8}$, $\frac{1}{4}$, $\frac{1}{2}$ and $\frac{3}{4}$, or similar, sectors have been cut are then inserted in turn in the cell holder below the sensitive surface and the readings noted. The readings should be proportional to the sector openings. The second test involves the use of a series of accurately calibrated neutral density filters, which are placed in front of the lamp holder after the galvanometer has been set up to read 100% with no filter in position. The filters used should be uniform, perfectly clean and inserted normal to the light beam. The third method of testing for linearity is to use one neutral density filter only with a transmission factor of about 75%. As before, the galvanometer is set to read 100% without the filter; the filter is then inserted, the reading noted, and the galvanometer re-set to this value without the filter in position. Thus if the first reading with the filter is 75%, the galvanometer is set to 75% without the filter and the filter re-inserted, when the new values should be 56·25%. This procedure is repeated, yielding successively lower values. A filter with a fairly high transmission factor should be chosen to give information on the upper portion of the calibration curve.

## E. CHOICE OF REFLECTIVITY STANDARDS

Because of the variations in reflectivity due to microscope conditions already discussed, it is necessary for accurate work to have a series of reflectivity standards to cover the range of reflectivity found in ore minerals. Ideally, standards should be isotropic, have low dispersion and be free from inclusions, cleavage and cracks. They should also be hard and resistant to tarnish and should preferably be a plane crystal face not requiring polishing. Two types of standard are used at present, one being transparent and the other absorbing. Transparent materials have a reflectivity up to about 20% and, if it is assumed that absorption is negligible, the reflectivity value can be calculated from the refractive index. Diamond, specially cut and mounted so as to eliminate reflections from the lower surface, has been used successfully as a standard for reflectivity measurements at low magnifications. However, with high-power objectives it is difficult to exclude reflections from the lower surface, and the presence of local internal reflections, caused by minute flaws and incipient cleavage, results in variations in reflectivity over small areas. The most satisfactory standard so far available is carborundum. This grows in plates parallel to the basal plane, and, as the crystals are uni-axial, the surface is mono-reflecting. Reflectivity measurements made by the direct method at the National Physical Laboratory (N.P.L.) are given in Table 6.

TABLE 6

*Spectral reflectivity of basal plate of carborundum*

| nm | $R\%$ | nm | $R\%$ |
|-----|------|-------|------|
| 400 | 21·8 | 546·1 | 20·4 |
| 420 | 21·5 | 589·3 | 20·2 |
| 440 | 21·2 | 620 | 20·0 |
| 460 | 21·0 | 660 | 19·9 |
| 480 | 20·8 | 700 | 19·8 |
| 500 | 20·6 | | |

Pyrite has been used for some time as a standard for "white" light measurements. The mean reflectivity value of eighteen specimens from various localities determined with a selenium-cell photometer and tungsten-filament lamp operating at 2850°K was shown to be 54·5 with a variation of less than 0·5 units about the mean; carborundum was used as the standard. All pyrite specimens, however, do not show the same dispersion of reflectivity. Hence it is necessary to have each specimen checked by the direct method if it is to be used as a reflectivity standard. Two specimens measured by N.P.L. gave the spectral values indicated in Table 7.

TABLE 7

*Spectral reflectivity of two specimens of pyrite*

| nm | $R\%$ | | nm | $R\%$ | |
|---|---|---|---|---|---|
| | (a) | (b) | | (a) | (b) |
| 400 | 37·6 | 37·8 | 546·1 | 53·7 | 53·2 |
| 420 | 39·5 | 39·6 | 589·3 | 55·0 | 54·4 |
| 440 | 42·0 | 42·1 | 620 | 55·2 | 54·7 |
| 460 | 45·0 | 45·1 | 660 | 56·1 | 55·6 |
| 480 | 48·2 | 48·1 | 700 | 56·5 | 55·8 |
| 500 | 50·8 | 50·4 | | | |

Good results have been obtained with specially prepared silver-backed cover-glass standards with a reflectivity in the range 87–96%. These have a uniform reflectivity over large areas and make a convenient standard for measurements with low-power objectives (n.a. < 0·30). Because of interference from the cover-glass, however, they are not suitable for use with high-power objectives. Values obtained by N.P.L. on two samples are given in Table 8.

TABLE 8

*Spectral reflectivity of silver-backed cover-glass standards (a) and (b)*

| nm | $R\%$ | | nm | $R\%$ | |
|---|---|---|---|---|---|
| | (a) | (b) | | (a) | (b) |
| 400 | 89·5 | 87·1 | 546·1 | 95·3 | 95·3 |
| 420 | 91·3 | 90·2 | 589·3 | 95·7 | 95·7 |
| 440 | 92·5 | 91·9 | 620 | 96·0 | 96·0 |
| 460 | 93·3 | 92·9 | 660 | 96·2 | 96·2 |
| 480 | 94·1 | 93·8 | 700 | 96·5 | 96·5 |
| 500 | 94·5 | 94·5 | | | |

### F. MEASUREMENT OF REFLECTIVITY IN AIR

The quality of polish is particularly important if reflectivity measurements are to be of diagnostic value. Quality can be assessed by measuring the reflectivity at successive stages before polishing is complete (Hallimond and Bowie, 1964). However, care has to be taken to avoid "over-polishing", which tends to produce a hardened layer with a reflectivity lower than that of the mineral. The surface of a polished section should be as near as possible to an optical flat (less than 5 fringes per centimetre), since unevenness can cause errors of a few per cent in the reflectivity value. It is also important that the surface is normal to the axis of the microscope. This is best checked before measurements are made by inserting the Bertrand lens, or else removing the ocular, and rotating the stage. If the specimen is level the image of the lamp filament will remain stationary when the stage is rotated. The microscope should be adjusted as indicated in Section IVB.

In making a series of readings, the standard is first placed on the stage and a flaw-free area brought into focus. The field and ocular iris diaphragms are adjusted so that, with no more than a quarter of the field illuminated, the galvanometer reading is approximately equal to the standard's reflectivity. A small box painted black internally and with an aperture on top is then substituted for the standard and a second reading obtained on the galvano-meter. This reading, which is due to primary glare, is off-set by adjusting the zero-setting control until the galvanometer reads zero. The standard is then replaced and the irises adjusted slightly so that the galvanometer reading is exactly equal to the reflectivity of the standard. All subsequent galvanometer readings obtained on selected areas of unknown minerals give reflectivities directly, so long as no adjustments, other than refocusing, are made to the equipment.

Measurement of reflectivity by this method does not allow for secondary glare. This can be minimized by selecting a standard not too dissimilar from the unknown in reflectivity and by observing the other precautions described in Section IVB. Alternatively, secondary glare can be eliminated by drawing up a calibration curve for any particular wavelength of light by using a series of not less than five standards covering a wide range of reflectivity (Fig. 13). The curve illustrated was based on pyrite as the primary standard because its reflectivity is approximately half way between 0 and 100. It will be noted that in adjusting the microscope field and ocular irises to allow pyrite to read its "true" value, the microscope value for carborundum is too low (too much has been subtracted for secondary glare); whereas the value for silver is too high (too little has been subtracted for secondary glare).

### G. SPECTRAL REFLECTIVITY VALUES

Many ore minerals show considerable dispersion of the reflectivity so that, if values for light of different wavelengths are required, it is necessary to have a monochromator with a fairly narrow pass band. However, the pass band should not be so narrow that the amount of light falling on the specimen is insufficient for reliable readings. One of the most satisfactory types of mono-chromator is an interference filter of the "continuous-band" type described by Harrison and Day (1963). Alternatively a series of single interference filters of suitable wavelengths can be used. These interference-type filters have a half-height band width of about 25 nm and a transmission factor of nearly 40%. The continuous-band filter covers wavelengths from 400 to 700 nm.

Band widths centred on any wavelength can be selected by sliding the filter along in front of a slit. Moreover, the band width changes little with increase in slit width from 0 to 6 mm. At 4 mm, which is a convenient slit width for most microscopes, the half-height value is 25·5 nm. Several manufacturers produce interference-type filters, but some have second-order transmission peaks requiring an auxiliary filter for their elimination.

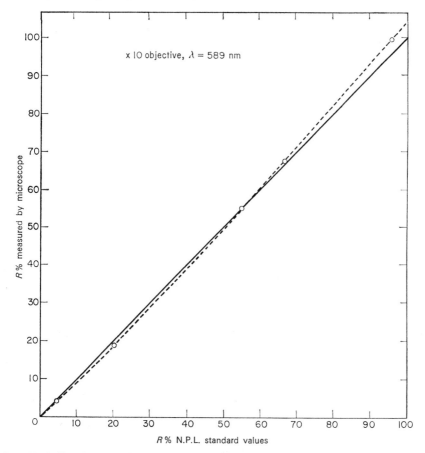

FIG. 13. Calibration curve based on pyrite as the primary standard. Note that the microscope value for carborundum is too low and that the value for silver is too high owing to the secondary glare component being "correct" for pyrite only.

Having selected a standard of reflectivity near to the unknown, the simplest procedure in measuring spectral reflectivity is to set the galvanometer so that it reads nearly 100 for the highest reflectivity recorded when the "continuous-band" filter is run through, first with the standard and then with the unknown on the stage. Values for primary glare $(C)$ are measured at the required wavelengths and a series of readings are taken on the standard $(G_{st})$ and finally on the unknown $(G_{sp})$. The reflectivity at each wavelength is then evaluated from the expression:

$$\frac{R_{sp}}{R_{st}} = \frac{G_{sp} - C}{G_{st} - C}$$

and the spectral curve plotted (Fig. 14). In this way the variable sensitivity of the cell is automatically allowed for. If a standard of similar reflectivity to the unknown is not available, secondary glare can be eliminated by preparing a calibration curve for each wavelength, as mentioned in the preceding section.

### H. MEASUREMENT OF REFLECTIVITY IN OIL

Data on the reflectivity of mineral species in oil are as useful in identification as measurements in air. Measurements in oil can be made by using oil-immersion objectives in a similar manner to making measurements in air. Care must be taken to ensure that the oil recommended by the manufacturer of the microscope is used and that its refractive index and dispersion characteristics are noted when results are given. An oil with $n = 1.515$ at a wavelength of 589 nm and a temperature of 20°C is commonly used.

An alternative method is that of Cambon (1949) in which a cover-glass of the same refractive index as the oil is placed on top of a drop of oil, allowing dry objectives to be used. Corrections necessary when this technique is used are discussed by Piller and von Gehlen (1964).

## V. Measurement of Rotation Properties

For quantitative measurement of the rotation properties of ore minerals, it is necessary to have a microscope fitted with an analyser, with vernier adjustment, that can be rotated through at least 90° and can be read to ± 0.1°. A rotatable elliptical compensator, with a vernier scale of similar accuracy is also required for the measurement of phase difference. A convenient compensator is a rotating mica plate with an optical thickness of $1/6\lambda$ to $1/20\lambda$. A third accessory required is a Nakamura plate, used for the accurate setting of the analyser.

The analyser and elliptical compensator can be built into the microscope or, alternatively, a Wright slotted ocular with rotatable analyser and facilities for taking either a Nakamura plate or rotary elliptical compensator can be used.

### A. ADJUSTMENT OF THE MICROSCOPE

Proper adjustment of the microscope is particularly important if accurate measurements of rotation properties are to be obtained. Hence it is necessary to carry out the following tests: (1) Ensure that the polarizer is oriented accurately with the vibration plane parallel to the rotation axis of the reflecting plate. The crossed position of the analyser is then determined and recorded as $A_0$. (2) Each objective to be used must be set at a position of extinction and accurately centred. (3) The levelling of the specimen must be tested by observing the image of the aperture diaphragm with the Bertrand

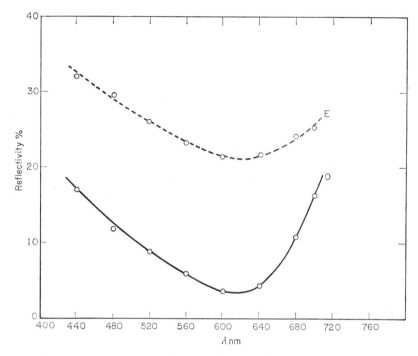

FIG. 14. Spectral reflectivity of covellite: O is the curve for the ordinary ray and E that for the extraordinary ray.

lens. If these adjustments are satisfactory, a section of covellite when viewed in light of $\lambda = 589$ nm with the polars crossed will show extinction positions $90° \pm 0.5°$ apart. (4) With the polars accurately crossed, and observing the polarization figure of magnetite, the zero position of the mica compensator is now determined for sodium light. This position is recorded as $G_0$. (5) The orientation of fast and slow rays of the mica plate is checked by using a gypsum (full-wave) plate. (6) The correction factor for the reflecting plate at wavelengths from 470 to 650 nm must be ascertained. To do this, a polarizer is placed on the microscope stage and illuminated from below. The polarizer is exactly crossed with the analyser, then rotated successively through 5°, 10°,

15°, 20° and 25° and the analyser rotated in the same direction until extinction is produced. With the polarizer in the 5° position the analyser setting will be about 6° and this additional rotation due to the glass reflector must be allowed for in obtaining correct values for the rotation angle, $A_r$.

## B. PROCEDURE FOR MEASUREMENT OF $A_r$

(1) Place a polished section, say of pyrrhotine, on the stage and examine under exactly crossed polars with a 16 mm objective. Choose a grain showing maximum anisotropism and insert an interference filter to give light of $\lambda = 589$ nm. (2) Determine the four extinction positions of the grain. These should be 90° apart. If not, alternate pairs of positions should show the same degree-interval. Calculate the bisectrix between two adjacent extinction positions. This gives one the 45° position. (3) Turn the stage to bring the grain to the 45° position. (4) Insert the Nakamura plate in the Wright ocular with the plate boundary orientated at 45° to the ocular crosshairs and passing through the crosshairs junction. (5) Rotate the analyser until the two halves of the Nakamura plate exactly match. Record the setting of the analyser and call it $A_1$. (6) Rotate the stage 90°. The mineral is now in the second 45° position. (7) Rotate the analyser until the Nakamura plate again shows a perfect match of the two halves. Record the setting of the analyser and call it $A_2$. (8) Calculate $A_r = \frac{1}{2}(A_1 - A_2)$. This is the observed $A_r$, which must be multiplied by the correction factor (approximately 5/6) to determine the true value of $A_r$.

## C. DETERMINATION OF SIGN OF PHASE DIFFERENCE, $\Delta_{x,y}$

(1) Using a low-power objective, focus on a grain showing maximum anisotropism. Determine the extinction directions of the grain. Rotate the stage to bring the grain into the 45° position. (2) Insert an interference filter and adjust to give light of $\lambda = 589$ nm. (3) Rotate the analyser anti-clockwise, to give extinction. If extinction is not obtained rotate the stage through 90°. (4) Remove the interference filter. (5) Insert a gypsum plate. If the colour of the grain appears yellow it indicates that the sign of $\Delta_{x,y}$ is negative, as for example in the case of covellite.

## D. MEASUREMENT OF PHASE DIFFERENCE, $2\delta$, WITH A MICA COMPENSATOR

(1) Repeat procedures (1), (2) and (3) above. (4) Attach a high-power objective, focus, and centre if necessary. Remove the eyepiece and substitute a pinhole eyepiece, or insert a Bertrand lens and focus on the polarization figure, which will now be seen. It will consist of two isogyres, in the NW and SE quadrants. (5) Insert the mica compensator into the tube slot. (6) Rotate the mica compensator and analyser simultaneously until the isogyres combine to form a cross. The centre of the cross must be black. This is the position

of compensation. (7) Record the settings of the analyser and mica compensator as $A_1$ and $G_1$, respectively. (8) Rotate the stage through 90° and repeat step (6), recording the setting of the analyser and mica compensator as $A_2$ and $G_2$, respectively. (9) From the value of $A_r$ and the values of $G_1$ and $G_2$ calculate $\beta_G$, which is the rotation of the mica compensator from its zero position coincident with the analyser to its position of compensation. The following possible cases may arise depending on the amount and direction of rotation of the compensator.

(a) The angular rotation of the compensator from $G_0$ is in the same direction as that of the analyser, but larger:

$$\beta_G = \frac{G_1 - G_2}{2} - A_r$$

(b) The angular rotation of the compensator is in the same direction as that of the analyser, but smaller:

$$\beta_G = A_r - \frac{G_1 - G_2}{2}$$

(c) The compensator and analyser are rotated in opposite directions:

$$\beta_G = A_r + \frac{G_1 - G_2}{2}$$

If the mica compensator is rotated clockwise and the analyser anti-clockwise, the sign of $\beta_G$ is positive; if the mica compensator is rotated anti-clockwise and the analyser clockwise, the sign of $\beta_G$ is negative. (10) Calculate the phase difference $2\delta$* from the formula:

$$\sin 2\delta = 2\beta_G \sin \Delta_G$$

where $\Delta_G$ is the phase difference for the mica compensator for the wavelength of light in use and $\beta_G$ is the corrected angle for compensation.

## VI. Micro-indentation Hardness

The hardness of minerals is a complex property dependent mainly on such factors as the nature of the chemical bond, inter-atomic distances, valency, density of packing of the atoms and their co-ordination number. Much of the fundamental work on hardness in the past has been based on Mohs scale and it seems likely that further work will prove much more fruitful now that more precise data on hardness can be obtained with low-load indenters.

* The phase difference $2\delta$ is related to the phase difference $\Delta_{x,y}$ by the equation:

$$\tan 2\delta = \tan \Delta_{x,y} \sin 2\alpha,$$

where $\alpha = 45 - A_r$. As $A_r$ seldom exceeds a few degrees, $\sin 2\alpha$ will usually be nearly unity, and the difference between $2\delta$ and $\Delta_{x,y}$ will be imperceptible.

Ore mineralogists have used three basically different ways of measuring hardness. One method is that of scratch hardness as used by Mohs; the second is to estimate the resistance of the mineral to abrasion (polishing hardness); and the third is to apply a steady load with an indenter and to calculate the hardness from the area or depth of the impression produced.

An attempt was made to refine scratch hardness by using a machine to measure the load necessary for a diamond point to scratch the mineral and promising results were obtained (Talmage, 1925). However, the method did not become standard practice, mainly because of the delicate nature of the instrument used and because of the difficulty in distinguishing scratching from fracturing in hard minerals. Van der Veen (1925), and more recently Uyten-bogaardt (1951) and Schneiderhöhn (1952), used polishing hardness to compare the relative hardness of two adjacent minerals. With the use of modern polishing techniques, however, particularly the use of diamond abrasive on lead laps, this property is rarely observable.

Static indenters were first used for determining the hardness of minerals during the Second World War, but indentation hardness was not used systematically as an aid to ore mineral identification until 1958 (Bowie and Taylor). The original and most frequently used static instruments are the Vickers indenter and the Knoop indenter, and it would appear that these are as good as the triangular- and cone-type indenters introduced later. The Knoop indenter is an elongated pyramid with two faces with an included angle of 130° between them and two with an included angle of 172° 30′. The shape of the perfect impression is that of a parallelogram in which one diagonal is about seven times the length of the other and thirty times the depth of the impression. The Vickers indenter is a square pyramid with a 136° included angle between opposite faces, so that a perfect indentation is a square with equal diagonals seven times the depth of penetration.

### A. LOW-LOAD MICRO-INDENTATION HARDNESS TESTERS

Several low-load micro-indentation hardness testers are available on the market; some are complete instruments, whereas others can be attached to a microscope. The method of operating both is similar. The diamond is brought into contact with the mineral surface and a known load, which varies between 0·1 g and 500 g, is applied. The apparatus used by the author is a GKN instrument, with Vickers indenter, which is designed for use with any good microscope (Fig. 15). It consists of an upper mounting plate with a threaded adaptor to secure it to the base of the vertical illuminator in place of the normal objective holder. Below this plate is a rotatable plate, which carries the diamond pyramid and two centralizing objectives. The indenter is mounted on a counterpoised beam suspended by two beryllium-bronze leaf springs, set at right angles to ensure rigidity to horizontal and vertical deflections, but with freedom to rotate (Fig. 16).

To test the hardness of a mineral grain, the indenter is turned into position in place of an objective and the sub-stage of the microscope raised until the polished section is just clear of the diamond pyramid. The fine-focusing control

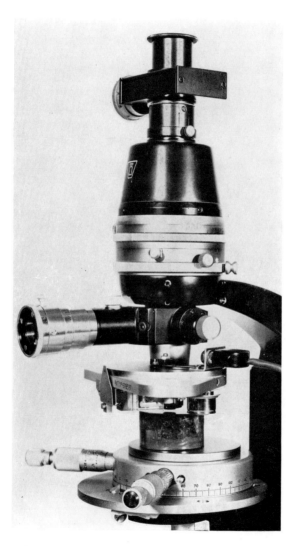

Fig. 15. Vickers microscope with GKN micro-indentation hardness tester, Smith reflector unit and micrometer eyepiece. Reproduced with permission from Hallimond and Payne (1964).

of the microscope is then used to lower the indenter, or alternatively to raise the specimen, with a slow steady movement until the indenter makes contact with the mineral surface. The application of the full load is indicated electrically by a miniature neon lamp which is arranged to be in circuit when the beam is in balance. When the specimen takes the load, the circuit is broken and the neon lamp glows with reduced intensity. The load is left applied for 15 sec, when the indenter is removed, first by using the fine focusing control and then by the coarse adjustment on the sub-stage. The objective is then

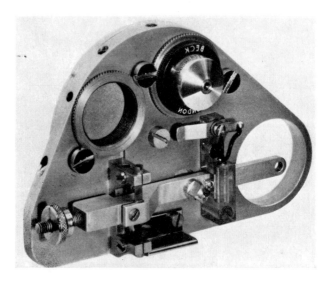

Fig. 16. Lower rotating plate of GKN micro-indentation hardness tester showing beam with diamond pyramid indenter and objective holders. Reproduced with permission from Bowie and Taylor (1958).

brought back into position and the indentation orientated by means of a mechanical stage, the distance across the diagonals being measured by means of a screw-micrometer eyepiece.

Indentation hardness is defined as the ratio of the load applied to the surface area of the indentation; thus for a Vickers indenter:

$$H = \frac{2\,L\,\sin\,\theta}{d^2}$$

where $H$ is the diamond pyramid hardness number in kg/mm$^2$, $L$ is the load in kg, $d$ is the length of the diagonal in mm and $\theta$ is half the included angle between opposite pyramid faces. For a 136° indenter, $2 \sin \theta = 1{\cdot}8544$, so that the Vickers hardness number ($VHN$) is equal to $1{\cdot}8544\,L/d^2$. For convenience, hardness numbers are normally read off tables that give solutions to the above equation for standard loads.

### B. INSTRUMENTAL FACTORS AFFECTING HARDNESS-NUMBER DETERMINATIONS

It is relatively easy to determine the precision with which micro-indentation hardness measurements can be made with a particular instrument. However, the accuracy of the results is more difficult to establish, and it is therefore recommended that polished blocks of stainless steel of standard hardness, say of *VHN* 300 and *VHN* 800, be used to facilitate comparison of results with other laboratories.

### 1. Instrumental Errors

Four main conditions relating to the instrument may be sources of error. These are: (1) the load applied is not known accurately; (2) the angle $\theta$ of the pyramid is not accurately machined; (3) there is lateral movement of the indenter; (4) the measuring device is incorrectly calibrated.

Inaccuracies in applying the load commonly result from the faulty operation of the instrument, from inertia or vibration effects and from variations in the time over which the load is applied. Friction in moving parts may also be a source of error.

### 2. Vibration Effects

Perhaps the most common source of error in low-load hardness testing is vibration. Every precaution must therefore be taken to reduce this effect as much as possible. A convenient method of overcoming vibration errors is to mount the hardness tester on a heavy stone slab which is supported on a sorbo-rubber mat and six shock absorbers. With this arrangement, the reproducibility of results over the range of hardness encountered in ore minerals is better than 5% of the measured value.

### 3. Time of Indentation

The exact time of indentation is not so important when the hardness of minerals is measured as for measurements on such materials as plastics and lacquers, in which penetration by the indenter increases markedly with time. However, a certain amount of creep occurs in minerals so that the load must be applied for several seconds before static conditions are reached. Vibrational effects are more likely to produce erroneous results the longer the period of loading, hence a compromise has to be reached. It is now generally accepted that a loading time of 15 sec is satisfactory and that the loading speed should be of the order of 10–20 $\mu$m/sec.

G*

## 4. Calibration of Instrument and Measurement of Indentations

The normal procedure in measuring indentations is to use a rotating-drum eyepiece micrometer on an optical microscope. In these circumstances the accuracy of the calibration of the equipment is dependent on that of the stage micrometer used and on the resolving power of the objective. For calibration purposes, optical gratings are much to be preferred to engraved glass or metal micrometers.

The resolving power of an objective is given by the expression, 0·5 $\lambda$/n.a., where $\lambda$ is the wavelength of light used and n.a. is the numerical aperture. With an oil-immersion lens of n.a. 1·30 and white light of average $\lambda$ = 550 nm, the resolving power is about 0·2 $\mu$m. Thus it is possible to measure the calibration graticule with an accuracy of $\pm$ 0·4 $\mu$m. If greater precision is required, it is preferable to use monochromatic light at the blue end of the spectrum, say at a wavelength of 400 to 500 nm. This would improve the accuracy of the calibration measurement to about $\pm$ 0·3 $\mu$m. The microscope field is rarely perfectly flat over its whole area, therefore only the central portion should be used both when calibrating the eyepiece micrometer and when making measurements of impressions. For the same reason the graticule distance used for calibration should be approximately the same as the diagonal lengths to be measured.

Consideration of what can and cannot be measured by the microscope is complicated by the combined resolving power of the objective and that of the "normal" human eye. Theory indicates, and observation confirms, that the apparent distance between two points decreases when an objective of large numerical aperture is replaced by one of smaller numerical aperture and thus of lower resolving power. Brown and Ineson (1951) show that for green-light measurements, the diagonal length of a Vickers indentation will appear 0·56 $\mu$m short when measured with an objective of n.a. 0·45 as compared with the measurement with an objective of n.a. 1·40. To this there is a further apparent shortening of approximately 0·2 $\mu$m due to the primary magnification of the former objective being only about a quarter of that of the latter.

In addition to the measurement errors discussed above, the precise measurement of micro-indentation hardness is also rendered difficult by uncertainty in locating the ends of the diagonals. This results mainly from hair fracturing and from the piling up or depression of material around the impression. It is thus doubtful if there is any advantage in using oil-immersion objectives of high n.a. for mineralogical purposes. In the author's experience a dry objective of n.a. 0·85 and light of $\lambda$ = 500 nm is to be recommended as giving optimum results from the point of view both of precision and the time taken to measure the diagonals of the impression.

## C. MINERAL PROPERTIES AFFECTING HARDNESS-NUMBER VALUES

The main factors affecting the hardness value determined for particular mineral species are: (1) method of preparing the polished surface; (2) orientation of mineral grain; (3) degree of crystallinity; (4) variations in chemical composition; (5) elastic recovery of the impression.

### 1. Effect of Polishing Method

The hard layer produced by some polishing methods may result in the hardness value appearing greater at low loads, but since this layer is usually no more than 10–20 $\mu$m thick, the hardness value is not likely to be affected when loads of 100 g are used. However, for loads of 10 g or less the effect of the surface-layer may become important. Further tests are required to establish the effect of surface-layer formation, but tests carried out with a 100 g load on the same sections of each of five common ore minerals polished by four different techniques show excellent agreement (Table 9).

TABLE 9

*Hardness values (VHN 100 g) for minerals polished by different methods*

|  | Galena | Sphalerite | Chalcopyrite | Löllingite | Magnetite |
|---|---|---|---|---|---|
| Diamond on lead lap | 72 | 202 | 206 | 437 | 554 |
| MgO on lead lap | 75 | 185 | 193 | 473 | 554 |
| $\gamma$-Alumina on lead lap | 76 | 191 | 193 | 433 | 560 |
| $\gamma$-Alumina on cloth | 77 | 206 | 197 | 437 | 554 |

Reproduced with permission from Bowie and Taylor (1958).

The results indicate that over the range of hardness *VHN* 70–600 no appreciable difference can be attributed to polishing whether by diamond dust, magnesium oxide or gamma-alumina on lead laps or by gamma-alumina on a cloth lap. The mean values are all within, or in the case of sphalerite and löllingite just outside, the standard deviation limits of the measurements. Polishing by the "rough" methods mentioned earlier may result in Beilby layers of 100 $\mu$m or more, and in this case hardness values may be anomalously high even with 100 g loads.

### 2. Orientation of mineral grain

Most crystals show some degree of hardness anisotropy, but generally hardness differences are not appreciable. However, in the case of fibrous and prismatic crystals of minerals, such as löllingite, pyrolusite, enargite, zincite, millerite and niccolite, hardness anisotropy is quite marked. The range of hardness in löllingite, for example, is from *VHN* 421–920 depending on orientation alone. The property of hardness anisotropy is not confined to anisotropic minerals, but also occurs in cubic species, such as galena, in which it can be related to glide and slip planes (Young and Millman, 1964).

### 3. Degree of crystallinity

It is important when making measurements of hardness to note whether the results are obtained on crystals that are large compared with the indentation, or whether aggregate hardness is measured, as in the case of microcrystalline or cryptocrystalline masses. The hardness numbers of microcrystalline hematite and goethite, for example, are only about 70% of those obtained on coarsely crystalline material.

Metamictization also tends to reduce hardness values, so that a wide range of results may be encountered in radioactive minerals, such as betafite or euxenite.

### 4. Chemical Composition

The variation in hardness number with chemical composition of solid-solution series has so far been little studied. Recent work (Young and Millman, 1964) shows that for (110) faces of members of the sphalerite–marmatite series the hardness increases rapidly with iron content 0·07–1·2% w/w, but with iron contents of from 1·2–10·7% w/w the hardness progressively decreases. In the hübnerite–wolframite–ferberite series, hübernite is consistently softer than ferberite in similar orientation. However, when the proportions of FeO and MnO are approximately equal, the hardness values are appreciably higher than for the two end members.

### 5. Effect of Elastic Recovery

Several workers including Nakhla (1956) have demonstrated a tendency for the measured hardness number to increase with low loads. Theoretically, however, the hardness value obtained with a 136° pyramid indenter is independent of depth, hence of load, which implies that the contraction of the diagonals is proportional to the unrecovered diagonal length. Further work is necessary to clarify the reasons for the apparent increase in hardness with decreasing load. At present, however, there seem to be three possible causes: (1) "rough" polishing methods produce a hardened layer that would result in hardness values appearing to be higher when the depth of penetration is small compared with the thickness of the layer; (2) the slip mechanism producing deformation is not the same for small indentations as it is for larger ones; (3) inertia in beam-type testers becomes important when the mass of the beam is greater than that of the applied load. This results in the hardness number appearing higher for low loads on equipment not designed for low-load testing. Instruments designed for test loads of 10–100 g should therefore not be used for test loads of 0·1–1 g if useful results are to be obtained.

## D. SHAPE AND QUALITY OF INDENTATIONS IN MINERALS

It is rarely found even with low-load testers that perfectly square impressions are made in mineral grains. Often the shape of the impression is a parallelogram with concave or convex sides, or one pair of sides may be concave and the other two convex. These features, though making accurate

FIG. 17. Indentations in bornite (grey) and chalcocite (light grey). Note development of parting in bornite more than one diagonal length away from the impression. (× 390)

determinations more difficult, are nevertheless of considerable diagnostic value. The shape and quality of the indentation can conveniently be described as: perfect (p), slightly fractured (sf), fractured (f), concave sided (cc) or convex sided (cv). The pile-up or sinking-in of material adjacent to the indentation may cause fracturing in some crystal orientations, but not in others. These fractures are often shell-like or radiate from the corners of the impression but may also be at right angles to the sides of the impression. In minerals with good cleavage, parting or twinning, characteristic patterns are developed which are also of diagnostic value (Fig. 17). An example of two minerals with similar hardness numbers, but with quite different shapes of indentation is provided by chalcopyrite and pentlandite. The former mineral produces indentations with nearly straight edges, whereas in the latter the impressions are markedly concave.

G**

The shape of the indentations produced in prismatic sections of chalcocite, covellite, ilmenite and hausmannite vary considerably with orientation. When the diagonals of the pyramid are orientated parallel to and at right angles to the prism faces, the indentations have two opposite sides concave and two convex; on rotation into the 45° position, however, the indentations are elongate or kite-shaped with slightly concave edges. Where there is obvious hardness anisotropy, it is recommended that separate readings be noted for each diagonal rather than to attempt to obtain an average value by taking the mean of the hardness value in two directions (Fig. 18).

FIG. 18. Indentation in zincite showing marked directional hardness. The long diagonals are parallel to the (001) cleavage direction. (×330)

In a few minerals, such as graphite and molybdenite, the impressions are ill defined and prove extremely difficult to measure accurately. Surface flaking or cracking adjacent to indentations is rarely serious even in brittle minerals such as pyrite, but with loads of more than 100 g can be a handicap in obtaining precise results.

### E. PRECISION AND ACCURACY OF RESULTS

It is normally easier to produce an average hardness number for a particular mineral than it is to establish a true hardness range. An approximation to the former can be obtained by making several indentations on randomly orientated grains if impressions associated with bad fracturing or cracking are ignored. Tests carried out (Winchell, 1945; Bowie and Taylor, 1958) show that individual determinations of hardness are within 5% of the average. The

precision with which measurements can be made can be tested as follows:
If $H_1, H_2, H_3 \ldots \ldots H_n$ are the hardness numbers obtained in n tests on
the same specimen and the respective differences from the mean $\bar{H}$ are
$(H_1 - \bar{H}), (H_2 - \bar{H}), (H_3 - \bar{H}) \ldots \ldots (H_n - \bar{H})$, then the standard error of
any individual observation is expressed by:

$$\sqrt{\frac{\sum (H - \bar{H})^2}{n}}$$

and the standard error of the mean by:

$$\sqrt{\frac{\sum (H - \bar{H})^2}{n^2}}$$

Fairly large variations in hardness have been reported (Brown and Ineson,
1951) on results obtained on different types of indenter and it seems doubtful
if an accuracy of better than $\pm 0.5$ $\mu$m in the measurement of diagonals can
generally be obtained unless calibration is carried out carefully on metal
standards. However, a precision, or reproducibility of measurement, on the
same apparatus is probably about $\pm 0.1$ $\mu$m.

The variation in $VHN$ assuming an accuracy of $\pm 0.5$ $\mu$m, and an accuracy
of $\pm 0.1$ $\mu$m, in the measurement of diagonals is shown in Fig. 19. This
illustrates the comparative inaccuracy resulting from using a 10 g load as
compared with 100 g load from errors due to measurement alone. The choice
of load is a compromise, since heavy loads shatter brittle minerals and no
useful hardness data are obtained. On the other hand, some mineral grains
are so small that low loads have to be used in order to comply with the
requirement that precise results will only be obtained if the diameter of the
grain is more than three times the length of the diagonals of the indentation.
As a general rule a load of at least 100 g should be used whenever possible
as this minimizes measurement errors and at the same time avoids differences
in hardness due to Beilby-layer formation.

### F. COMPARISON OF VICKERS HARDNESS NUMBERS WITH MOHS SCALE

Several authors have compared micro-indentation hardness values for the
minerals selected by Mohs for his standards of hardness and there is some
disagreement as to whether the relationship approximates to log/linear or
log/log. Which relationship one chooses largely depends on the orientation of
the standard hardness minerals when measured by the indentation method.
Experiments have shown (Tabor, 1954) that for one mineral of hardness $H_p$ to
scratch another of hardness $H_s$, $H_p$ must be $\geqslant 1.2 H_s$. Assuming that $H_p =$

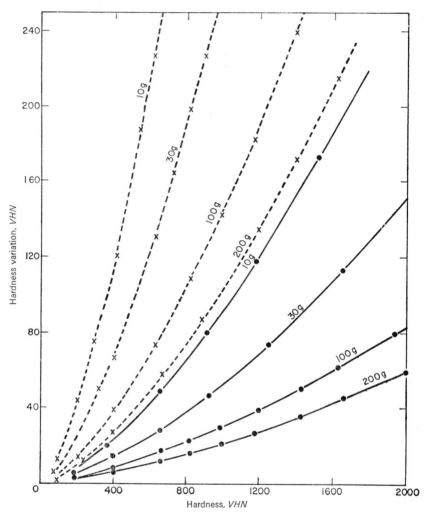

FIG. 19. Variation in hardness for 200, 100, 30 and 10 g loads on the assumption that measurements are made with an accuracy of ±0·5 μm (dotted curves) and of ±0·1 μm (full curves).

$1·2\ H_s$ then the Mohs hardness number, $M$, is related to the Vickers hardness number by the expression:

$$VHN = K\,(1·2)^M \text{ or}$$
$$\log VHN = M \log 1·2 + C$$
$$\text{where } C = \log K$$

A graph of Mohs hardness values against the log of Vickers numbers should therefore be a straight line of slope log 1·2. In practice, however, the gap between hardness standards must be somewhat greater in order to avoid ambiguities in scratch-hardness determinations, and, in the case of the Mohs scale, the slope of the line approximates to 1·77, which corresponds to an increment of 77% for each standard (Fig. 20).

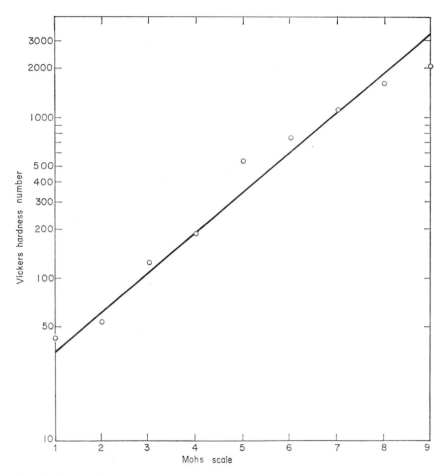

FIG. 20. Graph of Mohs hardness values against log of Vickers hardness numbers showing relationship approximating to a straight line of slope log 1·77.

## VII. Systematic Scheme of Ore Mineral Identification

When mean reflectivity value is plotted against mean Vickers hardness number, five main groups of similar composition, structure and bonding are apparent (Bowie and Taylor, 1958). The metals and metalloids with metallic bonding are relatively soft, but have a high reflectivity. At the other extreme, the metallic oxides with ionic bonds are relatively hard and have a low re-flectivity. Between these two groups are a wide range of sulphides, sulpho-salts and cobalt–nickel–iron sulphides, with covalent and ionic bonding, which have medium reflectivity and a wide range of hardness (Fig. 21). Some

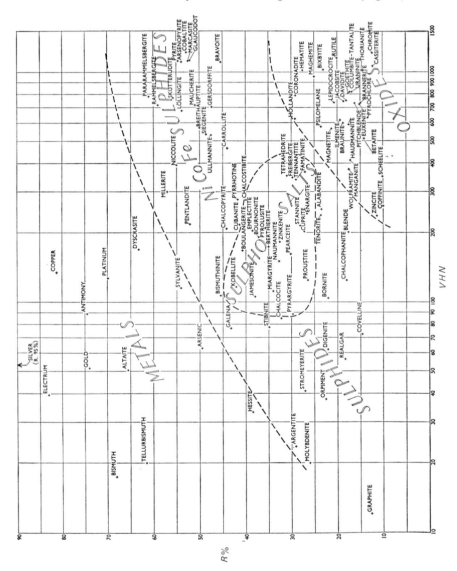

FIG. 21. Distribution of the more common ore minerals on the basis of mean reflectivity and Vickers hardness number. Reproduced with permission from Bowie and Taylor (1958).

anomalies occur within these broad groups—for example, cuprite $Cu_2O$, pyrolusite $MnO_2$ and naumannite $Ag_2Se$ occur among the sulpho-salts—but generally any unknown mineral can be assigned to one of the five main groups on the basis of hardness and reflectivity alone.

## A. REFLECTIVITY

The spectral reflectivity of only a few minerals has been determined with narrow-band filters of the type described in Section IVG. Thus until more spectral data are available reflectivity values for "white" light must be used. Values obtained with a selenium-cell photometer with correction filter (Fig.

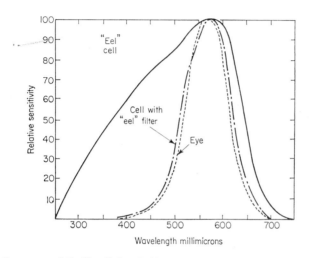

Fig. 22. Response of "eel" cell for "white" light, with and without correction filter, compared with that of the human eye. Reproduced by permission of Evans Electro-selenium Ltd.

22) can be compared directly with the values available in the literature for visual photometer determinations. The values quoted in Table 10 are for "white" plane polarized light (3050°K) in air relative to a pyrite standard of 54·5 and should be reproducible if a selenium-cell photometer is used, the light source being approximately of the same colour temperature.

## B. MICRO-INDENTATION HARDNESS

Further work is required before the accuracy of micro-indentation hardness values on the various indentation hardness testers is available. Also, it is desirable that there should be some degree of standardization of the load, the numerical aperture of the objective and the wavelength of the light used.

TABLE 10

*Minerals arranged in order of increasing reflectivity*

The values are for white plane polarized light (3050°K) in air relative to a pyrite standard of 54·5

| Number of localities | Mineral species | Mean | Range | Difference, mainly due to bireflection |
|---|---|---|---|---|
| 1 | Coffinite | 9·9 | — | — |
| 1 | Scheelite | 10·0 | — | — |
| 1 | Zincite | 11·2 | — | — |
| 2 | Cassiterite | 12·0 | 11·2–12·8 | 1·6 |
| 2 | Chromite | 12·1 | — | — |
| 2 | Graphite | 12·5 | 6·0–17·0 | 11·0 |
| 2 | Pyrochlore | 13·4 | 13·0–13·8 | 0·8 |
| 3 | Covellite | 14·5 | 7·0–22·0 | 15·0 |
| 3 | Betafite | 14·5 | 14·0–14·9 | 0·9 |
| 3 | Thorianite | 14·6 | 14·0–15·3 | 1·3 |
| 4 | Brannerite | 14·8 | 13·7–16·0 | 2·3 |
| 3 | Euxenite | 15·0 | — | — |
| 5 | Pitchblende | 16·0 | — | — |
| 1 | Uraninite | 16·8 | — | — |
| 2 | Manganite | 17·0 | 14·0–20·0 | 6·0 |
| 3 | Columbite-tantalite | 17·1 | 16·3–18·0 | 1·7 |
| 5 | Goethite | 17·3 | 16·1–18·5 | 2·4 |
| 3 | Wolframite | 17·3 | 16·2–18·5 | 2·3 |
| 5 | Sphalerite | 17·5 | — | — |
| 1 | Hausmannite | 17·5 | 16·0–19·0 | 3·0 |
| 3 | Davidite | 17·8 | — | — |
| 1 | Chalcophanite | 18·1 | 10·2–26·0 | 15·8 |
| 1 | Realgar | 18·5 | — | — |
| 1 | Jacobsite | 18·5 | — | — |
| 2 | Braunite | 18·8 | 17·8–19·8 | 2·0 |
| 4 | Ilmenite | 19·4 | 17·8–21·1 | 3·3 |
| 2 | Rutile | 20·2 | — | — |
| 1 | Lepidocrocite | 20·4 | 15·8–25·0 | 9·2 |
| 4 | Magnetite | 21·1 | — | — |
| 4 | Bornite | 21·9 | — | — |
| 1 | Digenite | 22·0 | — | — |
| 1 | Orpiment | 22·6 | 20·3–25·0 | 4·7 |
| 1 | Bixbyite | 23·0 | — | — |
| 1 | Alabandite | 23·4 | — | — |
| 1 | Tenorite | 23·4 | 20·0–26·9 | 6·9 |
| 1 | Psilomelane | 23·5 | 23·0–24·0 | 1·0 |
| 1 | Maghemite | 25·0 | — | — |
| 3 | Molybdenite | 26·0 | 15·0–37·0 | 22·0 |
| 1 | Proustite | 26·4 | 25·0–27·7 | 2·7 |
| 3 | Enargite | 26·5 | 25·0–28·1 | 3·1 |
| 1 | Famatinite | 26·9 | 25·1–28·7 | 3·6 |
| 2 | Cuprite | 27·1 | — | — |
| 2 | Stromeyerite | 27·1 | 25·5–28·7 | 3·2 |
| 5 | Hematite | 27·5 | 25·0–30·0 | 5·0 |
| 2 | Stannite | 28·0 | — | — |
| 1 | Tennantite | 28·9 | — | — |
| 2 | Argentite | 29·0 | — | — |
| 1 | Coronadite | 29·0 | 26·0–32·0 | 6·0 |
| 2 | Hollandite | 29·2 | 26·0–32·5 | 6·5 |
| 3 | Pyrargyrite | 29·6 | 28·4–30·8 | 2·4 |

TABLE 10 (*continued*)

| Number of localities | Mineral species | Mean | Range | Difference, mainly due to bireflection |
|---|---|---|---|---|
| 1 | Freibergite | 29·8 | — | — |
| 1 | Pearceite | 30·1 | — | — |
| 4 | Tetrahedrite | 30·7 | — | — |
| 4 | Chalcocite | 32·2 | — | — |
| 1 | Zinkenite | 32·3 | — | — |
| 1 | Naumannite | 32·6 | 31·0–34·2 | 3·2 |
| 2 | Miargyrite | 33·9 | 31·8–36·0 | 4·2 |
| 1 | Berthierite | 35·0 | 30·0–40·0 | 10·0 |
| 2 | Stibnite | 35·1 | 30·2–40·0 | 9·8 |
| 2 | Pyrolusite | 35·8 | 30·0–41·5 | 11·5 |
| 1 | Bournonite | 37·1 | 36·0–38·2 | 2·2 |
| 1 | Jamesonite | 38·0 | 36·0–40·0 | 4·0 |
| 1 | Hessite | 38·5 | — | — |
| 1 | Emplectite | 38·5 | 36·0–41·0 | 5·0 |
| 2 | Chalcostibite | 40·0 | 37·1–43·0 | 5·9 |
| 2 | Boulangerite | 40·6 | 37·0–44·1 | 7·1 |
| 2 | Cubanite | 41·2 | 40·0–42·5 | 2·5 |
| 5 | Pyrrhotine | 41·6 | 38·0–45·2 | 7·2 |
| 1 | Kobellite | 42·1 | 40·9–43·2 | 2·3 |
| 4 | Galena | 43·2 | — | — |
| 1 | Carrollite | 44·0 | — | — |
| 7 | Chalcopyrite | 44·0 | 42·0–46·1 | 4·1 |
| 2 | Bismuthinite | 45·4 | 42·0–48·7 | 6·7 |
| 1 | Bravoite | 45·5 | — | — |
| 1 | Gersdorffite | 47·5 | — | — |
| 1 | Ullmannite | 47·5 | — | — |
| 3 | Siegenite | 48·6 | 47·3–49·8 | 2·5 |
| 1 | Arsenic | 49·5 | 48·0–51·0 | 3·0 |
| 1 | Breithauptite | 49·9 | 45·3–54·6 | 9·3 |
| 2 | Maucherite | 51·2 | — | — |
| 3 | Pentlandite | 52·0 | — | — |
| 4 | Marcasite | 52·2 | 48·9–55·5 | 6·6 |
| 1 | Glaucodot | 52·5 | — | — |
| 2 | Cobaltite | 52·7 | — | — |
| 3 | Arsenopyrite | 53·7 | 51·7–55·7 | 4·0 |
| 6 | Löllingite | 53·8 | 53·0–54·7 | 1·7 |
| 1 | Sylvanite | 54·0 | 48·0–60·0 | 12·0 |
| 5 | Pyrite | 54·5 | — | — |
| 3 | Niccolite | 55·1 | 52·0–58·3 | 6·3 |
| 5 | Skutterudite | 55·8 | — | — |
| 1 | Millerite | 57·0 | 54·0–60·0 | 6·0 |
| 2 | Rammelsbergite | 59·0 | 58·0–60·0 | 2·0 |
| 1 | Pararammelsbergite | 61·0 | 60·5–62·0 | 1·5 |
| 1 | Tellurbismuth | 61·5 | 60·5–62·5 | 2·0 |
| 1 | Dyscrasite | 63·2 | 62·0–64·5 | 2·5 |
| 1 | Altaite | 65·5 | — | — |
| 1 | Bismuth | 67·9 | — | — |
| 1 | Platinum | 70·0 | — | — |
| 2 | Gold | 74·0 | — | — |
| 1 | Antimony | 74·5 | 72·0–77·1 | 5·1 |
| 1 | Copper | 81·2 | — | — |
| 1 | Electrum | 83·0 | — | — |
| 2 | Silver | 95·0 | — | — |

Reproduced with permission from Bowie and Taylor (1958).

The data quoted in Table 11 are for "white" light, of colour temperature 3050°K, with a 100 g load and objective of n.a. 0·45.

## C. IDENTIFICATION PROCEDURE

After some experience, most common ore minerals can be identified in polished section in much the same way as quartz, muscovite, biotite, sphene and other minerals can be recognized in thin section without the necessity for carrying out any tests. The recognition of ore minerals is probably somewhat more difficult because a large number are white or exhibit various shades and tints of grey and have a reflectivity between 20 and 50%.

TABLE 11

*Minerals arranged in order of increasing Vickers hardness numbers*

| Number of localities | Mineral species | Mean | Range | Quality of indentation* | Remarks |
|---|---|---|---|---|---|
| 2 | Graphite | 12 | 12 | p. | |
| 3 | Molybdenite | 17 / 23 | 16–19 / 21–28 | p. | ⊥ to cleavage / ‖ to cleavage |
| 1 | Bismuth | 18 | 16–19 | p.cv. | |
| 1 | Tellurbismuth | 21 | 20–21 | sf.cv. | |
| 2 | Argentite | 24 | 20–30 | p.cc. | |
| 1 | Hessite | 33 | 28–41 | p.cc. | |
| 1 | Orpiment | 38 | 23–52 | sf.cc. | |
| 1 | Electrum | 40 | 34–44 | p.cc. | |
| 2 | Stromeyerite | 41 | 38–44 | p. | |
| 1 | Altaite | 51 | 48–57 | p. | |
| 2 | Gold | 51 | 50–52 | p.cv. | |
| 2 | Silver | 53 | 48–63 | p.cv. | |
| 1 | Realgar | 56 | 53–60 | p. | |
| 1 | Digenite | 61 | 56–67 | p.cc. | |
| 1 | Arsenic | 63 | 57–69 | p. | |
| 3 | Pyrargyrite | 71 / 106 | 50–97 / 98–126 | sf.cc. | ⊥ to cleavage / ‖ to cleavage |
| 3 | Covellite | 72 | 69–78 | sf.cc., cv. | |
| 4 | Galena | 76 | 71–84 | p. | |
| 2 | Pyrolusite | 76 / 252 / 279 / 292 | 76 / 252 / 256–346 / 225–405 | sf.cc. | Average hardness ⊥ to fibres / Average hardness ‖ to fibres / Isotropic sections / Microcrystalline |
| 2 | Stibnite | 77 | 42–109 | p. | |
| 1 | Chalcophanite | 81 / 124 / 133 | 71–85 / 103–165 / 110–178 | sf. | ⊥ to cleavage / ‖ to cleavage / Isotropic sections |
| 4 | Chalcocite | 84 | 68–98 | p.cc., cv. | |
| 1 | Antimony | 89 | 83–99 | p. | |
| 1 | Jamesonite | 99 / 113 | 96–105 / 105–121 | p.cc. | Granular allotriomorphic sections / Prismatic sections |

Table 11 *(continued)*

| Number of localities | Mineral species | Mean | Range | Quality of indentation* | Remarks |
|---|---|---|---|---|---|
| 4 | Bornite | 103 | 97–105 | p.cc. | |
| 2 | Bismuthinite | 107 | 92–119 | p. | |
| 2 | Miargyrite | 110 | 104–123 | sf.cc. | |
| 1 | Sylvanite | 110 | 102–125 | p. | |
| 1 | Kobellite | 116 | 69–173 | p. | |
| 1 | Proustite | 123 | 109–135 | p. | |
| 1 | Platinum | 126 | 125–127 | | |
| 1 | Copper | 134 | 120–143 | p.cv. | |
| 1 | Naumannite | 148 | 115–185 | p.cc. | |
| 1 | Zincite | {154 / 304 | 150–157 / 295–318} | p.cc. | ⊥ to cleavage / ‖ to cleavage |
| 1 | Pearceite | 160 | 153–164 | sf.cv. | |
| 3 | Enargite | {160 / 272 | 133–185 / 245–346} | sf. | ⊥ to cleavage / ‖ to cleavage |
| 2 | Boulangerite | 166 | 157–183 | p. | |
| 1 | Dyscrasite | 167 | 162–178 | p. | |
| 1 | Berthierite | 171 | 155–185 | sf. | |
| 1 | Zinkenite | 178 | 162–207 | sf. | |
| 1 | Emplectite | {191 / 222 | 168–213 / 197–238} | p.cc. | ‖ to elongation / ⊥ to elongation |
| 1 | Bournonite | 192 | 185–199 | sf. | |
| 7 | Chalcopyrite | 194 | 186–219 | p. | |
| 5 | Sphalerite | 198 | 186–209 | p.cc. | |
| 2 | Cuprite | 199 | 192–218 | p. | |
| 2 | Stannite | 210 | 197–221 | sf. | |
| 2 | Cubanite | 213 | 199–228 | sf. | |
| 3 | Pentlandite | 215 | 202–230 | p.cc. | |
| 1 | Tenorite | 236 | 209–254 | p.cc. | |
| 1 | Millerite | {236 / 254 / 348 | 225–256 / 235–280 / 318–376} | p. | Isotropic sections / ‖ to elongation / ⊥ to elongation |
| 5 | Pyrrhotine | {248 / 303 | 230–259 / 280–318} | p. | Anisotropic sections / Isotropic sections |
| 1 | Alabandite | 251 | 240–266 | p.cc. | |
| 1 | Coffinite | 258 | 236–333 | p. | |
| 2 | Chalcostibite | 276 | 264–285 | sf. | |
| 3 | Niccolite | {336 / 446 | 328–348 / 433–455} | p. | Anisotropic sections / Isotropic sections |
| 1 | Tennantite | 338 | 320–361 | p. | |
| 1 | Freibergite | 345 | 317–375 | sf. | |
| 2 | Scheelite | 348 | 285–429 | f. | |
| 4 | Tetrahedrite | 351 | 328–367 | sf. | |
| 1 | Famatinite | 363 | 333–397 | sf. | |
| 3 | Wolframite | 373 | 357–394 | p.cc. | |
| 2 | Manganite | 410 | 367–459 | p. | |
| 1 | Carrollite | 463 | 351–566 | p. | |
| 6 | Löllingite | {486 / 825 | 421–556 / 739–920} | p. | ⊥ to elongation / ‖ to elongation |

Table 11 (continued)

| Number of localities | Mineral species | Mean | Range | Quality of indentation* | Remarks |
|---|---|---|---|---|---|
| 3 | Siegenite | 524 | 503–533 | sf. | |
| 1 | Ullmannite | 525 | 498–542 | sf. | |
| 3 | Betafite | 525 | 503–560 | p. | |
| 4 | Ilmenite | {536 / 681 | 519–553 / 659–703} | sf.cc., cv. | Possible differences in composition |
| 5 | Goethite | {554 / 803 | 525–620 / 772–824} | sf. | Microcrystalline / Coarsely crystalline |
| 6 | Magnetite | 560 | 530–599 | p. | |
| 1 | Breithauptite | 563 | 542–584 | p. | |
| 1 | Psilomelane | 572 | 503–627 | p.cc. | |
| 1 | Hausmannite | 587 | 541–613 | sf.cc., cv. | |
| 2 | Braunite | 595 | 584–605 | p. | |
| 2 | Pyrochlore | 613 | 572–665 | sf. | |
| 2 | Hollandite | 620 | 560–724 | p. | |
| 5 | Skutterudite | 653 | 589–724 | p. | |
| 1 | Gersdorffite | 698 | 665–743 | sf. | |
| 2 | Maucherite | 704 | 685–724 | p. | |
| 3 | Euxenite | 707 | 599–782 | p. | |
| 2 | Rammelsbergite | 712 | 687–778 | sf. | |
| 4 | Brannerite | 720 | 710–730 | p. | |
| 5 | Pitchblende | 720 | 673–803 | f. | Fresh specimens, oxidation produces marked decrease in hardness. |
| 1 | Lepidocrocite | 724 | 690–782 | sf. | |
| 1 | Jacobsite | 734 | 724–745 | p. | |
| 3 | Davidite | 745 | 707–803 | p. | |
| 5 | Hematite | {755 / 1,009 | 739–822 / 920–1,062} | sf. | Microcrystalline / Coarsely crystalline |
| 1 | Pararammelsbergite | 772 | 762–803 | sf. | |
| 1 | Coronadite | 784 | 767–813 | f. | |
| 3 | Columbite-tantalite | 803 | 724–882 | p. | |
| 1 | Uraninite | 808 | 782–839 | f. | |
| 1 | Maghemite | 946 | 894–988 | p. | |
| 1 | Bixbyite | 1,018 | 1,003–1,033 | p. | |
| 3 | Thorianite | 1,918 | 988–1,115 | f. | |
| 2 | Cassiterite | 1,053 | 1,027–1,075 | p. | |
| 3 | Arsenopyrite | 1,094 | 1,048–1,127 | sf. | |
| 1 | Bravoite | 1,097 | 1,003–1,288 | sf. | |
| 4 | Marcasite | 1,113 | 941–1,288 | f. | |
| 1 | Glaucodot | 1,124 | 1,071–1,166 | sf. | |
| 2 | Rutile | 1,139 | 1,074–1,210 | p. | |
| 5 | Pyrite | 1,165 | 1,027–1,240 | f. | |
| 2 | Cobaltite | 1,200 | 1,176–1,226 | sf. | |
| 2 | Chromite | 1,206 | 1,195–1,210 | p. | |

* p = perfect, sf. = slightly fractured, f. = fractured; cc. = concave and cv. = convex refer to the shape of the edges of the indentation where curvature is marked.

Reproduced with permission from Bowie and Taylor (1958).

Reflectivity is possibly the most important aid to ore mineral identification, and often all that is required to confirm deductions made from other visual observations is to check the reflectivity value. However, if there is still any uncertainty the micro-indentation hardness can be measured. No further confirmation may be necessary, but if it is, the next step is generally to observe such properties as bireflection, hardness anisotropy and easily observable properties such as colour, anisotropism, polarization colours, or the presence or absence of internal reflections. All these characteristics can be studied without damage to the polished surface and without complicated and time-consuming measurements. It is therefore recommended that such observations precede any additional measurements or tests that may be necessary.

### D. COLOUR

Colour was recognized by Murdoch (1916) as a useful aid in ore mineral identification. But as few show distinct colours it is not easy to recognize the delicate differences in shade and colour that exist. The main difficulty in the use of colour, however, is one of definition. Available colour charts all give much stronger colours than are observed in ore minerals and, until suitable standards are available, full use of the property of colour in opaque mineral identification can only be made by using a comparison eyepiece (Fig. 23). In this equipment two microscopes are bridged by a special eyepiece which permits the unknown to be compared directly with the standard by placing one on each microscope stage and examining the two half fields in juxta-position. The light sources used should be identical. However, if the colour temperature of the two lamps is not exactly the same this can be adjusted by fitting a rheostat in the lamp circuit.

Some minerals show distinctly different colours in air and oil. For example a basal section of covellite is deep blue in air, but red–violet in cedar oil ($n = 1\cdot515$).

### E. BIREFLECTION

Isotropic minerals in polished section normally give one reflectivity value in the same way as transparent isotropic minerals give one refractive index in transmitted light. Similarly, most sections of anisotropic minerals have two principal directions that differ in their reflecting power for rays vibrating parallel to them. The bireflection of any mineral grain can readily be measured by rotating the stage to one extinction and noting the reflectivity value, and repeating this for the second extinction position. The difference between the two readings will give a measure of the bireflection of the section. The value will not necessarily be the maximum bireflection for the mineral, but if a number of grains of the same mineral showing the most marked anisotropism are examined, the maximum bireflection can usually be obtained.

Bireflection can be observed in oil as well as in air; some minerals show appreciable enhancement of bireflection in oil and others little change.

### F. ANISOTROPISM AND POLARIZATION COLOURS

The reflectivities corresponding to the two principal directions in the surface of an anisotropic mineral often vary independently of the wavelength of the light employed. This property is known as dispersion of the bireflection and is the main cause of the colours seen under crossed polars. Polarization colours can be observed both in air and in oil, but, as they are more pronounced in oil, they are normally given for oil immersion. It has been shown

FIG. 23. Comparison eyepiece fitted to two identical ore microscopes. One lamp is in circuit with a variable resistance to control any difference in light intensity between the two lamps. Reproduced with permission from Bowie and Taylor (1958).

(Galopin, 1947; Hallimond, 1953) that if the microscope is accurately adjusted to give extinction, by using a Nakamura plate, sections in the 45° position often exhibit characteristic polarization colours. Strain in objectives may cause difficulty in the accurate crossing of the polars; however, this can be minimized by using objectives that can be rotated in their holder, if they are not entirely strain free. Polarization colours for most anisotropic minerals in the 45° position display one or two predominant colours or shades of these colours (Bowie and Taylor, 1958). The colours differ from those quoted in standard text-books on ore minerals. For example, niccolite, which has a characteristic polarization colour in shades of blue-green, is usually given in the literature as being anisotropic with strong colours variously described as yellowish, greyish-green, bluish-green or violet-green.

The relation of the positions of extinction to the crystallographic directions, such as crystal outlines of cleavage, should always be noted as such information can be of diagnostic value.

### G. INTERNAL REFLECTIONS

A number of ore minerals are sufficiently transparent for some of the light incident upon them to penetrate well into the crystal. Some of this light may be reflected back from cleavage planes, crystal boundaries or from cracks and other flaws. Sphalerite, for example, often shows pale yellow to brown internal reflections and zincite red internal reflections. Usually a number of grains of any particular mineral have to be examined before it can be decided whether or not it shows internal reflections. Occasionally, the presence or absence of internal reflections is a guide to chemical composition, as in the case of sphalerite which exhibits abundant internal reflection if it is low in iron and no internal reflections if it is rich in iron. The presence of internal reflections can result in erroneous reflectivity values being obtained, and areas showing internal reflections should therefore be avoided when measurements are made.

## VIII. Determinative Tables

In cases where mineral species cannot be distinguished on the basis of reflectivity and micro-indentation hardness; colour, bireflection, ansotropism, polarization colours and presence or absence of internal reflections can be used to assist in their identification. Tables 12–15 (reproduced with permission from Bowie and Taylor, 1958) incorporating these data are given under the following headings: oxides; isotropic cobalt, nickel and iron sulphides and arsenides; anisotropic cobalt, nickel and iron sulphides and arsenides; and sulpho-salts.

## IX. Conclusions

Quantitative methods in the field of ore mineralogy are undergoing rapid development following the introduction of sensitive photometers and "continuous band" interference filters that enable reflectivity dispersion curves to be prepared for different mineral species. Low-load indentation instruments suitable for use in determining the hardness of mineral grains have also become available. It is to be anticipated that within a few years accurate quantitative data for micro-indentation hardness and for spectral reflectivity both in air and oil will be available. Use of these constants will greatly facilitate opaque-mineral identification and is likely to eliminate etch

## TABLE 12

*Oxides*

| Mineral species | Reflectivity in white light (mean and range) | Hardness (mean and range) | Colour | Anisotropism | Internal reflections |
|---|---|---|---|---|---|
| *Oxides of low reflectivity* | | | | | |
| Scheelite $CaWO_4$ | 10·0 | 348 / 285–429 | dark grey | masked by internal reflections | white |
| Zincite $ZnO$ | 11·2 | 234 / 150–318 | dark grey | masked by internal reflections | red |
| Cassiterite $SnO_2$ | 12·0 / 11·2–12·8 | 1,053 / 1,027–1,075 | dark grey | moderate | yellow-brown |
| Chromite $FeCr_2O_4$ | 12·1 | 1,206 / 1,195–1,210 | dark grey | isotropic | red |

| Mineral species | Reflectivity in white light (mean and range) | Hardness (mean and range) | Colour | Anisotropism | Predominant polarization colours in 45° position |
|---|---|---|---|---|---|
| *Oxides of medium reflectivity* | | | | | |
| Manganite $MnO.OH$ | 17·0 / 14·0–20·0 | 410 / 367–459 | light grey to dark brownish grey | strong | yellowish white |
| Goethite $FeO.OH$ | 17·3 / 16·1–18·5 | 803* / 772–824 | grey | moderate | blue-grey mouse-grey |
| Wolframite $(Fe,Mn)WO_4$ | 17·3 / 16·2–18·5 | 373 / 357–394 | brownish grey | weak | greenish grey |
| Hausmannite $Mn_3O_4$ | 17·5 / 16·0–19·0 | 587 / 541–613 | dark grey | strong | olive-buff with numerous fine scratches |
| Jacobsite $MnFe_2O_4$ | 18·5 | 734 / 724–745 | olive-grey | isotropic | |
| Braunite $(Mn, Si)_2O_3$ | 18·8 / 17·8–19·8 | 595 / 584–605 | dark grey | weak | grey |
| Ilmenite $FeTiO_3$ | 19·4 / 17·8–21·1 | 611 / 519–703 | light brown | moderate | pale greenish grey |
| Rutile $TiO_2$ | 20·2 | 1,139 / 1,074–1,210 | grey | strong, usually masked by internal reflections | |
| Lepidocrocite $FeO.OH$ | 20·4 / 15·8–25·0 | 724 / 690–782 | grey-white | strong | light greenish yellow |
| Magnetite $Fe_3O_4$ | 21·1 | 560 / 530–599 | grey | isotropic | |
| *Oxides of high reflectivity* | | | | | |
| Bixbyite $(Mn, Fe)_2O_3$ | 23·0 | 1,018 / 1,003–1,033 | white with yellow tint | isotropic | |

TABLE 12 (*continued*)

| Mineral species | Reflectivity in white light (mean and range) | Hardness (mean and range) | Colour | Anisotropism | Predominant polarization colours in 45° position |
|---|---|---|---|---|---|
| *Oxides of high reflectivity* | | | | | |
| Psilomelane $BaMnMn_8O_{16}$ $(OH)_4$ | 23·5 23·0–24·0 | 572 503–627 | white | isotropic (poorly crystalline material) | |
| Maghemite $Fe_2O_3$ | 25·0 | 946 894–988 | white | isotropic | |
| Hematite $Fe_2O_3$ | 27·5 25·0–30·0 | 1,009* 920–1,062 | white | strong | pale green |
| Coronadite $MnPbMn_6O_{14}$ | 29·0 26·0–32·0 | 784 767–813 | white with yellow tint | strong | pale green |
| Hollandite $MnBaMn_6O_{14}$ | 29·2 26·0–32·5 | 620 560–724 | white with yellow tint | strong | pale green |

| Mineral species | Reflectivity in white light (mean and range) | Hardness (mean and range) | Colour | Anisotropism | Internal reflections |
|---|---|---|---|---|---|
| *Oxides of uranium and thorium (including coffinite)* | | | | | |
| Coffinite $U(SiO_4)_{1-x}$ $(OH)_{4x}$ | 9·9 | 258 236–333 | grey-brown | anisotropic | |
| Pyrochlore $NaCaNb_2O_6F$ | 13·4 13·0–13·8 | 613 572–665 | light grey | isotropic | greenish yellow to brown |
| Betafite $(U, Ca)(Nb, Ti, Ta)_3O_9$ | 14·5 14·0–14·9 | 525† 503–560 | light grey | isotropic (metamict) | brown |
| Thorianite $ThO_2$ | 14·6 14·0–15·3 | 1,018 988–1,115 | light grey | isotropic | |
| Brannerite $(U, Ca, Fe, Y)_3Ti_5O_{16}$ | 14·8 13·7–16·0 | 720 710–730 | light grey | isotropic (metamict) | reddish brown |
| Euxenite $(Y, Er..., U)(Nb, Ti, Ta)_2O_6$ | 15·0 | 707 599–782 | light grey | isotropic (metamict) | brown |
| Uraninite $UO_2$ | 16·8 | 808 782–839 | light grey | isotropic | |
| Pitchblende $UO_2$ | 16·0 | 720 673–803 | light grey | isotropic | |
| Columbite-tantalite $(Fe, Mn)(Nb, Ta)_2O_6$ | 17·1 16·3–18·0 | 803 724–882 | greyish white | anisotropic | reddish |
| Davidite $(Fe, U, Ce)(Ti, Fe)_3O_7$ | 17·8 | 745 707–803 | light grey | isotropic (metamict) | reddish brown |

\* Coarsely crystalline material.
† Betafite from Madagascar has a mean hardness of 642.

TABLE 13

*Isotropic cobalt, nickel, and iron sulphides and arsenides*

| Mineral species | Reflectivity (white light) | Hardness (mean and range) | Colour |
|---|---|---|---|
| Carrollite $CuCo_2S_4$ | 44·0 | 463 351–566 | white |
| Bravoite $(Ni, Fe)S_2$ | 45·5 | 1,097 1,003–1,288 | pale violet |
| Gersdorffite NiAsS | 47·5 | 698 665–743 | white |
| Ullmannite NiSbS | 47·5 | 525 498–542 | white |
| Siegenite $(Co, Ni)_3S_4$ | 48·6 | 524 503–553 | pinkish cream |
| Cobaltite CoAsS | 52·7 | 1,200 1,176–1,226 | pinkish white |
| Pyrite $FeS_2$ | 54·5 | 1,165 1,027–1,240 | yellowish white |
| Skutterudite $(Co, Ni)As_3$ | 55·8 | 653 589–724 | white |

TABLE 14

*Anisotropic cobalt, nickel, and iron sulphides and arsenides*

| Mineral species | Reflectivity in white light (mean and range) | Hardness (mean and range) | Colour | Anisotropism | Predominant polarization colours in 45° position |
|---|---|---|---|---|---|
| Breithauptite NiSb | 49·9 45·3–54·6 | 563 542–584 | coppery pink | strong | bright yellowish green |
| Maucherite $Ni_3As_2$ | 51·2 | 704 685–724 | pinkish white | weak | grey |
| Marcasite $FeS_2$ | 52·2 48·9–55·5 | 1,113 941–1,288 | pale yellowish white | strong | green, pinkish buff |
| Glaucodot $(Co, Fe)AsS$ | 52·5 | 1,124 1,071–1,166 | creamy white | weak | grey, almost isotropic |
| Arsenopyrite FeAsS | 53·7 51·7–55·7 | 1,094 1,048–1,127 | creamy white | strong | reddish brown, greenish blue |
| Löllingite $FeAs_2$ | 53·8 53·0–54·7 | 671 421–920 | white | strong | pale greenish yellow in prismatic sections |
| Niccolite NiAs | 55·1 52·0–58·3 | 392 328–455 | coppery pink | strong | pale green, greenish blue |
| Millerite NiS | 57·0 54·0–60·0 | 301 225–376 | yellow | strong | pale yellow |
| Rammelsbergite $NiAs_2$ | 59·0 58·0–60·0 | 712 687–778 | white | strong | light brownish grey, bluish grey |
| Pararammelsbergite $NiAs_2$ | 61·0 60·5–62·0 | 772 762–803 | white | strong | pinkish buff |

TABLE 15

*Sulpho-salts*

| Mineral species | Reflectivity in white light (mean and range) | Hardness (mean and range) | Colour | Anisotropism | Predominant polarization colours in 45° position (prismatic sections) |
|---|---|---|---|---|---|
| Proustite $Ag_3AsS_3$ | 26·4 25·0–27·7 | 123 109–135 | bluish grey | strong | greenish grey |
| Enargite $Cu_3AsS_4$ | 26·5 25·0–28·1 | 239 133–346 | pinkish grey | strong | reddish brown, pinkish buff |
| Famatinite $Cu_3SbS_4$ | 26·9 25·1–28·7 | 363 333–397 | pinkish brown | strong | pinkish buff, yellowish olive |
| Tennantite $(Cu, Fe)_{12}As_4S_{13}$ | 28·9 | 338 320–361 | pale greenish grey | isotropic | |
| Pyrargyrite $Ag_3SbS_3$ | 29·6 28·4–30·8 | 88 50–126 | bluish grey | moderate, often isotropic | shades of grey |
| Freibergite $(Cu, Ag)_{12}Sb_4S_{13}$ | 29·8 | 345 317–375 | yellowish grey | isotropic | |
| Pearceite $(AgCu)_{16}As_2S_{11}$ | 30·1 | 160 153–164 | pale greenish grey | isotropic | |
| Tetrahedrite $(Cu, Fe)_{12}Sb_4S_{13}$ | 30·7 | 351 328–367 | brownish grey | isotropic | |
| Zinkenite $Pb_6Sb_{14}S_{27}$ | 32·3 | 178 162–207 | white | weak, almost isotropic | shades of grey |
| Miargyrite $AgSbS_2$ | 33·9 31·8–36·0 | 110 104–123 | bluish grey | strong | pale yellow |
| Berthierite $FeSb_2S_4$ | 35·0 30·0–40·0 | 171 155–185 | white | strong | bluish grey |
| Stibnite $Sb_2S_3$ | 35·1 30·2–40·0 | 77 42–109 | white | strong | pale green |
| Bournonite $CuPbSbS_3$ | 37·1 36·0–38·2 | 192 185–199 | white | strong | greenish blue, greenish grey |
| Jamesonite $Pb_4FeSb_6S_{14}$ | 38·0 36·0–40·0 | 109 96–121 | white | strong | greyish green |
| Emplectite $CuBiS_2$ | 38·5 36·0–41·0 | 203 168–238 | creamy white | strong | yellowish green, green |
| Chalcostibite $CuSbS_2$ | 40·0 37·1–43·0 | 276 264–285 | white | strong | bluish green |
| Boulangerite $Pb_5Sb_4S_{11}$ | 40·6 37·0–44·1 | 166 157–183 | white | strong | greenish grey |
| Kobellite $Pb_6FeBi_4Sb_2S_{16}$ | 42·1 40·9–43·2 | 116 69–173 | white | strong | greyish brown |
| Bismuthinite $Bi_2S_3$ | 45·4 42·0–48·7 | 107 92–119 | white | strong | yellowish green |

and microchemical tests. Etching techniques, however, are still necessary to bring out textural features and internal structures in mineral grains such as twinning, zonal growth and "etch parting". Microchemical tests for specific elements are rarely employed as an aid to opaque-mineral determination, and quantitative measurement of the elements comprising mineral species by electron probe is rapidly replacing the qualitative technique of assessing the distribution of elements by chromatographic contact printing (Evans, 1966).

## ACKNOWLEDGEMENTS

The writer is indebted to Professor E. N. Cameron for permission to publish Section V of this Chapter which is based on instructions given at the International Summer School for Quantitative Methods in Reflected-light Microscopy held at Cambridge, England, in 1963. He is also grateful to the Director, National Physical Laboratory, for permission to publish the data contained in Tables 6, 7 and 8.

## REFERENCES

Berek, M. (1937). Optische Messmethoden im polarisierten Auflicht. *Fortschr. Miner. Kristallogr. Petrogr.* 22, 1.
Bowie, S. H. U. (1954). *Mem. geol. Surv. Summ. Prog. 1953*, p. 71.
Bowie, S. H. U. (1957). The photoelectric measurement of reflectivity. *Mineralog. Mag.* 31, 476.
Bowie, S. H. U. and Henry, N. F. M. (1964). *Trans. Instn Min. Metall.* 73, 467.
Bowie, S. H. U. and Taylor, K. (1958). A system of ore mineral identification. *Mining Mag., Lond.* 99, 265, 337.
Brown, A. R. G. and Ineson, E. (1951). Experimental survey of low-load hardness testing instruments. *J. Iron Steel Inst.* 169, 376.
Cambon, T. (1949). Service de documentation de l'aeronautique. *Publs. scient. tech. Minist. Air.* No. 236.
Cameron, E. N. (1957). Apparatus and techniques for measurement of certain optical properties of ore minerals in reflected light. *Econ. Geol.* 52, 252.
Cameron, E. N. (1961). "Ore Microscopy." John Wiley and Sons, New York.
Campbell, W. (1906). The microscopic examination of opaque minerals. *Econ. Geol.* 1, 751.
Davy, W. M. and Farnham, C. M. (1920). "Microscopic Examination of the Ore Minerals." McGraw-Hill, New York.
Donnay, J. D. H. (1930). Thinned polished sections. *Econ. Geol.* 25, 270.
Drude, P. (1888). Beobachtungen über die Reflexion des Lichtes am Antimonglanz. *Annln. Phys.* 34, 489.
Evans, W. D. (1966). Membrane colorimetry. *Trans. Instn Min. Metall.* 75B, 165.
Försterling, C. (1908). Die optische Konstanten von Eisenglanz. *Neues Jb. Miner. Geol. Paläont. BeilBd.* 25, 344.
Galopin, R. (1947). Observations sur la dispersion du plan de polarisation après réflexion normale sur les minéraux métalliques anisotropes. Avec un exposé général sur la réflexion normale. *Schweiz. miner. petrogr. Mitt.* 27, 190.
Hallimond, A. F. (1953). "Manual of the Polarizing Microscope." (2nd edn.) Cooke, Troughton and Simms Ltd., York, England.
Hallimond, A. F. (1963). Polishing mineral specimens. *Mining Mag., Lond.* 108, 197.

Hallimond, A. F. and Bowie, S. H. U. (1964). On the reflectivity of pyrite. *Mining Mag.*, *Lond.* **111**, 385.

Hallimond, A. F. and Payne, B. O. (1964). New British polarizing microscope. *Mining Mag.*, *Lond.* **111**, 162, 247.

Harrison, R. K. and Day, G. (1963). A continuous monochromatic interference filter. *Mineralog. Mag.* **33**, 517.

Kennedy, G. C. (1945). The preparation of polished thin sections. *Econ. Geol.* **40**, 355.

Murchison, D. G. (1964). Reflectance techniques in coal petrology and their possible application in ore mineralogy. *Trans. Instn Min. Metall.* **73**, 479.

Murdoch, J. (1916). "Microscopical Determination of the Opaque Minerals." John Wiley and Sons, New York.

Nakhla, F. M. (1956). The hardness of metallic minerals in polished sections. *Econ. Geol.* **51**, 811.

Orcel, J. (1927). Sur l'emploi de la pile photo-électrique pour la mesure de pouvoir réflecteur des minéraux opaques. *C. r. hebd. Séanc. Acad. Sci., Paris* **185**, 1141.

Phillips, F. C. (1937). A universal ore-polishing machine. *Mineralog. Mag.* **24**, 595.

Piller, H. and von Gehlen, K. (1964). On errors of reflectivity measurements and of calculations of refractive index n and absorption coefficient k. *Am. Miner.* **49**, 867.

Schneiderhöhn, H. (1928). Ein neues Photometerokular zur Messung des Reflexionsvermögens in Erzanschliffen. *Centbl. Miner. Geol. Palaont.* A, 394.

Schneiderhöhn, H. (1952). "Erzmikroskopisches Praktitkum." E. Schweizerbart'sche Verl., Stuttgart.

Schneiderhöhn, H. and Ramdohr, P. (1931). "Lehrbuch der Erzmikroskopie 2." Gebrüder Bornträger, Berlin.

Short, M. N. (1931). Microscopic determination of the ore minerals. *Bull. U.S. geol. Surv.* **825**; 2nd edn (1940), **914**.

Smith, F. H. (1964). A new incident illuminator for polarizing microscopes. *Mineralog. Mag.* **33**, 725.

Tabor, D. (1954). Mohs's Hardness Scale—A physical interpretation. *Proc. phys. Soc.* **67B**, 249.

Talmage, S. B. (1925). Quantitative standards for hardness of the ore minerals. *Econ. Geol.* **20**, 531.

Uytenbogaardt, W. (1951). "Tables for Microscopic Identification of Ore Minerals." Princeton University Press.

van der Veen, R. W. (1925). "Mineragraphy and Ore Deposition." 1. G. Naeff, The Hague.

Winchell, H. (1945). The Knoop microhardness tester as a mineralogical tool. *Am. Miner.* **30**, 583.

Young, B. B. and Millman, A. P. (1964). Microhardness and deformation characteristics of ore minerals. *Trans. Instn Min. Metall.* **73**, 437.

H

CHAPTER 4

# X-ray Fluorescence Spectrography

## K. NORRISH

*C.S.I.R.O. Division of Soils, Adelaide, South Australia*

## B. W. CHAPPELL

*Geology Department, Australian National University, Canberra, A.C.T.*

## I. Introduction

Just as in emission spectrography (Chapter 10) elements can be excited to cause them to emit light that can be used for analytical purposes, so also with appropriate excitation techniques elements can be made to emit in the

161

X-ray wavelength region, and these radiations can be used to identify and estimate the concentrations of elements in samples.

In X-ray fluorescence spectrography, as the name implies, excitation of characteristic X-rays is by means of a primary X-ray beam. Alternatively, excitation may be effected by an electron beam, as in the method of electron probe microanalysis (Chapter 5). There is much in common between the two techniques, so that many features of the present Chapter can be usefully referred to for both.

X-ray spectrography originated over 50 years ago with Moseley's study of the X-ray spectra of the elements, and in 1932 von Hevesy (1932) produced the first text book on the subject. However, the method had to await the technological advances of the last two decades before it could be taken from the physics laboratory and used as a simple and reliable method of chemical analysis.

Present commercially available X-ray spectrographs are capable of detecting the spectra of at least eighty elements (atomic number $> 10$) and analyses can be made on a wide variety of samples over a concentration range from 100% to several parts per million. With correct procedures, very precise analyses can be made in a short time. The method is essentially non-destructive, although some form of sample preparation is often required. The usefulness of such an analytical tool is obvious.

This Chapter is too brief to cover the whole field that has grown out of X-ray spectrography, and so the authors have concentrated on those analytical methods that are likely to be of most use in a mineralogical laboratory equipped with a commercial X-ray spectrograph. Most of the discussion is concerned with the various analytical methods that can be used and as far as possible each method is related to the fundamental properties of X-rays. Although successful analyses can be made by using the equipment in a completely empirical manner, some understanding of the theory underlying the various methods enables the analyst to choose the most appropriate method in a particular case. At the same time the limitations of each method will be appreciated. Several methods which in the authors' experience are applicable to a great variety of mineral analyses are given in some detail.

The nature of this book pre-supposes the use of equipment designed for the laboratory and there is no discussion of field instruments in which isotopes are used for excitation. Likewise, consideration of those instruments and techniques, such as on-stream analyzers, designed for industrial use is omitted. Non-dispersive methods of analysis are not considered, since although they can be applied successfully to particular samples, they would not meet the requirements of most mineralogical laboratories. Non-dispersive methods will find wider applicability as solid-state detectors are improved (Bowan *et al.*, 1966). The analysis of such samples as liquids or alloys is not covered fully. However, some of the methods described are applicable to such samples. Absorptiometry as a method of chemical analysis and X-ray absorptiometry and emission as applied to crystal chemistry (determination of valence state and co-ordination number) are subjects not treated. All of the above omissions are dealt with in the books and journals devoted to X-ray spectrography.

A book which deals fully with the nature, production and properties of X-rays is *X-rays in Theory and Experiment* (Compton and Allison, 1935). Books on X-ray spectrography giving a broader treatment of the subject than this Chapter have been written by Birks (1959), Liebhafsky *et al.* (1960) and Blokhin (1962). A recent book dealing with the more practical aspects is that of Jenkins and de Vries (1967).* Amongst the journals which should be consulted are *Advances in X-ray Analysis, Analytical Chemistry* and *Applied Spectroscopy*. The Analytical Review published as part of *Analytical Chemistry* every second year is particularly valuable (last issue April 1966). In addition to summarizing recent developments these reviews list references under element and type of sample. Manufacturers of X-ray spectrographic equipment also publish various application reports, newsletters and bibliographies.

## II. The Nature and Production of X-rays

### A. NATURE OF X-RAYS

X-rays form that portion of the spectrum of electromagnetic radiation that, for the purposes of this discussion, has wavelengths in the range from 0·1 to 50 Å.

X-rays may also be regarded as consisting of individual photons of energy, the energy of each photon being given by the relationship:

$$E = h\nu = \frac{hc}{\lambda} \tag{1}$$

where $h$ is Planck's constant, $\nu$ is the frequency of the radiation, $c$ is the velocity of electromagnetic radiation and $\lambda$ is the wavelength of the radiation. If the values of the constants $h$ and $c$ are substituted in equation (1) and if $\lambda$ is expressed in Ångström units and $E$ in electron volts, this expression may be written:

$$E = \frac{12,400}{\lambda} \tag{2}$$

It can be seen that X-rays with wavelengths between 0·1 and 50 Å consist of photons with energies in the range from 120 keV to 0·25 keV.

### B. PRODUCTION OF X-RAYS

X-rays are produced when a beam of electrons of sufficient energy strike any matter. X-rays may also be produced by irradiating matter with primary

* Also published recently in "X-ray Emission Spectography in Geology", by I. Adler (Elsevic, 1966).

X-rays produced by electron bombardment, the term "fluorescent" generally being applied to these secondary X-rays.

## C. X-RAY SPECTRA

The X-ray spectrum consists of two parts.

### 1. The Continuous Spectrum

When electrons of sufficient energy strike any matter, X-rays with a continuous spectrum of energies (and wavelengths) are produced. These energies range from that of the incident electrons down beyond the lower limit of X-ray photon energies. The maximum photon energy and the corresponding minimum wavelength depends only on the energy of the incident electrons and is independent of the nature of the material emitting the X-rays. Most diagrams of the continuous spectrum show intensity as a function of wavelength; when plotted in this way the curves have a maximum at approximately one and a half the minimum wavelength. The intensity (energy) produced per unit energy interval rises from zero at an energy corresponding to that of the incident electrons and in a linear fashion with decreasing photon energy (Cosslett and Dyson, 1957).

The total energy represented by the continuous spectrum, for a particular element, may be expressed as a proportion ($\epsilon$) of the energy of the incident electrons in the following approximate empirical relationship (Compton and Allison, 1935, p. 90):

$$\epsilon = 1 \cdot 1 \times 10^{-9} ZV \qquad (3)$$

where $Z$ is the atomic number of the element and $V$ is the accelerating potential for the electrons, expressed in volts. Thus, for example, with tungsten ($Z = 74$) and an accelerating potential of 50 kV, 0·4% of the energy of the incident electrons is converted into radiation of the continuous spectrum. The energy of the incident electrons is also proportional to $V$ and hence the energy of the continuous radiation for a given electron current is proportional to $ZV^2$.

The continuous spectrum obtained from an X-ray tube differs from that produced within the anode of the tube, since the lower energy radiation is selectively filtered out by both the anode (in which X-rays are produced to a depth depending on the energy of the incident electrons) and the window of the tube. An increase in the accelerating voltage results in greater absorption of X-rays in the anode, with the result that the energy outputs from X-ray tubes do not increase as rapidly as $V^2$. Likewise, the increased absorption in higher-atomic-number anodes (see equation 11) causes the emission from the X-ray tube to increase with $Z$ at a lesser rate than predicted by equation (3).

## 2. The Characteristic Spectrum

Characteristic or line spectra are produced when the incident electrons possess sufficient energy to remove electrons from the inner shells of an atom. The X-ray photons that result when outer electrons fall into the vacancy have an energy that is characteristic of a particular element. These characteristic spectral lines are superimposed on the continuous spectrum.

Incident electrons can expel electrons from the inner shells of atoms provided that their kinetic energy exceeds certain values, which are a physical characteristic of individual elements. The accelerating potentials required to give electrons these energies are referred to as *excitation potentials*. Characteristic spectra are also produced when matter is irradiated with X-rays. The analysis of these secondary X-rays forms the basis of X-ray fluorescence spectrography.

X-ray excitation differs from electron excitation in that no continuous spectrum is produced; in other respects, however, the principles are the same. X-ray photons of sufficient energy can remove an electron, known as a photoelectron, from the inner shells of an atom. The X-ray photons must possess an energy corresponding to the relevant excitation potential to remove a photoelectron.

When an electron is expelled by either process the atom is said to be excited and the vacancy is rapidly occupied by an electron falling from an outer shell. The energy of an electron in its initial state in an outer shell $(E_i)$ is higher than when it is in its final state in an inner shell $(E_f)$. As a result, an X-ray photon is emitted with an energy equal to the difference $(E_i - E_f)$. If the energies $E_i$ and $E_f$ are expressed in electron volts, then from equation (2) the wavelength of the emitted X-ray will be given by:

$$E_i - E_f = \frac{12,400}{\lambda} \tag{4}$$

The energies of electrons in the atoms of a particular element are fixed, so the difference $(E_i - E_f)$ can only take a limited number of values which are characteristic for each element.

The characteristic spectrum is divided into different series, the $K$, $L$, $M$, ... series, depending on whether the spectrum originated from a vacant space in the $K$, $L$, $M$, ... shell, respectively. Electron transitions into these spaces are governed by selection rules and some transitions are not possible. For those that are, different probabilities exist and these determine the relative intensities of the emission lines in a given spectral series. Spectral lines that arise from electron transitions predicted by the quantum theory are known as diagram lines, since the electron transitions can be represented by lines on an energy-level diagram. Some lines cannot be represented in this way. These are known as non-diagram lines or satellites (Compton and Allison, 1935, p. 654).

The atomic energy level for barium is shown in Fig. 1. The energies necessary to expel an electron from the different energy levels, i.e. the excitation potentials, are plotted along the vertical axis on a logarithmic scale. The differences between the various levels correspond to the possible values of $(E_i - E_f)$ in equation (4). Some of the various possible electron transitions between the levels are represented diagrammatically, the symbols for the radiation produced by these transitions being shown. They include all the characteristic X-rays likely to be of important analytical interest.

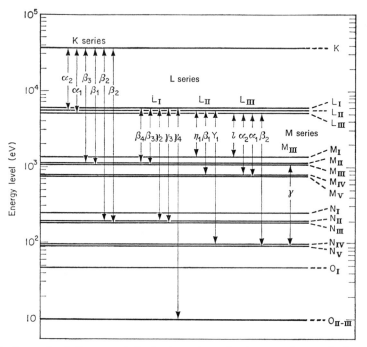

FIG. 1. Energy-level diagram of the barium atom showing energies required to expel electrons from a particular energy level. Data from Hill *et al.* (1952).

When the accelerating potential of the incident electrons or X-rays has an energy exceeding the excitation potential for a particular energy level, all the lines in the corresponding series of the characteristic spectrum are emitted simultaneously. So, for example, if electrons or X-rays with an energy of 5·25 keV strike a barium atom, the $L_{III}$ series lines will all be emitted, in addition to the M series, but no lines of the $K$, $L_I$ or $L_{II}$ series will be produced. The excitation potential for the $K$ electrons in barium is 37·4 keV and so an energy in excess of this is required to produce the complete characteristic spectrum of the barium atom. To excite the $L_{III}$ spectrum of barium, X-rays must have a wavelength shorter than 2·36 Å. This means that the $K_\alpha$ radiation of elements with the atomic number of chromium or greater will produce this series ($CrK_\alpha = 2·29$ Å, 5·41 keV). The portion of the associated continuous spectrum with wavelengths greater than 2·36 Å will not contribute. Chromium $K_\alpha$ radiation will not produce the $L_I$ and $L_{II}$ series of barium since energies of 5·99 and 5·63 keV, respectively, are required; however, the appropriate portion of the continuous spectrum will, and the associated chromium $K_\beta$ radiation will excite the $L_{II}$ spectrum but not the $L_I$ spectrum ($CrK_\beta = 5·96$ keV).

For any element, the energy differences ($E_i - E_f$) are greatest when the $K$ shell is concerned, and the $K$ spectrum X-rays thus have the highest energies and the shortest wavelengths. The $L$ spectral lines have lower energies, and so on. An idea of the relative energies of the various characteristic radiations can be obtained from Fig. 1.

In 1913, Moseley showed that the wavelength of a particular characteristic line changed in a systematic fashion with the atomic number, $Z$, the two being related by the equation:

$$\frac{1}{\lambda} = K(Z - \sigma)^2 \tag{5}$$

where $K$ and $\sigma$ are constants whose values depend on the particular series. For the $K_\alpha$ lines, equation (5) takes the form:

$$\frac{1}{\lambda_{K_\alpha}} = 8·2303 \times 10^4(Z-1)^2 \tag{6}$$

Equation (5) is called Moseley's Law. Its discovery provided a simple test of the order of the elements according to atomic number, showed where elements were missing from the periodic table and led to the discovery of some of these elements.

In the $K$ spectrum, the $\alpha_1$ and $\alpha_2$ lines are generally unresolved in commercial spectrographs and $\beta_1$ is the only other line with appreciable intensity, so that this spectrum consists effectively of two main lines, the $K_\alpha$ and the $K_\beta$. The latter line has about one-tenth the intensity of the $K_\alpha$ line, which is the one normally used analytically. The $L$ spectrum is more complex. The $\alpha_1$ line is the most intense but $\beta_1$, $\beta_2$, $\gamma_1$ and others have appreciable intensities so that a wider selection is available in choosing an analytical line.

Table 1 shows the absorption edges and the main lines associated with the $K$ and $L$ spectra. Cauchois and Hulubei (1947) and Bearden (1964) have compiled tables of X-ray wavelengths. Tables giving the spectra of elements as energies (keV) are also available (Fine and Hendee, 1956).

H*

In an X-ray tube, the intensity (rate of production of photons per unit time) of any characteristic line is given by the relationship:

$$I = Ci\,(V - V_0)^p \tag{7}$$

where $C$ is a constant, $i$ is the electron current, $V$ is the accelerating potential for the electrons, $V_0$ is the appropriate excitation potential and $p$ is an exponent whose value is about $1\cdot7$ for potentials $V$ less than about $3V_0$. For higher voltages, $p$ has smaller values.

### D. THE PASSAGE OF X-RAYS THROUGH MATTER

X-rays are both scattered and absorbed in passing through matter. Except in the case of short wavelength X-rays and low-atomic-number elements, for which combination Compton scattering (see p. 182) is comparatively high (Compton and Allison, 1935, p. 535), the process of photoelectric absorption is the dominant one. If a parallel beam of monochromatic X-rays of intensity $I_0$ passes through a material of thickness $x$, then the intensity $I$ of the radiation that emerges, after losses by both scattering and absorption, is related to $I_0$ by the following equation:

$$I = I_0\,e^{-\mu x} \tag{8}$$

$\mu$ being known as the *linear absorption coefficient*.

A more useful coefficient, which is independent of the physical state of the absorbing medium, can be derived by introducing the density $\rho$ of the medium into the above equation, so that we have:

$$I = I_0\,e^{-\mu/\rho\;\rho x} \tag{9}$$

where $\mu/\rho$ is known as the *mass-absorption coefficient* and $\rho x$ is the mass per unit area of the absorber.

Mass-absorption coefficients have a fixed value for a given element for a particular wavelength of X-rays. Tables of these coefficients may be found in various books (Compton and Allison, 1935; Allen, 1956; Birks, 1963; Theisen, 1965), but the best data is probably that of Heinrich (1966).

If the weight fractions of the elements in a sample are $p_1, p_2, p_3, \ldots$ and the corresponding mass-absorption coefficients for a particular wavelength are $(\mu/\rho)_1, (\mu/\rho)_2, (\mu/\rho)_3, \ldots$ then the mass-absorption coefficient of the sample is given by:

$$(\mu/\rho) = p_1(\mu/\rho)_1 + p_2(\mu/\rho)_2 + p_3(\mu/\rho)_3 + \cdots \tag{10}$$

As an example, we may calculate the mass-absorption coefficient of diopside,

$CaMgSi_2O_6$, for chromium $K_\alpha$ radiation. The following data are available:

$$CrK_\alpha \; \lambda = 2 \cdot 291 \text{ Å}$$

| Element | Percentage present | $\mu/\rho$ for $\lambda = 2 \cdot 291$ Å |
|---------|--------------------|-------------------------------------------|
| O | 44·32 | 39 |
| Mg | 11·23 | 119 |
| Si | 25·94 | 184 |
| Ca | 18·51 | 469 |

The mass-absorption coefficient of the mineral is given by:

$$\mu/\rho = 0 \cdot 443 \times 39 + 0 \cdot 112 \times 119 + 0 \cdot 259 \times 184 + 0 \cdot 185 \times 469 = 165$$

For a given element, the mass-absorption coefficient increases with increasing wavelength, except for discontinuities at particular wavelengths, known as *absorption edges*. Between these absorption edges, the absorption coefficients increase with wavelength approximately according to the following empirical relationship (Compton and Allison, 1935, p. 534)

$$\mu/\rho = K \lambda^u Z^v \tag{11}$$

where $K$, $u$ and $v$ are constants. The value of $K$ changes at each absorption edge. The exponents $u$ and $v$ are to some extent a function of $\lambda$ and $Z$, having a value of about 3.

These characteristics of absorption are illustrated in Fig. 2, in which the mass-absorption coefficient of barium carbonate is shown for various wavelengths. In this figure, the value of $\mu/\rho$ increases according to the requirements of equation (11) until the wavelength reaches a value of $0 \cdot 332$ Å, at which point an absorption edge is reached and the magnitude of the absorption coefficient drops sharply. At longer wavelengths, the absorption rises again until three more absorption edges are encountered at wavelengths of $2 \cdot 068$, $2 \cdot 204$ and $2 \cdot 363$ Å. The wavelengths at which these four absorption edges occur correspond to energies of $37 \cdot 43$, $5 \cdot 99$, $5 \cdot 63$ and $5 \cdot 25$ keV. Reference to Fig. 1 shows that it is these energies that are required to expel $K$, $L_I$, $L_{II}$, and $L_{III}$ electrons from a barium atom. It can be seen that these absorption edges occur because X-ray quanta with wavelengths shorter than that corresponding to an absorption edge of the absorbing element have enough energy to eject electrons from a particular shell of that absorber. The absorption edge corresponding to the ejection of $K$ shell electrons is known as the $K$ absorption edge, those corresponding to the ejection of the $L_I$, $L_{II}$ and $L_{III}$ electrons are known as the $L_I$, $L_{II}$ and $L_{III}$ absorption edges, and so on.

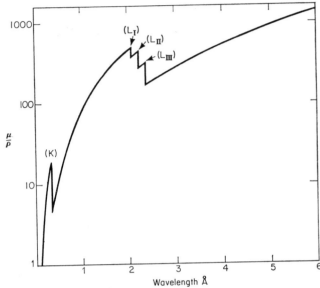

FIG. 2. Mass-absorption coefficient of barium carbonate as a function of wavelength.

Values of the wavelengths of absorption edges are readily available and these are extremely useful in selecting the optimum conditions of excitation in X-ray fluorescence analysis (see Table 1).

# III. Equipment

### A. COMPONENTS OF SPECTROGRAPHS

An X-ray spectrograph is an instrument for dispersing X-rays and for measuring their intensity as a means of making qualitative or quantitative analyses. The arrangement of the components of an X-ray spectrograph is shown in Fig. 3.

The fluorescent radiation is produced by directing X-rays from an X-ray tube operating with a stabilized power supply on to the sample to be analyzed.

The various components of X-ray spectrographs will be briefly discussed below. Many aspects of these subjects are treated more fully in the publications of manufacturers of commercial equipment.

### 1. Geometry

Many different geometrical arrangements have been used in X-ray spectrographs, but the one most commonly used is the flat crystal non-focusing type.

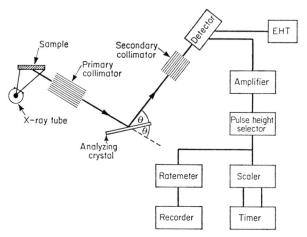

FIG. 3. Components of an X-ray spectrograph.

In this, a small proportion of the fluorescent radiation emitted by the sample is selected by a primary collimator and allowed to pass on to the flat surface of a crystal that is cleaved in a direction parallel to the diffracting planes being utilized. These X-rays are diffracted by the crystal according to the *Bragg equation*:

$$n\lambda = 2\, d\, \sin\theta \tag{12}$$

where $n$ is an integer and $\lambda$ is the wavelength of X-rays diffracted through an angle $2\theta$ by diffracting planes in the crystal spaced a distance $d$ apart. A secondary collimator is placed at an angle $2\theta$ to the primary collimator and the X-rays pass through this to the detector. A goniometer enables the crystal and detector to be moved together so that they are always at angles of $\theta$ and $2\theta$, respectively, to the primary collimator. The detailed geometry of the non-focusing spectrograph has been discussed by Spielberg *et al.* (1959). Spectrographs may have a focusing geometry and these types have a better resolution. However, since these geometries are discussed by Long elsewhere in this book (Chapter 5) they will not be treated here.

### 2. Generator and X-ray Tube

The function of the high-voltage generator is to supply a stabilized high voltage which can be applied between the cathode and anode of the X-ray tube and to stabilize the electron current through the tube, at selected values of tube voltage and current.

X-ray tubes are available with various anode materials, including: gold, platinum, tungsten, silver, molybdenum, chromium and titanium. The choice of X-ray tube in a particular case is governed by the following factors:

(a) A tube clearly cannot be used for the analysis of small amounts of an element if the anode material corresponds to the element being analyzed or if the characteristic radiation from the anode or its Compton-scattered radiation interferes with the measurement of the element being sought. For example, the characteristic radiation from a Cr tube interferes with determination of chromium ($CrK_\alpha$ line) and manganese ($CrK_\beta$ line). The $L$ spectrum from a W tube interferes with determinations of nickel, copper and zinc.

(b) The continuous spectrum produced by an X-ray tube generally has many times more energy than the superimposed characteristic spectrum. Thus for the majority of elements, most of the fluorescent radiation produced in a sample is excited by the continuous spectrum whose energy, according to (3), increases with atomic number of the anode; i.e. for general use, gold, platinum and tungsten are the best targets. For elements having absorption edges at a slightly longer wavelength than the characteristic lines from the anode, these characteristic lines can be used to good effect. The use of Cr radiation in this way to excite Ba$L$ radiation has been mentioned above. Likewise, the Au characteristic spectrum is suitable for exciting nickel, copper, zinc and other elements.

(c) X-ray tubes with chromium, titanium and some other anode materials are manufactured with thin beryllium windows ($\sim 0\cdot3$ mm as against the 1 mm normally used). These windows transmit a greater proportion of the long wavelengths that give efficient excitation of the light elements ($Z < 20$), and so such tubes should be used for the analysis of light elements. For example, a Cr tube with a thin window gives about six times the intensity for Si$K_\alpha$ and Al$K_\alpha$ as a W tube operated at the same power.

Of the common target materials, chromium, gold and molybdenum are the most useful for the trace and major analysis of silicates.

For the analysis of very light elements ($Z = 5$–$9$) demountable, windowless or very thin windowed, X-ray tubes have been constructed (Henke, 1963; Wyckoff and Davidson, 1965). A variety of anodes can be used in these tubes, which are generally operated at a comparatively low voltage and high current. Such X-ray tubes require considerably more maintenance than sealed tubes, and are not commercially available at present.

## 3. Analyzing Crystals

The function of the analyzing crystal is to disperse the X-rays being analyzed, according to the Bragg equation (12). A wide variety of analyzing crystals is available, including the following:

| Crystal | | Reflection plane | 2d Spacing (Å) |
|---|---|---|---|
| Topaz | | (303) | 2·712 |
| Lithium fluoride | LiF | (220) | 2·848 |
| Lithium fluoride | LiF | (200) | 4·028 |
| Sodium chloride | NaCl | (200) | 5·641 |
| Silicon | Si | (111) | 6·271 |
| Germanium | Ge | (111) | 6·532 |
| Quartz | | (10$\bar{1}$1) | 6·686 |
| Pentaerythritol | PE | (002) | 8·742 |
| Ethylenediamine-d-tartrate | EDDT | (020) | 8·808 |
| Ammonium dihydrogen phosphate | ADP | (110) | 10·648 |
| Gypsum | | (020) | 15·185 |
| Potassium acid phthalate | KAP | (001) | 26·63 |

Various multilayer crystals (e.g. lead stearate, $2d \sim 100$ Å) are available for use in the very long wavelength region ($\lambda > 15$ Å, $Z < 9$).

The most useful crystals are the two varieties of LiF and the Ge, PE, ADP and KAP crystals. Tables giving $2\theta$ values of the spectra of the elements for various analyzing crystals have been published by Powers (1957) and White et al. (1965). A selection of analyzing crystals is required for the following reasons:

(a) It is not possible to use a single crystal to disperse satisfactorily the complete range of X-ray wavelengths used in spectrographic analysis. A crystal clearly cannot diffract radiation whose wavelength exceeds the $2d$ value for the crystal. In practice, the wavelength limit is lower than this, since the mechanical requirements of a spectrometer prevent the crystal being used at $2\theta$ values higher than about 145°. The lower angular limit is set by the fact that a crystal will not intercept all of the radiation passing through the primary collimator at low $2\theta$ values, and by the fact that lines are too close together at low angles.

(b) Analyzing crystals vary in their reflection efficiency, and one with a strong reflection is used if possible. It may, however, be desirable to sacrifice some intensity and use a crystal that gives greater dispersion in order to improve the resolution of the spectrograph. An expression for the dispersion of an analyzing crystal can be derived by differentiating equation (12), so that:

$$\frac{d\theta}{d\lambda} = \frac{n}{2d} \cdot \frac{1}{\cos\theta} \tag{13}$$

To increase dispersion, the (220) LiF crystal may often be used in preference to the (200) LiF crystal although it has a lower reflection efficiency. The analysis of rare earths and the determination of trace amounts of vanadium in the presence of larger amounts of titanium are instances in which high dispersion is desirable.

(c) Sometimes the second-order reflection of an element may coincide with the first-order reflection of the element being analyzed. If the element giving the second-order reflection is present in high concentration then the interference may be serious, even if pulse-height selection is used to reduce the unwanted second order. Such interference is not uncommon and examples are given below. In these cases a crystal with a weak or absent second-order reflection can be used to advantage. The Si and Ge crystals do not give second-order reflections (Lublin, 1960).

(d) Crystals should be chosen so that as far as possible any fluorescent radiation generated within the crystal is removed in wavelength from that being measured. For example, the ADP and gypsum crystals will often give high backgrounds if they are used for determining sulphur, phosphorus, silicon or aluminium. NaCl, Si and quartz crystals suffer from the same defect when used for chlorine, phosphorus and sulphur radiations, and it is better to use a Ge crystal that has a similar spacing.

### 4. Collimators

The two collimators in a non-focusing spectrograph are normally of the Soller type, consisting of a series of closely spaced parallel plates. The angular resolution of the collimators is determined by the length of the plates and the distance between them. A choice of primary collimators is normally available and the one with the narrower plate spacing is used if high resolution is required, which is generally the case when analyses at low $\theta$ values are made. It can be seen that both the primary collimator and analyzing crystal can be varied to alter the spectrograph resolution.

### 5. Spectrometer Path

Between the shortest wavelengths used and about 3 Å, the spectrograph can be operated in air, but appreciable increases in intensity will be obtained if a vacuum is used for wavelengths above 1·5 Å, and above 3 Å a vacuum is necessary. For samples that cannot be subjected to a vacuum, the air can be replaced by helium.

### 6. Detectors

The function of an X-ray detector is to convert the energy of individual X-ray photons into pulses of electrical energy that can then be measured. The simplest type of detector is the Geiger counter. However, since this produces pulses with an energy independent of the energy of the incident X-ray photons, and since it has a long dead time (of the order of hundreds of microseconds), it has been generally superseded. The two types of detector in common use are the scintillation counter and proportional counter. These both produce pulses with an energy proportional to the energy of the incident photons and have dead times of less than one microsecond.

The scintillation counters used to detect X-ray photons normally consist of a thallium-activated sodium iodide phosphor, the scintillations of which are detected by a photomultiplier. Such a counter has a high counting efficiency for short-wavelength radiation and is therefore used for detecting wavelengths from the shortest encountered in X-ray spectrographic analysis up to about 3 Å. Proportional counters are gas filled, and each photon that interacts with a gas atom produces a pulse on a centrally situated wire anode. Since the proportional counters are used in conjunction with scintillation counters, they are filled with a gas which is readily ionized by long-wavelength X-rays; most counters use P10 gas (argon 90%, methane 10%) or an argon–carbon dioxide mixture since argon is the most suitable gas for detecting the lower energy X-rays. For very long wavelengths ($\lambda > 10$ Å) methane is a suitable gas. Proportional counters must be fitted with a thin window (often beryllium, aluminium, Mylar or polypropylene) in order to transmit long-wavelength X-rays. The counter gas may diffuse through these thin windows and consequently it is necessary to maintain a constant flow of gas through the counter. Counters of this type are referred to as flow-proportional counters. Argon-filled flow-proportional counters are highly efficient in detecting long-wavelength X-rays and may be used in the range from 1·5 to 12 Å. In the range from 1·5 to 3 Å for which either the scintillation counter or the flow-proportional counter may be used, the choice in a particular case must depend on other factors.

Polyethylene tetraphthalate (Mylar or Melinex), 4–6 $\mu$m thick, is the most common window material of commercial spectrographs. Such a film, 5·5 $\mu$m thick, absorbs about 50% of Al$K_\alpha$, 70% of Mg$K_\alpha$, 85% of Na$K_\alpha$, and 98% of F$K_\alpha$ radiations, so that Na$K_\alpha$ count-rates are lowered considerably and fluorine cannot be determined. Polypropylene has about 60% of the absorption coefficient, and about 70% of the density, of Mylar, so the use of an equally thin polypropylene window will increase intensities. Unoriented polypropylene film can be stretched to a thickness of the order of 1 $\mu$m, and if supported by the Soller slits in front of the counter, or by a high transmission grid, it will withstand the vacuum of the spectrograph (Henke, 1964, 1965). Such windows supported by the Soller slits have a life of 2–4 months, and this is satisfactory, in view of the improved performance, if the instrument is used for magnesium, sodium or fluorine determinations.

For incident monochromatic radiation, the output pulses from both counters have a spread of voltages. The distribution of these pulses is approximately Gaussian and the relative standard deviation of this distribution is given by:

$$\sigma\% = \frac{K}{\sqrt{E}} \tag{14}$$

where $K$ is a constant and $E$ is the energy of the incident photons. Values of $E$ can be obtained by using equation (2). It can be seen that the pulse-energy distribution is broader for longer wavelength X-rays. If $E$ is expressed in keV, then for an argon-filled proportional counter $K$ will have a value close to 16. For a scintillation counter, the spread in voltages is much greater for photons of the same energy (see Fig. 4). This is due to the low efficiency of the phosphor and of the photocathode in the photomultiplier. For these counters, $K$ will have a value of about 50.

For both counters, in addition to the single ideal energy peak there may be a smaller peak at a lower energy (see Fig. 4). This is known as an *escape peak* and is characterized by being at a fixed energy difference below the main

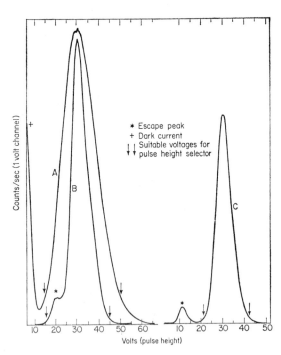

FIG. 4. Pulse-height distributions from proportional and scintillation counters. (A) $CuK_\alpha$. Scintillation counter. Scale $10^4$ counts/sec. (B) $CuK_\alpha$. Proportional counter. Scale $3 \times 10^4$ counts/sec. Apart from the detector the conditions for A and B were identical. (C) $TiK_\alpha$. Proportional counter.

peak. This difference corresponds to the energy of the X-ray photons emitted by the counter gas in the case of a proportional counter and elements such as iodine in the phosphor in the case of a scintillation counter. Escape peaks are only produced by X-rays whose energy is greater than the excitation potential of the counter material emitting the X-rays.

## 7. Amplifier and EHT supply

The pulses produced by both counters are very small and need to be amplified to an energy of a volt or more before they can be measured. The amplitude of the pulses out of the linear amplifier depends on three factors: (a) The energy of the X-ray photons incident on the detector. (b) The detector gain. The voltage amplification of the detector is markedly dependent on the detector voltage $V$ (EHT), and is typically a function of $V^{14}$ for the proportional counter and of $V^{6.3}$ for the scintillation counter. (c) The gain of the amplifier.

The second two factors can be varied, so that for X-rays of a particular energy a voltage pulse of a particular amplitude may be obtained by using either a high EHT and low amplifier gain or vice versa.

Certain counting characteristics deteriorate if high values of EHT are used. For scintillation counters, the dark-current count increases at higher voltages. The dark current is the count-rate observed in the absence of a phosphor and is due to thermal emission from the photocathode in the photomultiplier. As can be seen from Fig. 4 the dark current consists of low-energy pulses, and the scintillation counter should only be used to detect X-rays whose energies can be resolved from the dark current. For proportional counters, the effects of using a high EHT are much more serious. It results in a drift in pulse height as the counting rate increases, as illustrated in Fig. 5. At the same time the pulse-height distribution broadens, e.g. in Fig. 4 the value of $\sigma$ for the proportional counter is greater than that predicted by equation (14) owing to the high input into the counter (approximately $10^5$ counts/sec). These effects will be aggravated if the anode wire of the counter has deteriorated. The shift of pulse height with rate is often considerably more than that of Fig. 5 (Birks, 1963, p. 104; Bender and Rapperport, 1966; Beaman, 1966). It can be seen that both detectors are best used at low values of EHT with a high amplifier gain, i.e. low attenuation. However, as amplifier gain is increased to compensate for low amplification in the detector, amplifier noise may be registered. The optimum operating conditions are those in which the lowest possible EHT is used consistent with the exclusion of amplifier noise.

## 8. Pulse-height Selector

Pulses from the linear amplifier can be passed through a pulse-height selector to reject pulses with an unwanted energy. The use of a pulse-height

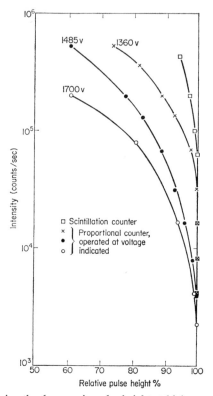

FIG. 5. Showing the decrease in pulse height at high count-rate. $ZnK_\alpha$.

selector becomes increasingly necessary as lower concentrations are measured. The lower and upper level settings are normally chosen so that the window between them passes more than 90% of the monochromatic X-ray beam. This is illustrated in Fig. 4. Pulse-height selection thus improves the peak-to-background ratio by rejecting harmonics (for which $n$ equals 2 or more in the Bragg equation) and spurious pulses as well as those due to natural radiation, and by removing most pulses due to scattering and fluorescence from the analyzing crystal. In using pulse-height selection with a proportional counter the drift in pulse height with counting rate must be considered, and if this effect is present then the counting intensities must be restricted or a wide window must be used. If escape peaks are present, these must be either completely included or excluded by the pulse-height selector in the interests of stability.

It can be seen that the settings of the pulse-height selector and the three factors listed above that determine the amplitude of the pulses entering the selector are all inter-related. In order to keep the EHT as low as possible, the settings on the selector should be kept at a low level. Having set a lower and upper level on the pulse-height selector, pulses from any particular radiation can be brought into the pre-set window by varying the EHT. It is only necessary to alter the settings on the pulse-height selector when the width of the window relative to the height of the lower level has to be changed. It is necessary to use different settings of this type in changing from the proportional counter to the scintillation counter because of the different pulse distributions produced by these two counters. Even with a single detector it may be desirable to alter the settings when measurements of widely different wavelengths are being made (see equation 14). In practice, the maximum safe amplifier gain may be chosen as follows. Set the lower level of the pulse-height selector to a low value, ensuring that the selector itself generates no spurious pulses. Increase the amplifier gain (or decrease attenuation) until amplifier noise is just registered in the counting equipment (this is best done with the EHT turned off). Operate with the amplifier gain reduced (or attenuation increased) by a factor of 4 below the value at which noise appears. Under these conditions amplifier noise will not give rise to spurious counts and the noise will not cause any appreciable broadening of the pulse-height distribution.

When the proportional counter is used, a convenient channel width for most wavelengths is that which is numerically equal to the lower level setting, i.e. ratio of window width to mean height (average of upper and lower levels) equals 66%. A narrower channel is theoretically possible with shorter wavelengths, in the range 1–2·5 Å, but pulse-height drift (see Fig. 5) will normally preclude its use.

With the scintillation counter, the same pulse-height selector settings can be used for the shorter wavelengths but the longer wavelengths ($\sim 1·5$ Å) may require the channel width to be doubled.

## 9. Ratemeter, Recorder and Scaler

The pulses passing through the window of the pulse-height selector may be recorded as an intensity on a ratemeter. The precision of ratemeters is generally no better than a few per cent and they are only used in making qualitative analyses. The ratemeter is usually connected to a pen recorder that can record the spectrum of a sample as the goniometer moves through a selected angular range. Scalers count the pulses arising from individual X-ray photons and are operated in conjunction with a timing device. The scaler and timer can be used to measure either the number of pulses in a given time (fixed time) or else the time taken for the scaler to accumulate a predetermined number of pulses (fixed count). The relative merits of these two methods of counting will be discussed later in the consideration of counting statistics.

## B. COUNTING LOSSES

Owing to the finite resolution time of the detectors and the following counting circuits, a pulse occurring shortly after another may be lost. The time interval for which the equipment is inoperative after registering a pulse is generally referred to as the "dead time" of the equipment. The dead time results in counting losses and if measurements of actual X-ray intensities are required then the observed intensity must be corrected for these losses.

If $C_0$ is the observed intensity and $C$ the actual intensity and if the dead time of the equipment is $t$ then we have:

$$C = \frac{C_0}{1 - C_0\, t} \tag{15}$$

If a fixed number of counts $N$ is measured in a time $T_0$ then this correction is more conveniently made as:

$$C = \frac{N}{T} \text{ where } T = T_0 - Nt \tag{16}$$

$T$ is the live time of the equipment. This means that the observed time is corrected simply by subtracting a constant time, e.g. $0.3$ sec for $t = 3\ \mu\text{sec}$ and $N = 10^5$.

Dead time is best measured by the single-foil method (Short, 1960), as follows. A thin aluminium foil is inserted in front of the detector, and the time taken to accumulate a fixed number of counts before $(T_1)$ and after $(T_2)$ it is inserted is measured for a range of count-rates. A value of $Nt$ (see equation 16) is found, for which the ratio $(T_1 - Nt)/(T_2 - Nt)$ is constant and independent of the count-rate. $t$ is then the dead time of the equipment. Above a certain intensity, the ratio $(T_1 - Nt)/(T_2 - Nt)$ can no longer be made constant and this intensity should never be exceeded in quantitative work. Dead time is not necessarily constant and may be changed by altering various instrumental settings, such as the levels on the pulse-height selector and the time constants on the amplifier, which alter the pulse shape (Heinrich, 1960). An apparent change in dead time can be caused by the pulse amplitudes moving out of the window in the pulse-height selector at high count-rates, as described above (Fig. 5).

## C. SAMPLE PREPARATION

For qualitative analysis, samples may be placed in an X-ray spectrograph in virtually any form. The simplest method is to place powdered samples in a sample holder with a plastic window. This method cannot be used for the very light elements, since their emitted radiation would be seriously attenuated by the plastic.

For quantitative analysis it is desirable to use compacted self supporting samples, since these can be prepared in a reproducible fashion and used without a supporting window. Powders may be pressed directly into suitable holders or made into pellets. Pellets require less sample and may be stored indefinitely. It is undesirable and unnecessary to mix a bonding agent with the powder. For most analyses, powdered samples must be finely ground; the reasons for this are discussed later. For the quantitative analysis of light elements it is generally desirable to dissolve samples into a lithium borate glass, and this may be cast into discs for direct insertion into the spectrograph. For the analysis of liquids, liquid cells are available.

Suitable methods of preparing powdered and fused samples are described in Section VIII.

# IV. Qualitative Analysis

## A. SENSITIVITY

Most commercially available X-ray spectrographs may be used to detect all the elements heavier than sodium. A vacuum is necessary when the elements between sodium and calcium are analyzed. If a very thin proportional counter window is used, as discussed earlier, detection can be extended to fluorine, oxygen, nitrogen and carbon. The use of special demountable X-ray tubes improves the sensitivity for these elements and enables boron and beryllium to be detected (Henke, 1965).

For all elements the limit of detection is determined by the size of the peak that can be distinguished from background variations (see later, equation 37). In the authors' laboratories, with powdered silicate samples, minerals, rocks and soils, the limits of detection, for counting times not exceeding 200 sec, were approximately as follows: carbon and oxygen $\sim 1\%$, fluorine $\sim 0.4\%$, sodium $\sim 200$ p.p.m., with methane counter gas and a 1 $\mu$m polypropylene counter window; magnesium $\sim 100$ p.p.m., aluminium $\sim 30$ p.p.m., sulphur $\sim 8$ p.p.m., scandium–molybdenum $\sim 3$ p.p.m., iodine–uranium $\sim 6$ p.p.m. These figures are intended only as a guide, and in favourable circumstances the sensitivity may be a little better, while in unfavourable circumstances it may be considerably worse.

## B. IDENTIFICATION OF ELEMENTS

Identification involves the measurement of $2\theta$ values for the radiation emitted from the sample. Knowing the $d$ spacing of the analyzing crystal, the wavelength $\lambda$ of the radiation can be calculated by using equation (12). Books of tables are available that enable one to proceed directly from the $2\theta$ values for various analyzing crystals to the element emitting the radiation (e.g.

Powers, 1957; White *et al.*, 1965). Compared with optical spectra, X-ray spectra are simple and lines can generally be identified without ambiguity.

The angular position of peaks can be measured on a chart recorder synchronized with the motor-driven goniometer. A quick manual scan is generally sufficient in checking for the presence of major elements. The quickest way of accurately determining the $2\theta$ position of a line is by averaging the goniometer settings that give the same intensity, 50–80% of the peak intensity, on either side of the peak. This method gives results that are reproducible to 0·01°. Locating the peak position by determining the angle of greatest intensity is slow, owing to the peaks being almost flat topped over a small angular interval.

If there is any possibility of confusion in identifying a particular line, this can easily be checked by looking for other lines that should occur in the spectrum of a particular element. Spectra can always be compared with those obtained from pure elements or compounds if there is any doubt about the nature of the element giving rise to a particular line. The pulse-height selector can be used to determine the order of the reflection giving a particular line.

Lines from the anode of the X-ray tube and from tube contaminants, as well as their higher order reflections, will appear on the spectrum. Compton-scatter peaks will also be present. These are broader peaks, originating from the anode, which lie alongside and are approximately 0·03 Å greater in wavelength than the parent peak. Compton scatter increases with decreasing wavelength and with decreasing atomic number of the scattering material (sample). In cases in which an X-ray tube emits short wavelength lines (e.g. $AgK_\alpha$ and $MoK_\alpha$), the Compton peak may be comparable to, or greater in intensity than, the coherently scattered peak.

Some tables do not list non-diagram lines, and one must be conscious of such lines in interpreting spectra. The $K_{\alpha_3}$ line, for example, a non-diagram line, generally appears in the spectrum of the light elements but is not listed in many tables. The same is true of various other satellites. The tables of White *et al.* (1965) are comprehensive in this respect.

Abnormal reflections may be produced by X-rays being diffracted from planes in the analyzing crystal other than the ones normally used (Spielberg and Ladell, 1960). This results in an incorrect wavelength being assigned to these lines on the basis of the angular setting of the goniometer. This effect is generally restricted to analyzing crystals of comparatively low symmetry, such as topaz (orthorhombic) and EDDT (monoclinic).

For various reasons the angular position of a peak may not correspond precisely to that listed in tables. The equipment may be slightly misaligned and also the spacing ($2d$) of the crystal may differ slightly from that used in computing the tables. Temperature changes will cause a variation in crystal spacing, particularly in the case of the PE crystal, which has a comparatively large coefficient of thermal expansion. Although for analytical purposes the X-ray spectrum of an element is taken as fixed and independent of its chemical state, this is not strictly true. Wavelengths do vary slightly, and these changes are the basis of various crystal-chemistry studies (Leonard *et al.*, 1964; White *et al.*, 1960; Baun and Fischer, 1965). For the heavier elements it is not possible to observe these changes with spectrographs designed for analytical use. However, as the atomic number decreases the changes become comparatively larger, so that for instance the $K_\alpha$ lines from the metal and oxide of aluminium occur at noticeably different angles, 0·05° $2\theta$, with a PE crystal.

## V. Theory of Quantitative Analysis

### A. FLUORESCENT GENERATION OF X-RAYS

The production of fluorescent X-rays within a sample in an X-ray spectrograph is illustrated in Fig. 6. The intensity of the primary radiation at a

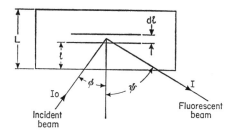

FIG. 6. Geometry of excitation and emission in the sample.

depth $l$ will be reduced by a factor $e^{-\mu_1 l \sec\phi}$ where $\mu_1$ is the linear absorption coefficient of the primary radiation and $\phi$ is the angle between the primary X-ray beam and the normal to the sample surface. This attenuation factor is only strictly valid if the primary X-rays are monochromatic. This assumption can reasonably be made, since much of the excitation from the continuous spectrum comes from a relatively narrow range of wavelengths on the immediate short-wavelength side of the appropriate absorption edge. In any case, characteristic X-rays often form an important part of the primary radiation.

The characteristic radiation emitted per unit area of the sample by an element, present in concentration $p$, in a thickness $dl$ will be proportional to $p \, \rho \, dl$, where $\rho$ is the density of the sample. If this fluorescent radiation emerges at an angle $\psi$ to the normal to the sample surface, it will be attenuated by a factor $\exp(-\mu_2 \, l \sec \psi)$ where $\mu_2$ is the linear absorption coefficient of the sample for the characteristic radiation.

If a primary beam of intensity $I_0$ is used, the intensity of the fluorescent radiation will be:

$$I = K I_0 p \rho \, e^{-l(\mu_1 \sec \phi + \mu_2 \sec \psi)} dl$$

where $K$ is a constant.

The total intensity per unit area from a sample of thickness $L$ is given by:

$$C = \int_0^L I = K \frac{I_0 p \rho [1 - e^{-L(\mu_1 \sec \phi + \mu_2 \sec \psi)}]}{\mu_1 \sec \phi + \mu_2 \sec \psi} \tag{17}$$

If $e^{-L(\mu_1 \sec \phi + \mu_2 \sec \psi)} \ll 1$ the sample is effectively infinitely thick and:

$$C = \frac{K p \rho I_0}{\mu_1 \sec \phi + \mu_2 \sec \psi} \tag{18}$$

If $A_1$ and $A_2$ are the mass-absorption coefficients of the incident and fluorescent radiations, then $A_1 = \mu_1/\rho$ and $A_2 = \mu_2/\rho$, so that:

$$C = \frac{K p I_0}{A_1 \sec \phi + A_2 \sec \psi} \tag{19}$$

In those samples that do not have a major element with an absorption edge between $\lambda_1$ and $\lambda_2$, $A_1$ and $A_2$ will be related so that $A_1 = k A_2$. The constant $k$ will have a value about $0 \cdot 3$ [i.e. $\sim (\lambda_1/\lambda_2)^3$] for many elements when the $K_\alpha$ line is used for analysis. For many spectrographs, $\sec \phi$ and $\sec \psi$ have values near to 1 and 2, respectively. Therefore the expression $(A_1 \sec \phi + A_2 \sec \psi)$ may be replaced by $2 k' A_2$, where $k'$ has a definite value, which will be not much greater than unity. In these particular samples then, not only is $A_1$ related to $A_2$, but the effect of $A_2$ predominates. Dropping the various constants from equation (19) we then have:

$$C \propto \frac{p}{A_2} \tag{20}$$

Equations (19) and (20) show that absorption coefficients influence fluorescent intensities, and this is the main matrix effect associated with fluorescent analysis. Quantitative analysis must, directly or indirectly, make allowance for the variations in mass-absorption coefficients of samples. Before considering the various methods of doing this, two effects that may invalidate (19) must be considered. These are microabsorption effects and enhancement effects.

### B. MICROABSORPTION EFFECTS

In the derivation of equation (19), it was assumed that the sample being analyzed has a uniform composition. Claisse and Samson (1962) and Bernstein

(1963), have studied the effect of particle size and have shown that for any X-ray wavelength a sample gives fluorescent intensities that are independent of particle size only for very fine or coarse particles. In between these extremes, which will vary with the sample and the wavelength under consideration, intensities depend in a complex way on the phases present and the shape of their particle-size-distribution curves, so that quantitative analysis becomes very difficult.

Under favourable circumstances, it may be possible to make quantitative analyses with comparatively coarse-grained samples (method 10, p. 201), and in such cases, where analysis depends on the samples having a certain minimum grain size, accuracy may suffer if the specimens are ground too finely. However, most methods of analysis assume a homogeneous sample and the requirements for such a sample will now be considered.

A specimen can be considered homogeneous if each ray that is completely absorbed has traversed an average section of the sample before being absorbed. The problem is basically one of contrast in absorption coefficients. If individual particles differ greatly in this respect, then each ray must traverse many particles, but if particles differ only a little, then passage through only a few will suffice. Brindley (1945) has treated the case of the effect of microabsorption on diffraction by powders, and this is essentially the same effect. He found the critical parameter to be $l(\mu - \mu_{av})$, so that a sample may be regarded as sufficiently uniform if:

$$|l(\mu - \mu_{av})| \ll 1 \qquad (21)$$

where $l$ is the linear dimension of the particles, $\mu$ is the linear absorption coefficient of a particular mineral and $\mu_{av}$ is the average linear absorption coefficient of all the particles.

It can be seen that as $\mu$ and $\mu_{av}$ approach one another, larger particle sizes can be tolerated and for monomineralic samples the effect is absent.

For equation (21) $\mu$ will have a value between those of $\mu_1$ and $\mu_2$ in equation (18). For mineral particles composed predominantly of the element being determined, $\mu_1$ may be several times as high as $\mu_2$. For other particles, $\mu_1$ is generally smaller than $\mu_2$.

Because $\mu$ varies as a power of wavelength (see equation 11), the actual particle sizes required for sample uniformity will not only vary with the phases present, but long wavelengths (light elements) will require very much finer particles than shorter wavelengths.

As the particle size of samples is generally unknown, a practical test to determine uniformity is to grind until further grinding gives no change in fluorescent intensity. Intensities generally increase as the particle size decreases if the grains containing the element being analyzed have $\mu > \mu_{av}$. If, however, $\mu_{av}$ is greater than the $\mu$ of the grains containing the element, intensities will decrease with grinding. Many grinding mills virtually cease grinding after a certain time owing to compaction of the sample. In such a case, of course, fluorescent intensities will change little with time of grinding, although the particle size may still be too coarse.

Equation (21) indicates that the required particle size is reciprocally related to $(\mu - \mu_{av})$. If a powdered sample is diluted it is best to choose a diluent whose absorption coefficient is near that of the sample. Quartz is a suitable diluent for many mineral samples. In addition, the hardness of quartz ensures fine grinding during the mixing and grinding. Low-atomic-number organic powders are often recommended as diluents. Incorporation of these may produce microabsorption effects and their softness may prevent further grinding of the sample.

For wavelengths less than 3 Å, the requisite particle sizes can be obtained by the very fine grinding of normal rocks. Powder samples are thus used for measuring the heavy trace elements in silicate samples. The use of a powder rather than a fused sample is to be preferred in measuring trace elements, since any dilution of the sample will result in a loss in sensitivity and may introduce contamination. Longer wavelengths require submicron particle sizes, and since this is difficult to achieve by grinding it is quicker and easier to render the sample uniform by fusion.

Fused samples are used for analyzing the major elements present in rock and mineral samples (see p. 190). Dissolving the sample in a glass overcomes particle-size problems and reduces matrix variations (Claisse, 1956; Andermann, 1961; Rose et al., 1963).

### C. ENHANCEMENT EFFECTS

The process of enhancement, or secondary fluorescence, was not considered in the derivation of equation (19). This is the process by which fluorescent X-rays generated within the sample are photoelectrically absorbed and as a result increase the emission of characteristic radiation of longer wavelengths. The maximum effect will occur when a major component of the sample emits at a wavelength that is just shorter than the absorption edge of another element. Such a case would occur, for example, in estimating iron in a zinc rich sample. The $ZnK$ radiation ($\alpha = 1\cdot437$ Å, $\beta = 1\cdot296$ Å) is shorter than the $FeK$ absorption edge ($\lambda = 1\cdot744$ Å) so that $ZnK$ radiation will be photoelectrically absorbed by the iron which will consequently emit more strongly.

Gillam and Heal (1952), Sherman (1955) and others have considered the problem theoretically. The problem is complex and involves many assumptions, and practical solutions have not been forthcoming. Souninen (1958) found that in metal and ore-type specimens the enhancement was no more than 10% even when studying the effects of arsenic in lead where the effect should be almost maximal. Data obtained in the authors' laboratories support these findings. In analyzing artificial mixtures of 1% iron (as $Fe_2O_3$) in ZnO and 1% strontium in $MoO_3$ secondary fluorescence increased the intensities by less than 15%.

Secondary fluorescence can be viewed to a first approximation as a negative absorption, and in some methods of analysis corrections for it can be applied in this form (Beattie and Brissey, 1954).

In practice it seems that the effect can be ignored in many mineral and rock analyses, since very many authors obtain satisfactory analyses taking no account of it.

# VI. Methods of Quantitative Analysis

## A. MEASUREMENTS OF X-RAY INTENSITIES

All methods of X-ray spectrographic analysis involve the measurement of the fluorescent intensity due to a particular element in a sample. At high or moderate concentrations this may be the only measurement required but for low concentrations a background correction must be applied. Background measurements are normally made at equal angular distances on either side of the peak position. However, the average value may not correspond to the background at the peak position. The background profile may be curved and it may be high at the peak position owing to characteristic radiation from impurities in the X-ray tube being scattered from the sample.

The background profile may be determined by measuring the intensity of fluorescent radiation at peak and background positions for a variety of pure substances. Reagents should be chosen the constituent elements of which have an average atomic number close to that of the unknown, since even for pure materials the peak/background ratio varies a little with composition. Powdered clear crystal quartz and spectroscopically pure $SiO_2$ and $Al_2O_3$ are useful. The ratio of peak-to-background intensity is taken, and this correction is applied to all subsequent analyses.

Adjacent spectral lines can interfere with peak and background measurements. The correction procedures adopted in such cases are illustrated by the following example. Rubidium and strontium are two elements commonly determined by X-ray spectrographic analysis. The radiation from these elements is not completely resolved by equipment with moderate resolution, so that in measuring the background intensities for Sr, both the $RbK_\alpha$ and $K_\beta$ peaks interfere at higher and lower angles, respectively. Below are some intensities, in counts per second, obtained at five angular positions for a quartz sample and for this sample with 500 p.p.m. of rubidium added. The data was obtained with a Philips PW 1540 spectrograph, equipped with a 160 μm primary collimator and a (200) LiF analyzing crystal.

| $2\theta =$ | 24·36° | 25·10° (Sr peak) | 25·84° | 26·58° (Rb peak) | 27·32° |
|---|---|---|---|---|---|
| Quartz | 2658 | 2487 | 2122 | 1951 | 1786 |
| +500 p.p.m. Rb | 2805 | 2551 | 2483 | 22497 | 2107 |

It can be seen that the presence of rubidium increases the count-rate at all positions. The average background, peak plus background and net peak intensities are as follows:

| | | Quartz | +500 p.p.m. Rb |
|---|---|---|---|
| Sr | average background | 2390 | 2644 |
| | peak + background | 2487 | 2551 |
| | net peak | 97 | −93 |
| Rb | average background | 1954 | 2295 |
| | peak + background | 1951 | 22497 |
| | net peak | −3 | 20202 |

An increase in the net peak intensity of Rb from −3 to 20202 counts/sec causes an apparent decrease in the Sr peak intensity from +97 to −93 counts/sec, a depression of 190 counts/sec. This depression is equal to 0·94% of the Rb peak intensity and any measurements made on strontium in natural samples will give intensities superimposed on this apparent depression. In making strontium analyses under the conditions applying when the above data were obtained, it would therefore be necessary to reduce the Sr background by 0·94% of the Rb peak intensity. In a similar way a correction for the effect of strontium on rubidium should be made. In this instance, the use of a lower $d$ spacing crystal, LiF (220), would reduce the magnitude of the interference.

A similar procedure can be applied with other cases of interference, irrespective of whether the interference occurs in a peak or background position. For example, with a (200) LiF analyzing crystal, the $SrK_\beta$ peak at 22·42° is close to the $ZrK_\alpha$ peak at 22·56° $2\theta$, and the correction that has to be made to the $ZrK_\alpha$ intensity is 15% of the intensity of the $SrK_\alpha$ line. Corrections such as this can be applied quantitatively to a wide range of elements. It is clear, however, that if the interfering element is present in a much greater abundance than the element being measured, serious limits may be placed on the accuracy of the measurement of the less abundant element. Detection limits are likewise affected. A good example is provided by rubidium and strontium in micas in which the content of these elements is commonly of the order of 1000 p.p.m. and less than 10 p.p.m., respectively. Problems arise when the error in the correction is of the same order as the net count-rate.

Apart from the interferences discussed above, high-order lines may cause interference. The correct use of the pulse-height selector can minimize this interference, but where it cannot be completely eliminated empirical correction factors must be determined and applied. In the authors' experience the factors can generally be determined accurately and simply on synthetic mixtures as above. When working at concentration levels below 100 p.p.m. corrections for some type of interference often become necessary.

Examples in which a first order line interferes with the analytical line include the following, where the analytical line is given, with the interfering line in brackets: $VK_\alpha$ ($TiK_{\beta1}$); $ScK_\alpha$ ($CaK_\beta^{III}$, $CaK_\beta^{IV}$); $TiK_\alpha$ ($BaL_{\alpha1}$); $BaL_{\alpha1}$ ($TiK_\alpha$); $CoK_\alpha$ ($FeK_{\beta0}$); $MnK_\alpha$ ($CrK_{\beta1}$); $RbK_\alpha$ ($UL_{\alpha2}$); $YK_\alpha$ ($RbK_{\beta1}$); $CeL_{\alpha1}$ ($BaL_{\beta1}$); $CuK_\alpha$ ($WL_\alpha$ Compton-scattered peak from X-ray target). Some instances of high order interference are: $PK_\alpha$ ($CaK_{\beta1}2$); $PK_\alpha$ ($CuK_\alpha4$); $HfL_{\alpha1}$ ($ZrK_\alpha2$); $SiK_\alpha$ ($FeK_{\beta1}4$); $AlK_\alpha$ ($CrK_{\beta1}4$); $ClK_\alpha$ ($CrK_{\alpha1}2$).

## B. METHODS OF ANALYSIS

Having measured a net count-rate, C, this must then be related to the concentration of the element in the sample. Some methods of doing this will now be discussed.

### 1. Linear Calibration Method

If $A$ (the mass-absorption coefficient) is constant, then a simple linear relationship will exist between fluorescent intensity and concentration. Normally for $A$ to be constant, the bulk composition of the samples must remain reasonably constant. The method will thus be applied directly to minerals and rocks where the composition varies only a little. In such cases satisfactory analyses can be made, the tolerable limit in rock variability being determined by the accuracy required.

Emerson (1959) obtained good results for $K_2O$ in potash feldspars using this method. Various analysts (Molloy and Kerr, 1960; Chodos and Engel, 1961; Volborth, 1963) have applied the method to rocks and minerals of limited variability to obtain results of moderate accuracy, generally better than 10% relative.

It must be remembered that $A$ ($A_1$ and $A_2$) of each element is normally approximately proportional to $Z^3$ so that relatively small variations in the concentrations of iron, calcium and potassium may cause significant changes in $A$ if heavier elements are being determined. For example, the addition of 20% $Fe_2O_3$ to $SiO_2$ doubles the value of $A$ for $SrK_\alpha$.

Under exceptional circumstances, $A$ may remain relatively constant even with major variations in the sample. Such a case of interest is in the determination of potassium, chlorine, sulphur and phosphorus in rocks and soils.

The heavier major oxides (CaO and $Fe_2O_3$) have much the same absorption coefficients as the lighter oxides (MgO, $Al_2O_3$ and $SiO_2$). In view of the difficulties in determining low amounts of chlorine and sulphur in silicates, the accuracy achieved assuming constant $A$ is acceptable.

If samples are sufficiently diluted, effects due to variations in absorption coefficients can be largely overcome. Dilution can be by fusion (Claisse, 1956; Andermann and Allen, 1961; Rose et al., 1963; Townsend, 1963; Welday et al., 1964), by fine grinding with a solid (Adler and Axelrod, 1954; Rose et al., 1965) or by a liquid after the sample has been taken into solution (Campbell and Thatcher, 1960). If a heavy absorber is added during dilution matrix effects can be further suppressed. Dilution, with or without a heavy absorber, reduces sensitivity approximately in proportion to the suppression of matrix effects, so it is necessary to compromise between the elimination of matrix effects and retention of adequate sensitivity.

For silicate analysis, fusion into a borate glass is a satisfactory method of dilution (Claisse, 1956). Lithium tetraborate is a suitable material, since it readily dissolves silicates and does not itself emit characteristic X-rays in the wavelength region used for analyses. Additional lithium in the form of lithium carbonate may be added to give a glass of composition $Li_6B_4O_9$, which has a lower melting point and is more fluid than $Li_2B_4O_7$. If a heavy element is added, it should be one that does not introduce spectral lines that interfere with the elements being determined. For the analysis of elements commonly determined in silicates, lanthanum is a suitable heavy absorber (Rose et al., 1962, 1963). A method that produces samples in the form of a glass disc has been described by Norrish and Hutton (1964) and is given in Section VIII.

The use of greater dilutions can achieve a somewhat similar result as the addition of a heavy absorber to the glass, and it is useful to compare the advantages of each method. The use of a heavy absorber can result in the production of fluorescent radiation in the analyzing crystal, which may increase the background intensity (Welday et al., 1964). However, pulse-height selection can be used to reduce this effect. Greater dilutions increase the relative importance of the X-ray background, and contamination of the reagents becomes more serious. In reducing matrix effects and sensitivity, the addition of a heavy absorber does not have the same effect as increased dilution for all elements. For iron, a dilution of 1/50 (in $Li_6B_4O_9$) would be required to produce the same sensitivity (counts/sec/%) as for a 1/6 dilution with 1/6 lanthanum oxide also added. For potassium a dilution of 1/11 and for magnesium a dilution of 1/8 would give the same effect. Lanthanum thus has its biggest effect on the heaviest elements, and this differential behaviour is desired, since matrix effects are biggest and sensitivities greatest for the heavy elements.

Many authors have established that provided analysis times are not too long and sensitivity is adequate, dilution is a very satisfactory method. If a variety of elements are being measured with different concentration levels, in order to retain adequate sensitivity, the degree of dilution may be limited to that which does not completely remove matrix effects. In such cases calibration should be made with samples of a similar composition, or, if variable samples are to be analyzed, matrix corrections should be applied (see method 7, p. 196).

### 2. Two-component Systems

This method relies on the fact that one component is measured (depending on the application a component may be regarded as an element, oxide or mineral). The value of $A$ for the second component (the rest of the sample) may be fixed or may vary in some pre-determined way with the measured component.

The analysis of binary mineral mixtures, or of single minerals subject to solid solution, are obvious applications of this method. Such systems will give a calibration which is a smooth curve. Although the full calibration curve (0–100%) will depart from linearity if the analyses are made on samples of limited variability, the relevant part of the curve may be approximately linear. This method has been successfully applied to the analysis of bauxite and clays (Sun, 1959), rock phosphates (Norrish and Sweatman, 1962) and some sulphide ores (Sweatman, 1963).

### 3. Internal Standard

In this method an element, which has a line and absorption edge near those of the element being measured, is added to the sample in a known concentration. Since absorption effects will be similar for the two elements, the ratio of the concentration of the unknown to the standard element will be related to the ratio of their intensities by a constant factor. This method is most satisfactorily applied to the heavier elements of low and moderate concentration, and when only one or two elements are being determined. Sometimes one addition may serve as an internal standard for several elements. If the internal standard is added to powdered samples in the form of an element, oxide or salt, then its particles will have a comparatively large linear absorption coefficient, and in the moderate wavelength range $\sim 1 - 3$ Å, this may lead to errors arising from microabsorption effects (see equation 21). These errors can be minimized by using a compound in which the concentration of the internal standard is not very high. Organic compounds can often be used and these have the additional advantage of grinding very finely. Addition of the internal standard as an aqueous solution does not necessarily eliminate microabsorption effects, since the salt may crystallize into large particles on

I

evaporation of the water. Precipitation of the internal standard, after addition in solution, can be satisfactory.

Microabsorption effects generally limit the method to those elements with $\lambda < 3$ Å. Within the above limitations, however, the method is a well established one that can give accurate results. It has been used to overcome matrix effects in the determination of rubidium and caesium (Axelrod and Adler, 1957), uranium (Smithson et al., 1960), thorium (Adler and Axelrod, 1955), strontium (Lucchesi, 1957), scandium (Heidel and Fassel, 1961) and barium, titanium, zinc (Lewis and Goldberg, 1956). In the analysis of solutions it is a very convenient method of allowing for matrix variations (Mitchell and O'Hear, 1962).

## 4. Spiking

Calibration can be made by adding known amounts of the element being measured to the sample and measuring the increase in intensity. From equation (20) we have $C_p = Kp/A$ before the addition, and if a weight fraction $S$ is added, $C_{p+S} = K(p + S)/A$ after the addition, so that:

$$\frac{p}{S} = \frac{C_p}{C_{p+S} - C_p} \tag{22}$$

For precision, the addition should be comparable to the concentration originally present, and since the addition should be small enough not to alter the absorption coefficient of the sample, the method is not normally applicable to the measurement of major constituents. Abdel-Gawad (1966) used this method for estimating hafnium in zircon. The limitations imposed by microabsorption effects are the same as in the previous method.

## 5. Measurement of Scattered Radiation

Under certain circumstances scattered radiation will depend on the absorption coefficient of the sample, and so it can be used to correct for variations in $A_2$.

If the intensity of the scattered radiation is $B$, then (Andermann and Kemp, 1958):

$$B \propto \frac{f(\text{composition})}{A_2 \sec \phi + A_2 \sec \psi} \tag{23}$$

Because of their proximity in wavelength, it is assumed that $A$ for the coherently and incoherently scattered radiations is the same. If the compositional variations in the sample are not large, then $f$ (composition) will be constant, so that:

$$A_2 = \frac{K}{B} \tag{24}$$

This equation may be used in making analyses when a measure of $A_2$ only is required (see equation 20), as then concentration will be linearly related to the ratio of line to scattered intensity.

Various measures of scattered radiation have been used in this connexion. Andermann and Kemp (1958) measured scattered radiation at 0·6 Å, this wavelength being in a region where scattered intensities are high so that it can be measured with precision. Background, which is normally measured near the analytical line, may also be used in the above manner (Kalman and Heller, 1962).

Other workers, for various reasons, have measured the intensity of Compton-scattered radiation from the anode (molybdenum) of the X-ray tube (Reynolds, 1963) or the coherently scattered radiation from the X-ray tube (Cullen, 1962).

Although there are limitations to the method (see below) this method of allowing for matrix is simple and quick, involving no additions to the sample. The authors have used the method successfully for many trace element determinations in soils, rocks and plants.

The limitations of this method include the following:

(a) The assumption that scattered background intensity is independent of compositional changes apart from the manner in which these are reflected in variations in $A_2$, is only approximate. (b) If the background is measured near the peak, it must be corrected for the tail of the line. (c) Background may be in error owing to interference from other lines, including non-diagram lines not always listed in tables. Errors of this type will be reduced if a strong line originating from the anode (Compton or coherently scattered) is measured. (d) Equation (24) is only valid if the background radiation originates in the X-ray tube, is scattered by the sample and is diffracted by the crystal. Background can, however, be due to other causes. It can arise from fluorescence in the analyzing crystal, those crystals that contain the heavier elements being troublesome in this respect, particularly if fluorescent radiation from the sample is of the correct wavelength to excite the crystal. To some extent, the analyzing crystal will Compton-scatter all wavelengths reaching it. Low-atomic-number crystals scatter most, particularly those containing hydrogen. The background intensity may also be increased by high-order diffraction from the crystal. Natural radiation and noise can also increase the background. In the above cases, unwanted background will often be of very different wavelength from the desired wavelength and this can be suppressed by pulse height selection.

The background at long wavelengths arises from these other causes and tends to be independent of $A$. This affects the application of the method in the analysis of light elements. This problem may be overcome by measuring the background at shorter wavelengths, provided no major absorption edges are crossed, since this assumes that equation (11) is valid.

## 6. Direct Measurement of Absorption Coefficient

Direct and reliable analyses can be made by the direct application of equation (19), by measuring the fluorescent intensity and the absorption co-efficients $A_1$ and $A_2$ of the sample to the exciting and fluorescent wavelengths. $A_1$ can be measured on spectrographs designed to do this (Norrish and Radoslovich, 1962). Direct measurement of $A_1$ is not easily made with commercial instruments but Carr-Brion (1964, 1965) has devised a method of measuring the effect of $A_1$ and $A_2$ on fluorescent intensities.

$A_2$ can be easily measured, and since for many samples $A_1$ and $A_2$ are related by a constant factor, this is sufficient. Salmon and Blackledge (1956) used $A_2$ to correct for matrix variations in mineral samples. If the element being measured is the only major element with an absorption edge between the exciting and fluorescent radiations then it is possible to write equation (19) as (Norrish, 1959):

$$p = \frac{k \, C \, A_2}{K - C} \tag{25}$$

The constants $k$ and $K$ may be determined empirically with analyzed samples. This equation is useful for analyzing ore-type specimens and Norrish (1959, 1960) has applied it to the analysis of lead and zinc ores. In equation (25), $K$ has a magnitude comparable to the fluorescent intensity observed from the pure element. When the fluorescent intensity from a sample is small compared to this, equation (25) reverts to the form of equation (20).

Mass-absorption coefficients can be measured directly on most samples for wavelengths shorter than about 3 Å. The method can thus be used for the analysis of elements heavier than calcium. This corresponds to the wavelength region for which pressed powder samples can be used to measure fluorescent intensity. The method of measuring $A_2$ is similar to that used in X-ray diffraction (Norrish and Taylor, 1962) and can be employed with commercial spectrographs by using a simple addition* (Sweatman et al., 1963). A method of preparing samples for absorption measurements is described in Section VIII.

Mass-absorption coefficients are measured by setting up the spectrograph to give a high intensity ($I_0$) of the required radiation. ($SrCO_3$ would be a suitable source of $SrK_\alpha$ radiation, for example.) The specimen, prepared as a parallel-sided slab of known weight per $cm^2$, $\rho x$, is then placed in front of the scintillation counter and the attenuated intensity $I$ is measured. Then, from equation (9):

$$A_2 = \frac{1}{\rho x} \ln \frac{I_0}{I} \tag{26}$$

Measurements made on duplicate samples generally agree within a few per cent. The measurements are quick and simple and one sample will suffice for

---

* Drawings of a suitable attachment can be obtained from the authors.

measuring the mass-absorption coefficients of a range of wavelengths. Normally, however, several samples of different thicknesses are required to cover the range $0.5 - 3$ Å.

In the absence of major absorption edges in the samples, the absorption coefficients for one wavelength may be used in analyzing several elements. Thus in normal rocks and soils the absorption coefficient for $ZnK_\alpha$ radiation could be used in the measurement of nickel, copper and gallium. The same coefficient could not be used for manganese analyses because of the major $FeK$ absorption edge between $ZnK_\alpha$ and $MnK_\alpha$ radiations.

In measuring absorption coefficients it is desirable to use pulse-height selection, since the attenuated beam may be high in second and third order harmonics and these could give rise to serious errors. Ratios of $I_0/I$ as high as $10^4$ can be used but when $I$ becomes small a background correction to it is necessary. If $I_0 \gg I$, the relative error in $A$ due to counting errors is:

$$\frac{\text{Counting error, \%}}{\ln (I_0/I)}$$

This shows that $A$ can be measured accurately with relatively few counts if $I_0/I$ is large. It also shows that it is difficult to measure $A$ accurately if $I_0/I$ is not much greater than unity. It is implicit in the above that count-rates have been corrected for dead time.

As an example of this method, data and calculations are given for strontium in the standard diabase W-1 and the standard sodium feldspar NBS 99. The concentration of strontium in W-1 (186 p.p.m.), obtained by isotope dilution*, is used to calibrate the method, and so the figure obtained for the strontium content of NBS 99 is compared with the isotope-dilution figure for this sample.

The net peak intensities for $SrK_\alpha$ radiation for 2 g pressed pellets of W-1 and NBS 99 were found to be 4118 and 5588 counts/sec, respectively. These figures correspond to values of $C$ in equation (20). The following data are used to calculate $A_2$.

| | W-1 | NBS 99 |
|---|---|---|
| Area of sample (diameter $= 1.27$ cm) | 1.267 sq. cm | 1.267 sq. cm |
| Mass of sample | 0.2973 g | 0.3427 g |
| $I_0$ | 105,600 counts/sec | 105,600 counts/sec |
| $I$ | 4,820 counts/sec | 18,240 counts/sec |
| $\frac{1}{\rho x} \ln I_0/I = A_2$ | 13.16 | 6.49 |

* Isotope-dilution figures for strontium in W-1 and NBS 99 were kindly made available by Dr W. Compston of the Department of Geophysics and Geochemistry at the Australian National University.

For W-1 we have a strontium content of 186 p.p.m. Thus in equation (20) we have $p = K A_2 C$ or $K = p/A_2 C = 0.00343$. For NBS 99 we then have $p = K A_2 C = 124.4$ p.p.m., which is in good agreement with the value of 125 p.p.m. obtained by isotope dilution.

Table 2 shows the power of the direct measurement of $A_2$ in overcoming matrix errors in such different materials as plants, plant ashes and mineral specimens. This method can be applied in determining most trace elements and has the advantage that additions to samples are not necessary, calibration being made with simple synthetic mixtures.

TABLE 2

*Strontium determinations using $A_2$*

| Sample | $A_2$ | Net line counts/sec | p.p.m. Sr X-rays | Atomic† absorption |
|--------|-------|---------------------|------------------|--------------------|
| Super phosphate | 13·6 | 237 | 320 | 311 |
| Phosphate rock | 16·7 | 202 | 330 | 341 |
| Powdered milk | 2·85 | 34·9 | 9·8 | 9·4 |
| Milk ash | 17·15 | 67 | 113 | 110‡ |
| Clover | 3·15 | 284 | 88 | 86 |
| Clover ash | 14·23 | 475 | 674 | 674‡ |
| Rye | 2·29 | 59·7 | 13·5 | 13·1 |
| Rye ash | 15·05 | 159 | 236 | 237‡ |

† Results from David (1962).
‡ Calculated from original Sr concentration and ash content.

### 7. Calculation of Absorption Coefficient from Measured Composition

Since the absorption coefficients of a sample depend only on the major elements present, if the fluorescent intensities of major elements are measured, then by appropriate calculations it is possible to determine absorption coefficients and actual concentrations. Beattie and Brissey (1954), Burnham *et al.* (1957), Mitchell (1958, 1960), Norrish (1959) and Lucas-Tooth and Pyne (1964) have considered the theoretical aspects of this method and have applied it to the analysis of alloys, mixed oxides, and lead and zinc ores.

This method can also be applied to the analysis of major elements in silicate samples (Norrish, 1965). We have seen that dilution of a sample by fusing it and the addition of a heavy absorber reduced matrix variations so that a linear calibration can be made if the samples do not vary greatly in composition. If the samples are of variable composition, then a linear calibration may be used to obtain an approximate composition, which may then be used to calculate absorption corrections to arrive at a final composition.

The presence of elements from the sample in the glass will alter the absorption coefficient of the borate glass. The changes in the coefficients will depend on the differences between the absorption coefficients of the elements (or oxides) and those of the glass. On this basis, correction coefficients can be calculated from tables of mass-absorption coefficients. Alternatively, the correction coefficients may be determined empirically with artificial samples. For a particular radiation we obtain an equation of the form:

$$A = X + \text{Fe.}\,Y_1 + \text{Mn.}\,Y_2 + \text{Ti.}\,Y_3 + \ldots + \text{loss.}\,Y_L \qquad (27)$$

for the absorption coefficient of the glass. Fe, Mn, Ti, ... represent the weight fractions of $Fe_2O_3$, MnO, $TiO_2$, ... in the sample and "loss" represents the weight fraction of volatile material in the sample, such as water, lost during fusion. The constant $X$ represents the mass-absorption coefficient of the glass for the radiation. $Y_1$, $Y_2$, $Y_3$, ... represent the differences between the mass-absorption coefficient of the glass and those of $Fe_2O_3$, MnO, $TiO_2$, ... multiplied by the dilution factor. $Y_L$ represents the corresponding difference for the material lost during the fusion. In practice it is convenient to normalize $X$, $Y_1$, $Y_2$, ... so that for an average rock $A$ is near unity.

If the line and background intensities are $C$ and $B$, the net fluorescent intensity for the glass is $(C - B)$. The concentration of the element in the sample is then given by:

$$p = K(C - B)A, \text{ where } K \text{ is a constant}$$

$$= KCA - KBA$$

From equation (23) we have $B = f\,(\text{sample})/A$, or $B.A = f\,(\text{sample})$. In the case of fusions the final samples are very similar in composition, and it follows that:

$$p = KCA - B'$$

where $B'$ is the background correction, which is a constant. It is the same for all samples and is in the form of a constant percentage correction. This method of treating the background has the considerable advantage that only the line intensity need be determined on each sample.

We can now express the above equation more fully as:

$$p = KC(X + \text{Fe.}\,Y_1 + \text{Mn.}\,Y_2 + \text{Ti.}\,Y_3 + \ldots + \text{loss.}\,Y_L) - B' \qquad (28)$$

The constants in this equation may be calculated or determined empirically (see Section VIII).

Calibration of the method, that is the conversion of fluorescent intensity to apparent percentage on a linear calibration, is achieved with artificial mixtures or analyzed samples (method 1). These values can then be inserted into the equations such as (28) to obtain the final values, which will normally be accurate to within a fraction of a per cent relative. If necessary, the calculations can be recycled several times until the concentrations reach a steady value. The calculations can easily be done on a desk calculator. If large numbers of analyses are to be processed, a simple computer programme can be used to make the corrections.

The background correction, $B'$, includes radiation scattered from the sample, as well as that due to impurities in the fusion mixture. $B'$ is determined by measuring the intensity of radiation from a sample free from the element being determined. This intensity is then inserted into equation (28) together with the appropriate values of Fe, Mn, Ti, . . . If $SiO_2$ is used to determine $B'$ for Fe, for example, we have $B' = KC(X + Y_{Si})$, which is in the form of a percentage.

Table 3 shows an example of the calculations for a sphene. The apparent

TABLE 3

*Refinement of the analysis of a sphene*

|  | $Fe_2O_3$ | $TiO_2$ | CaO | $SiO_2$ | $Al_2O_3$ | Total |
|---|---|---|---|---|---|---|
| Apparent % ($KC$) | 1·63 | 32·32 | 28·50 | 31·31 | 1·70 | |
| $A$ | 1·121 | 1·165 | 0·955 | 0·978 | 1·014 | |
| Background ($B'$) | 0·18 | 0·15 | 0·06 | 0·24 | 0·12 | |
| Actual % | 1·65 | 37·50 | 27·16 | 30·38 | 1·60 | 98·29 |

percentages are those obtained from intensities by using a linear calibration. These figures are used in equation (27) together with the coefficients of Table 6 to give $A$ for each element. The actual percentage is given by equation (28) as:

$$\text{Actual } \% = (\text{Apparent } \% \times A) - \text{background}$$

Sphene has an unusual composition and so the $A$ values show bigger than normal deviations from unity. In such cases the calculation can be recycled with the actual percentage in equation (27) to obtain better $A$ values for equation (28). However, even in the above case, several additional recyclings make negligible difference to the results. The final total of a little under 100% is acceptable since only five elements were determined.

Table 4 shows results of analyses of the standard samples G-1 and W-1 obtained by this method with artificial standards. These results are compared with the recommended values of Fleischer and Stevens (1962). Table 5 (from Norrish and Hutton, 1964) shows the instrumental conditions, count-rates and sensitivities for the various elements.

TABLE 4

*Analyses of two standard rocks*

|  | G-1 | | W-1 | |
|  | Accepted value | X-ray | Accepted value | X-ray |
|---|---|---|---|---|
| $SiO_2$ | 72·64 | 72·75 | 52·64 | 52·52 |
| $Al_2O_3$* | 14·04 | 14·24 | 14·85 | 14·93 |
| $Fe_2O_3$* | 1·96 | 1·89 | 11·17 | 11·14 |
| MgO | 0·33† | 0·41 | 6·62† | 6·71 |
| CaO | 1·36‡ | 1·31 | 10·94‡ | 11·07 |
| $Na_2O$ | 3·32 | 3·37§ | 2·07 | 2·16§ |
| $K_2O$ | 5·52‖ | 5·53 | 0·64 | 0·64 |
| $TiO_2$ | 0·26 | 0·26 | 1·07 | 1·10 |
| $P_2O_5$ | 0·09 | 0·08 | 0·14 | 0·14 |
| MnO | 0·03 | 0·032 | 0·16 | 0·18 |

* All Fe expressed as $Fe_2O_3$.
† Accepted value corrected by 0·63 BaO (Stevens and Niles, 1960).
‡ Accepted value corrected by 1·0 SrO (Stevens and Niles, 1960).
§ Determined by using a pressed powder sample.
‖ Recommended value of Fleischer (1965).
Unless otherwise indicated, the accepted values are those of Fleischer and Stevens (1962).

Analysis is not restricted to the elements shown in Table 5, but the same glass discs can be used for determining chromium, nickel, zinc, barium, etc.

This method, fusion into a glass coupled with matrix corrections, has been used in both authors' laboratories for several years to analyze major elements in a very wide range of samples, e.g. a variety of rocks, soils, minerals, bauxites, rock phosphates, metallic ores and plant ashes. It is the only method known to the authors that can be used on a wide variety of materials to give analyses of good accuracy and sensitivity. Synthetic standards may be used for calibration.

## 8. Calculation of Absorption Coefficient from Known Composition

This method may be used if the major element composition is known since the value of $A_2$ can then be calculated. The method can give good results for low concentrations (Hower, 1959), but is limited by the fact that the major element composition must be well known and it may be limited by the accuracy of published values of mass-absorption coefficients.

## 9. Thin Sample

An infinitely thin sample is one in which $L(\mu_1 \sec \phi + \mu_2 \sec \psi)$ is less than about 0·1 (see equation 17). If this is the case, then we have:

$$C \propto p\, \rho L \qquad (29)$$

I*

TABLE 5

*Instrumental conditions and sensitivity for silicate analysis using a lithium borate glass*

|  | Analyzing crystal | X-ray tube | kV | mA | counts/sec/%† | Background as percentage† | Counting time (sec) | Sensitivity‡ as percentage† |
|---|---|---|---|---|---|---|---|---|
| MgO | ADP | Cr | 44 | 20 | 7 | 0·36 | 100 | 0·07 |
| Al$_2$O$_3$ | PE | Cr | 44 | 20 | 70 | 0·12 | 50 | 0·015 |
| SiO$_2$ | PE | Cr | 44 | 20 | 67 | 0·24 | 50 | 0·025 |
| P$_2$O$_5$ | Ge | Cr | 44 | 20 | 200 | 0·06 | 50 | 0·008 |
| K$_2$O | PE | Cr | 44 | 20 | 2700 | 0·05 | 10 | 0·004 |
| CaO | LiF | Cr | 44 | 20 | 4200 | 0·06 | 10 | 0·004 |
| TiO$_2$ | LiF | Cr | 44 | 20 | 1430 | 0·15 | 10 | 0·01 |
| MnO | LiF | W | 50 | 20 | 1250 | 0·16 | 10 | 0·01 |
| Fe$_2$O$_3$ | LiF | W | 50 | 20 | 1600 | 0·18 | 10 | 0·01 |

$K_\alpha$ lines of all elements used.

Analyses made using a Philips PW1540 spectrograph with a 480μm primary collimator. A vacuum was used for all elements except Ti, Mn, Fe. A flow proportional counter with a 1 μm polypropylene window was used with a 5–10 volt channel on a pulse-height selector for all determinations. For Mg the amplifier gain was increased 25% above the symmetrical setting for the pulse-height analyzer, to decrease background originating from fluorescence of the ADP crystal.

† Percentage refers to oxide content of original unfused sample.
‡ Sensitivity amount which is significant above background at the $3\sigma$ level.

where $\rho L$ is the mass per unit area of the sample. $C$ therefore is simply proportional to the mass per unit area of the element in the sample. The method has been applied to the analysis of thin metal films to which it is ideally suited (Rhodin, 1955; Finnegan, 1962).

Solutions may be analyzed by evaporation of a small amount on to a supporting film (Gunn, 1961) or by impregnating a filter paper with the solution (Pfeiffer and Zemany, 1954). Ion-exchange papers form a suitable base, especially since the absorption may be coupled with a considerable concentration of the element to increase sensitivity (Zemany et al., 1958). This method is also applicable to mineral samples (Salmon, 1962), but is restricted to those elements giving short fluorescence wavelengths. To fulfil the above conditions for a thin sample, silicate specimens should have a mass per unit area of less than 0·04 g/sq. cm, if barium is to be determined with the $K_\alpha$ line ($\lambda = 0·387$ Å). With longer wavelengths the sample must be even thinner.

This method inevitably involves at least a ten-fold decrease in intensity. If the sample is physically thin any supporting medium should be of minimum thickness or else line/background contrast will suffer.

*10. Grain Count*

The previous methods have all been for uniform samples (i.e. for grains $\mu L \ll 1$). If $\mu L \gg 1$, then the X-rays are completely attenuated in the surface grains; this means that each ray has "seen" only one grain. If the element being analyzed is present in one phase only and if this phase has a constant composition, then the fluorescent intensity is proportional to the area of the grains exposed at the surface, since the mass-absorption coefficient for the radiation is constant. The method is equivalent to an accurate grain count in which the areas of grains are measured.

Where this method can be applied it will give a linear relationship between intensity and concentration. Because most powder particles are infinitely thick for the longer wavelengths, satisfactory results may be obtained for the lightest elements ($Z = 8, 11, 12$) by using powdered samples (Baird et al., 1963; Volborth, 1963). It must be remembered, however, that these elements should occur in the same mineralogical forms in the rocks being compared.

# VII. The Statistics of X-ray Measurement

## A. THE RANDOM NATURE OF X-RAY PRODUCTION

The production of X-ray photons by electron or X-ray excitation is a random process. It follows that although there is a true mean rate of emission of X-ray photons for a particular sample under particular instrumental conditions, various measurements of this rate, which can only be made over a finite time, will give results grouped about this mean value. These random fluctuations, commonly referred to as *counting error*, represent a fundamental restriction on the precision of intensity measurements in X-ray fluorescence and electron-probe analysis. It is therefore necessary to evaluate the effects of these fluctuations in performing analyses by these methods, since they impose an ultimate limit on the accuracy and sensitivity of concentration measurements.

### B. CALCULATION OF COUNTING ERROR

For X-ray emission, variations in $N$, the number of photons detected in a given time, are governed by the laws that apply to radioactive decay and are described by the *Poisson distribution*. For values of $N$ greater than about 50, which will always be the case in spectrographic analysis, the Poisson distribution is a close approximation to a *Normal* or *Gaussian distribution* of error. The standard deviation is given by $\sigma = \sqrt{N}$.

The nature of the Normal distribution is such that for an infinitely large number of measurements, 68·3% will lie within one standard deviation ($1\sigma$), 95·5% within $2\sigma$ and 99·7% within $3\sigma$ of the true mean value, $\bar{N}$. We are thus able to state the probability that $\bar{N}$ will lie within certain limits, for a particular value of $N$.

The relative standard deviation ($\epsilon$) is obtained by relating the absolute standard deviation $\sqrt{N}$ to $N$, so that:

$$\epsilon = \frac{\sqrt{N}}{N} = \frac{1}{\sqrt{N}} \tag{30}$$

The percentage relative standard deviation, $100\epsilon$, is known as the *coefficient of variation*.

### C. COUNTING ERRORS WHEN TWO OR MORE MEASUREMENTS ARE INVOLVED

The determination of the characteristic intensity due to a particular element normally involves more than one intensity measurement. These include measurements with the goniometer set on the peak position, generally one or two background measurements and measurements on a standard material to which all intensities are related. Each of these measurements is subject to counting error, so that error in the measurement of net line intensity is increased.

If several readings are averaged, i.e. if repetitive measurements are made, or if two background positions are averaged to obtain the value under the line, then:

$$\sigma = \frac{\sqrt{N_1 + N_2 + \ldots N_n}}{n} \tag{31}$$

If $N_p$ and $N_b$ are the number of quanta detected in $t_p$ and $t_b$ seconds, respectively, with the goniometer set in the peak and background positions, then the peak and background count-rates are given by:

$$C_p = \frac{N_p}{t_p} \text{ and } C_b = \frac{N_b}{t_b}$$

The relative standard deviation of the net count-rate will be:

$$\epsilon_n = \frac{\sqrt{\dfrac{C_p}{t_p} + \dfrac{C_b}{t_b}}}{C_p - C_b} \tag{32}$$

and if the counting times for the line and background are the same:

$$\epsilon_n = \frac{\sqrt{N_p + N_b}}{N_p - N_b} \tag{33}$$

If measurements are related to a standard, the error in the net line intensity will be given by:

$$\epsilon^2 = \epsilon_n{}^2 + \epsilon_s{}^2 \tag{34}$$

where $\epsilon_n$ is obtained from equation (33) and $\epsilon_s$ is the relative standard deviation for the standard.

### D. OBTAINING THE OPTIMUM COUNTING ERROR

In a given total analytical time, $T$, the standard deviation of the net count-rate will be a minimum if the times taken to measure peak and background counts are split according to the relationship:

$$\frac{t_p}{t_b} = \sqrt{\frac{C_p}{C_b}} \tag{35}$$

Under these conditions the relative standard deviation of the net count-rate is:

$$\epsilon_n = \frac{1}{\sqrt{T}(\sqrt{C_p} - \sqrt{C_b})} \tag{36}$$

The value of $(\sqrt{C_p} - \sqrt{C_b})$ can be used as a figure of merit, to select optimum conditions for operating the equipment.

I**

With variable samples of unknown composition it is often impractical to attempt to divide the counting time according to equation (35). Nevertheless, if the general level of concentration is known, a crude division between counting times of line and background can be made to achieve maximum counting precision in a short analytical time. If fixed time or fixed count techniques are available the former is to be preferred (Birks and Brown, 1962; Gaylor, 1962).

At low concentrations when $C_p$ and $C_b$ differ only a little the counting time should be divided equally between the line and background. Under these conditions $m/\sqrt{C_b}$ may be used as a figure of merit (Spielberg and Bradenstein, 1963), where $m$ is the number of counts per second obtained per unit of concentration for the element in the sample. The statistical lower limit of detection, i.e. peak minus background ($3\sigma$ confidence), may be taken as:

$$\text{lower limit of detection} = \frac{6}{m} \sqrt{\frac{C_b}{T}} \tag{37}$$

It must be emphasized that this gives only a statistical value for the lower limit of detection. The measuring time, $T$, cannot be extended indefinitely, since machine drift would result in variations in $C_p$ and $C_b$. (It should be noted that a fourfold increase in analytical time is needed to achieve a twofold increase in sensitivity.) In addition, the presence of characteristic radiation arising from contaminants in the X-ray tube may result in the detection limit being appreciably higher than would be predicted from equation (37). Spectral interference between two or more elements may also have a similar effect.

## VIII. Sample Preparation

### A. PRESSED POWDER SAMPLES

The procedure described by Norrish and Hutton (1964) produces pelleted samples with a boric acid back and edge designed to fit into Philips sample holders (PW 1527/20). The mass of powder used can vary between about 0·2 g and 2 g. The amount required depends on the composition of the sample, the elements to be analyzed and whether or not "infinitely thick" samples, as defined above after equation (17), are to be used.

There is no point in producing pellets containing much more than 2 g of sample since fluorescent radiation from the back of a very thick sample cannot reach the primary collimator owing to the mechanical apertures in the spectrograph. For the analysis of $K$ spectra, a 2 g sample will be infinitely thick for elements up to about atomic number 40 in most silicate samples. For heavier elements, the $K$ spectra can be used with infinitely thin samples or with samples of intermediate thickness. In the latter case, the mass-absorption coefficient must be measured and the fluorescent intensity corrected by using equation (17). The 2 g sample will be infinitely thick for all $L$ spectra.

The method described by Norrish and Hutton (1964) is essentially the same as that described by Baird (1961). Referring to Fig. 7, part (a) is placed

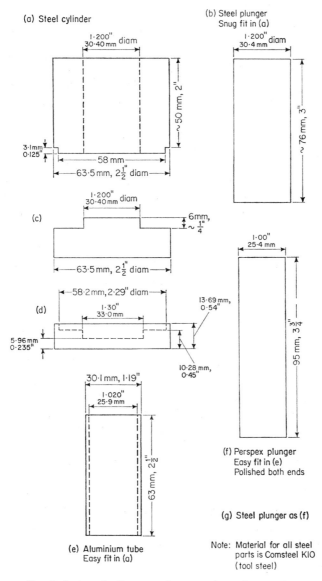

FIG. 7. Accessories for preparing pressed-powder samples.

on (c). (e) is inserted in (a) and the sample poured into it. The Perspex plunger (f) is then used to spread the sample evenly. The sample can be observed by looking through the polished end of the Perspex. With most mineral samples the gentle hand pressure applied in spreading the sample is sufficient. With other samples, such as fibrous minerals, considerable pressure must be applied with the steel plunger (g).

By pressing on the plunger (f) or (g) and rotating (e), and pressing on (e) and rotating the plunger, the sample is freed so that if the plunger and (e) are withdrawn slowly and carefully, the sample is left as a flat disc on (c). A teaspoon of boric acid (fine powder of B.P. grade) is poured into (a) to encase the sample. The amount of boric acid is not critical. (b) is inserted into (a) and a pressure is applied. The load required can vary between 1 and 10 tons. The pressure required to form a coherent sample varies with the nature of the sample and must be found by trial and error. Shearing of the sample is generally an indication of too high a pressure. If too low a pressure is used the sample will not be self supporting. If necessary, plastic can be used over (c) to prevent sample contamination and (d) can be used to press the sample from (a). Samples prepared in this way are quite robust and their identity can be written on the boric acid back with a soft pencil or a marking pen.

### B. SAMPLES FOR MEASURING MASS-ABSORPTION COEFFICIENTS

Samples used for measuring $A_2$ must be prepared with a uniform thickness and a known mass per unit area. Specimens that bond together under pressure can be made into small discs in a piston type die.* For $1 \cdot 27$ cm diameter specimens ($\frac{1}{2}$ in), the weight of powder should be between 20 and 500 mg, depending on the nature of the specimen and the wavelength being used. The thinner samples are difficult to make, particularly if they do not bond, and in these cases the sample can be diluted with filter-paper pulp. The pulp gives a bulkier robust sample of lower absorption coefficient. The amount of pulp is not critical, and 50% is a convenient amount. The measured absorption coefficient must be corrected for the presence of the pulp.

If for any reason, the dilution with pulp is undesirable, then friable samples can be mounted in a Perspex holder (Norrish and Taylor, 1962).

### C. SAMPLES DISSOLVED IN A LITHIUM BORATE GLASS

The fusion mixture described here is that proposed by Norrish and Hutton (1964), which is a modification of those described by Claisse (1956), Andermann (1961), Andermann and Allen (1961) and Rose *et al.* (1963).

* Drawings of this and other accessories are available from the authors on request.

To prepare the lithium borate–lanthanum oxide fusion mixture, the chemicals* are mixed in the proportions of 38·0 g anhydrous lithium tetraborate, 29·6 g lithium carbonate and 13·2 g lanthanum oxide. The lanthanum oxide should be ignited at 900°C before weighing, since this chemical absorbs $H_2O$ and $CO_2$. The other two chemicals should be dried at 550°C before weighing.

The mixture is heated in a large graphite crucible with a lid (1 in Fig. 8), at 1000°C for about 20 min, after which time effervescence should have ceased. The melt is poured on to a large polished aluminium sheet and after cooling is crushed and ground to a coarse powder. The powder will absorb moisture and if ground too finely this absorption may give rise to errors. The powdered mix is stored in air-tight containers.

The individual specimens are prepared in the following way. 1·50 g of borate mix, 0·02 g of sodium nitrate† and 0·28 g of sample are weighed, mixed and placed in a small crucible. Sodium nitrate is added to ensure oxidizing conditions during the fusion. The most satisfactory crucibles are the gold plated platinum variety.‡ These crucibles are not wetted by the melt and successive samples can be made without cleaning the crucible. Palau crucibles are also satisfactory in this respect, but they have a short life at the high temperatures of the fusion. Graphite crucibles are suitable for determining the heavier elements ($Z > 16$), but the glass discs then have some graphite powder on their surfaces, and this depresses count-rates from the lighter elements.

The mixture is fused at a temperature such that pure sodium fluoride salt just melts (980°C), for several minutes or until all the sample has dissolved and any reaction ceased. The melt is stirred (with a platinum-wire stirrer) or mixed thoroughly before pouring. A graphite disc (4 in Fig. 8) and an aluminium plunger (6) are kept on a hot plate at about 230°C. Immediately before pouring, a brass ring (5) is put over the graphite disc. The melt is then quickly poured on to the centre of the disc and the plunger is immediately brought down on to the melt. The plunger is then withdrawn and the brass ring removed. The glass disc is put between two asbestos mats on another hot plate at about 200°C. After several minutes the mats may be removed from the plate to allow the disc to cool slowly.

After cooling, any projecting edges are removed by rubbing with a file, and

* The recommended reagents are as follows:
  Lithium tetraborate  Code 1882 Baker and Adamson Products, Allied Chemical Corp., N.Y.
  Lithium carbonate  Extra pure DAB 6. E. Merck, Darmstadt.
  Lanthanum oxide  Code 528 Lindsay Rare Earth Chemicals, American Potash and Chemical Corp., N.Y.
† Sodium nitrate. Code 6537. E. Merck, Darmstadt.
‡ Supplied by Engelhard Industries Pty Ltd., Thomastown, Victoria.

the glass disc stored in a clean 2 × 2-in labelled envelope. The upper surface of the disc is used for analysis. Fusion discs produced in this way are infinitely thick for the $K$ spectra of all elements up to atomic number 40 and for all $L$ spectra. The discs have a slightly rough surface, but experience has shown that this does not seriously affect the precision.

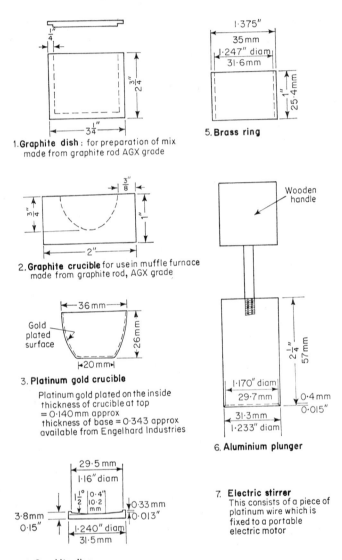

1.**Graphite dish**: for preparation of mix made from graphite rod AGX grade

5.**Brass ring**

2.**Graphite crucible** for use in muffle furnace made from graphite rod, AGX grade

3. **Platinum gold crucible**

Platinum gold plated on the inside thickness of crucible at top = 0·140 mm approx thickness of base = 0·343 approx available from Engelhard Industries

6. **Aluminium plunger**

4. **Graphite disc**

7. **Electric stirrer** This consists of a piece of platinum wire which is fixed to a portable electric motor

FIG. 8. Accessories for preparing lithium borate glass discs.

TABLE 6

Correction coefficients for major elements in silicate analysis

| X | | Y | | | | | | | | | | |
|---|---|---|---|---|---|---|---|---|---|---|---|---|
| | | Fe | Mn | Ti | Ca | K | P | Si | Al | Mg | Na | Loss |
| Fe | 1·075 | −0·043 | −0·065 | 0·138 | 0·122 | 0·127 | −0·072 | −0·099 | −0·106 | −0·113 | −0·118 | −0·167 |
| Mn | 1·035 | −0·021 | 0 | 0·203 | 0·197 | 0·181 | −0·045 | −0·060 | −0·068 | −0·083 | −0·083 | −0·160 |
| Ti | 0·960 | 0·058 | 0·030 | 0·104 | 0·617 | 0·625 | 0·010 | −0·015 | −0·042 | −0·070 | −0·096 | −0·149 |
| Ca | 0·857 | 0·086 | 0·082 | 0·049 | 0·122 | 0·737 | 0·160 | 0·140 | 0·103 | 0·086 | 0·069 | −0·133 |
| K | 0·867 | 0·099 | 0·081 | 0·013 | −0·012 | 0·049 | 0·180 | 0·159 | 0·134 | 0·113 | 0·080 | −0·134 |
| P | 0·903 | 0·054 | 0·041 | −0·036 | −0·072 | −0·072 | −0·100 | 0·126 | 0·109 | 0·091 | 0·063 | −0·138 |
| Si | 1·018 | 0·073 | 0·085 | −0·034 | −0·042 | −0·064 | −0·080 | −0·066 | 0·125 | 0·099 | 0·069 | −0·158 |
| Al | 1·008 | 0·121 | 0·113 | 0·044 | 0·010 | −0·010 | −0·028 | −0·040 | −0·030 | 0·181 | 0·167 | −0·157 |
| Mg | 1·021 | 0·105 | 0·096 | 0·043 | 0·017 | 0 | −0·026 | −0·037 | −0·055 | −0·071 | 0·154 | −0·160 |

In this Table the coefficients are for concentration (weight fraction) of the oxide, not concentration of the element. These coefficients apply only when the samples are prepared according to the recipe given in Section VIII C. Similar, but numerically different, coefficients can be derived for different sample preparations.

Many procedures for fusing samples into a borate glass are given in the literature. That given above was chosen as a reasonable compromise between sensitivity and the elimination of matrix effects. For most elements concentrations of 0·01% can be detected (Table 5), and the concentrations estimated from a linear calibration will generally be within several per cent relative of the correct value. For moderate accuracy these values can be used. For the highest accuracy interelement corrections must be applied and Table 6 shows the coefficients used for some elements (Norrish, 1965). It is to be noted that these coefficients apply only to the particular recipe given above.

## REFERENCES

Abdel-Gawad, A. M. (1966). X-ray spectrographic determination of hafnium–zirconium ratio in zirconium minerals. *Am. Miner.* **51**, 464.

Adler, I. and Axelrod, J. M. (1954). Aluminium powder as a binder in sample preparation for X-ray spectrometry. *Analyt. Chem.* **26**, 931.

Adler, I. and Axelrod, J. M. (1955). Determination of thorium by fluorescent X-ray spectrometry. *Analyt. Chem.* **27**, 1002.

Allen, S. J. M. (1956). *In* "Handbook of Chemistry and Physics." 38th Edition. Ed. C. D. Hodgman. Chemical Rubber Co., Cleveland.

Andermann, G. (1961). Improvements in the X-ray emission analysis of cement raw mix. *Analyt. Chem.* **33**, 1689.

Andermann, G. and Allen, J. D. (1961). The evaluation and improvement of X-ray emission analysis of raw-mix and finished cements. *Adv. X-ray Analysis* **4**, 414.

Andermann, G. and Kemp, J. W. (1958). Scattered X-rays as internal standards in X-ray emission spectroscopy. *Analyt. Chem.* **30**, 1306.

Axelrod, J. M. and Adler, I. (1957). X-ray spectrographic determination of cesium and rubidium. *Analyt. Chem.* **29**, 1280.

Baird, A. K. (1961). A pressed-specimen die for the Norelco vacuum-path X-ray spectrograph. *Norelco Reptr* **8**, 108.

Baird, A. K., McIntyre, D. B. and Welday, E. E. (1963). Sodium and magnesium fluorescence analysis – Part II: Application to silicates. *Adv. X-ray Analysis* **6**, 377.

Baun, W. L. and Fischer, D. W. (1965). The effect of valence and coordination on K series diagram and nondiagram lines of magnesium, aluminum, and silicon. *Adv. X-ray Analysis* **8**, 371.

Beaman, D. R. (1966). Effect of pulse amplitude shifts on electron probe intensity ratios. *Analyt. Chem.* **38**, 599.

Bearden, J. A. (1964). X-ray wavelengths. NYO 10586. U.S. Atomic Energy Commission.

Beattie, H. J. and Brissey, R. M. (1954). Calibration method for X-ray fluorescence spectrometry. *Analyt. Chem.* **26**, 980.

Bender, S. L. and Rapperport, E. J. (1966). *In* "The Electron Microprobe." Ed. T. D. McKinley, K. F. J. Heinrich and D. B. Wittry, pp. 405–414. Wiley, New York.

Bernstein, F. (1963). Particle size and mineralogical effects in mining applications. *Adv. X-ray Analysis* **6**, 436.

Birks, L. S. (1959). "X-ray Spectrochemical Analysis." Interscience, New York.

Birks, L. S. (1963). "Electron Probe Microanalysis." Interscience, New York.

Birks, L. S. and Brown, D. M. (1962). Precision in X-ray spectrochemical analysis. Fixed time vs. fixed count. *Analyt. Chem.* **34**, 240.

Blokhin, M. A. (1962). "X-Ray Spectroscopy." Hindustan Publishing Corporation, Delhi – 6.

Bowan, H. R., Hyde, E. K., Thompson, S. G. and Jared, A. C. (1966). X-ray unit uses solid-state detector. *Chem. Engng News* **44**, 42.

Brindley, G. W. (1945). The effect of grain or particle size on X-ray reflections from mixed powders and alloys, considered in relation to the quantitative determination of crystalline substances by X-ray methods. *Phil. Mag.* **36**, 347.

Burnham, H. D., Hower, J. and Jones, L. C. (1957). Generalized X-ray emission spectrographic calibration applicable to varying compositions and sample forms. *Analyt. Chem.* **29**, 1827.

Campbell, W. J. and Thatcher, J. W. (1960). Determination of calcium in wolframite concentrates by fluorescent X-ray spectrography. *Adv. X-ray Analysis* **2**, 313.

Carr-Brion, K. G. (1964). The X-ray fluorescence determination of zinc in samples of unknown composition. *Analyst, Lond.* **89**, 346.

Carr-Brion, K. G. (1965). The determination of tin in powder samples by X-ray fluorescence analysis. *Analyst, Lond.* **90**, 9.

Cauchois, Y. and Hulubei, H. (1947). "Constantes Selectionées Longeurs d'onde des Emissions X et des Discontinuities d'absorption X." Hermann et Cie. Editeurs. Paris.

Chodos, A. A. and Engel, C. G. (1961). Fluorescent X-ray spectrographic analyses of amphibolite rocks. *Am. Miner.* **46**, 120.

Claisse, F. (1956). Accurate X-ray fluorescence analysis without internal standard. *Quebec Dept. Mines P.R.* 327.

Claisse, F. and Samson, C. (1962). Heterogeneity effects in X-ray analysis. *Adv. X-ray Analysis* **5**, 335.

Compton, A. H. and Allison, S. K. (1935). "X-rays in Theory and Experiment." Van Nostrand, New York.

Cosslett, V. E. and Dyson, N. A. (1957). *In* "X-ray Microscopy and Microradiography." Ed. V. E. Cosslett, A. Engström and H. H. Pattee, pp. 405–413. Academic Press, New York.

Cullen, T. J. (1962). Coherent scattered radiation internal standardization in X-ray spectrometric analysis of solutions. *Analyt. Chem.* **34**, 812.

David, D. J. (1962). Determination of strontium in biological materials and exchangeable strontium in soils by atomic-absorption spectrophotometry. *Analyst, Lond.* **87**, 576.

Emerson, D. O. (1959). Correlation between X-ray emission and flame photometer determination of the $K_2O$ content of potash feldspars. *Am. Miner.* **44**, 661.

Fine, S. and Hendee, C. F. (1956). A table of X-ray $K$ and $L$ emission and critical absorption energies for all the elements. *Norelco Reptr* **3**, 113.

Finnegan, J. J. (1962). Thin-film X-ray spectroscopy. *Adv. X-ray Analysis* **5**, 500.

Fleischer, M. (1965). Summary of new data on rock samples G-1 and W-1, 1962–1965. *Geochim. cosmochim. Acta* **29**, 1263.

Fleischer, M. and Stevens, R. E. (1962). Summary of new data on rock samples G-1 and W-1. *Geochim. cosmochim. Acta* **26**, 525.

Gaylor, D. W. (1962). Precision of fixed-time vs. fixed-count measurements. *Analyt. Chem.* **34**, 1670.

Gillam, E. and Heal, H. T. (1952). Some problems in the analysis of steel by X-ray fluorescence. *Br. J. appl. Phys.* **3**, 353.

Gunn, E. L. (1961). X-ray fluorescent intensity of elements evaporated from solution onto thin film. *Analyt. Chem.* **33**, 921.

Heidel, R. H. and Fassel, V. A. (1961). Fluorescent X-ray spectrometric determination of scandium in ores and related materials. *Analyt. Chem.* **33**, 913.

Heinrich, K. F. J. (1960). Pulse-height selection in X-ray fluorescence. *Adv. X-ray Analysis* **4**, 370.

Heinrich, K. F. J. (1966). *In* "The Electron Microprobe." Ed. T. D. McKinley, K. F. J. Heinrich and D. B. Wittry, pp. 296–377. Wiley, New York.

Henke, B. L. (1963). Sodium and magnesium fluorescence analysis – Part I: Method. *Adv. X-ray Analysis* **6**, 361.

Henke, B. L. (1964). X-ray fluorescence analysis for sodium, fluorine, oxygen, nitrogen, carbon, and boron. *Adv. X-ray Analysis* **7**, 460.

Henke, B. L. (1965). Some notes on ultrasoft X-ray fluorescence analysis – 10 to 100 Å region. *Adv. X-ray Analysis* **8**, 269.

von Hevesy, G. (1932). "Chemical Analysis by X-rays and its Applications." McGraw-Hill, New York.

Hill, R. D., Church, E. L. and Mihelich, J. W. (1952). The determination of gamma-ray energies from beta-ray spectroscopy and a table of critical X-ray absorption energies. *Rev. scient. Instrum.* **23**, 523.

Hower, J. (1959). Matrix corrections in the X-ray spectrographic trace element analysis of rocks and minerals. *Am. Miner.* **44**, 19.

Jenkins, R. and de Vries, J. L. (1967). "Practical X-ray Spectrometry." Centrex, Eindhoven.

Kalman, Z. H. and Heller, L. (1962). Theoretical study of X-ray fluorescent determination of traces of heavy elements in a light matrix. *Analyt. Chem.* **34**, 946.

Leonard, A., Suzuki, S., Fripiat, J. J. and de Kimpe, C. (1964). Structure and properties of amorphous silicoaluminas. I. Structure from X-ray fluorescence spectroscopy and infrared spectroscopy. *J. phys. Chem., Wash.* **68**, 2608.

Lewis, G. J. and Goldberg, E. D. (1956). X-ray fluorescence determination of barium, titanium and zinc in sediments. *Analyt. Chem.* **28**, 1282.

Liebhafsky, H. A., Pfeiffer, H. G., Winslow, E. H. and Zemany, P. D. (1960). "X-ray Absorption and Emission in Analytical Chemistry." Wiley, New York.

Lublin, P. (1960). A novel approach to discrimination in X-ray spectrographic analysis. *Adv. X-ray Analysis* **2**, 229.

Lucas-Tooth, J. and Pyne, C. (1964). The accurate determination of major constituents by X-ray fluorescent analysis in the presence of large interelement effects. *Adv. X-ray Analysis* **7**, 523.

Lucchesi, C. A. (1957). Determination of strontium by X-ray fluorescence spectrometry. *Analyt. Chem.* **29**, 370.

Mitchell, B. J. (1958). X-ray spectrophotographic determination of tantalum, niobium, iron, and titanium oxide mixtures. *Analyt. Chem.* **30**, 1894.

Mitchell, B. J. (1960). X-ray spectrographic determination of zirconium, tungsten, vanadium, iron, titanium, tantalum and niobium oxides. *Analyt. Chem.* **32**, 1652.

Mitchell, B. J. and O'Hear, H. J. (1962). General X-ray spectrographic solution method for analysis of iron-, chromium-, and/or manganese-bearing materials. *Analyt. Chem.* **34**, 1620.

Molloy, M. W. and Kerr, P. F. (1960). X-ray spectrochemical analysis: an application to certain light elements in clay minerals and volcanic glass. *Am. Miner.* **45**, 911.

Norrish, K. (1959). The analysis of lead and zinc ores, and associated materials by X-ray fluorescent spectroscopy. 1. The determination of Mn, Fe, Cu, Zn, As, Pb and Sb using borax fusions. *Tech. Memo. Div. Soils CSIRO* **7/59**.

Norrish, K. (1960). The analysis of lead and zinc ores, and associated materials by X-ray fluorescent spectroscopy. 2. The determination of Zn and Pb using raw materials and borax fusions. *Tech. Memo Div. Soils CSIRO* **4/60**.

Norrish, K. (1965). Silicate Analysis by X-ray Spectrography. *Fifth Australian Spectroscopy Conference. Abstracts of Papers*, p. 23.

Norrish, K. and Hutton, J. T. (1964). Preparation of Samples for Analysis by X-ray fluorescent Spectrography. *Divl. Rep. Div. Soils CSIRO* **3/64**.

Norrish, K. and Radoslovich, E. W. (1962). A curved-crystal fluorescence X-ray spectrograph. *J. scient. Instrum.* **39**, 559.

Norrish, K. and Sweatman, T. R. (1962). Fluorescent X-ray Determination of Aluminium and Iron in Phosphate Rock. *Divl. Rep. Div. Soils CSIRO* **4/62**.

Norrish, K. and Taylor, R. M. (1962). Quantitative analysis by X-ray diffraction. *Clay Miner. Bull.* **5**, 98.

Pfeiffer, H. G. and Zemany, P. D. (1954). Trace analysis by X-ray emission spectrography. *Nature, Lond.* **174**, 397.

Powers, M. C. (1957). "X-ray Fluorescent Spectrometer Conversion Tables for Topax, LiF, NaCl, EDDT and ADP Crystals." Philips Electronics Instruments, New York.

Reynolds, R. C. (1963). Matrix corrections in trace element analysis by X-ray fluorescence: estimation of the mass absorption coefficient by Compton scattering. *Am. Miner.* **48**, 1133.

Rhodin, T. N. (1955). Chemical analysis of thin films by X-ray emission spectrography. *Analyt. Chem.* **27**, 1857.

Rose, H. J., Adler, I. and Flanagan, F. J. (1962). Use of $La_2O_3$ as a heavy absorber in the X-ray fluorescence analysis of silicate rocks. *Prof. Pap. U.S. geol. Surv.* **450–B**, 80.

Rose, H. J., Adler, I. and Flanagan, F. J. (1963). X-ray fluorescence analysis of the light elements in rocks and minerals. *Appl. Spectrosc.* **17**, 81.

Rose, H. J., Cuttitta, F. and Larsen, R. R. (1965). Use of X-ray fluorescence in determination of selected major constituents in silicates. *Prof. Pap. U.S. geol. Surv.* **525–B**, 155.

Salmon, M. L. (1962). A simple multielement-calibration system for analysis of minor and major elements in minerals by fluorescent X-ray spectrography. *Adv. X-ray Analysis* **5**, 389.

Salmon, M. L. and Blackledge, J. P. (1956). A review of improved mineral analyses by fluorescent X-ray spectrography. *Norelco Reptr* **3**, 68.

Sherman, J. (1955). The theoretical derivation of fluorescent X-ray intensities from mixtures. *Spectrochim. Acta* **7**, 283.

Short, M. A. (1960). Detection and correction of nonlinearity in X-ray proportional counters. *Rev. scient. Instrum.* **31**, 618.

Smithson, G. L., Eager, R. L. and van Cleave, A. B. (1960). Determination of uranium in flotation concentrates in leach liquors by X-ray fluorescence. *Adv. X-ray Analysis* **2**, 175.

Souninen, E. J. (1958). Influence of indirect excitation upon the intensity of a fluorescent X-ray line in two particular cases. *Suomal. Tiedeakat. Toim. Sarja A* **6**, 1.

Spielberg, N. and Bradenstein, M. (1963). Instrumental factors and figure of merit in the detection of low concentration by X-ray spectrochemical analysis. *Appl. Spectrosc.* **17**, 6.

Spielberg, N. and Ladell, J. (1960). Crystallographic aspects of extra reflections in X-ray spectrochemical analysis. *J. appl. Phys.* **31**, 1659.

Spielberg, N., Parrish, W. and Lowitzsch, K. (1959). Geometry of the non-focusing X-ray fluorescence spectrograph. *Spectrochim. Acta* **15**, 564.

Stevens, R. E. and Niles, W. W. (1960). Second report on a cooperative investigation of the composition of two silicate rocks. *Bull. U.S. geol. Surv.* **1113**, 3.

Sun, Shiou-Chuan. (1959). Fluorescent X-ray spectrometric estimation of aluminium, silicon and iron in the flotation products of clays and bauxites. *Analyt. Chem.* **31**, 1322.

Sweatman, T. R. (1963). The use of X-ray fluorescent spectroscopy for the rapid determination of sulphur in pyrite concentrates. *Divl. Rep. Div. Soils CSIRO* **7/63**.

Sweatman, T. R., Norrish, K. and Durie, R. A. (1963). An assessment of X-ray spectrometry for the determination of inorganic constituents in brown coals. *Misc. Rep. Div. Coal Research CSIRO* **177**.

Theisen, R. (1965). "Quantitative Electron Microprobe Analysis." Springer-Verlag, Berlin.

Townsend, J. E. (1963). X-ray spectrographic analysis of silica and alumina base catalyst by a fusion-cast disc technique. *Appl. Spectrosc.* **17**, 37.

Volborth, A. (1963). Total instrumental analysis of rocks. *Nevada Bureau of Mines Rept* **6**.

Welday, E. E., Baird, A. K., McIntyre, D. B. and Madlem, K. W. (1964). Silicate sample preparation for light-element analyses by X-ray spectrography. *Am. Miner.* **49**, 889.

White, E. W., McKinstry, H. A. and Bates, T. F. (1960). Crystal chemical studies by X-ray fluorescence. *Adv. X-ray Analysis* **2**, 239.

White, E. W., Gibbs, G. V., Johnson, G. G. and Zechman, G. R. (1965). X-ray emission line wavelength and two-theta tables. *A.S.T.M. Data Series*, 37.

Wyckoff, R. W. G. and Davidson, F. D. (1965). Windowless X-ray tube spectrometer for light element analysis. *Rev. scient. Instrum.* **35**, 381.

Zemany, P. D., Welbon, W. W. and Gaines, G. L. (1958). Determination of microgram quantities of potassium by X-ray emission spectrography of ion exchange membranes. *Analyt. Chem.* **30**, 299.

Zingaro, P. W. (1954). Principal emission lines of X-ray spectra. *Norelco Reptr* **1**, 67, 78.

# CHAPTER 5

# Electron Probe Microanalysis

## J. V. P. LONG

*Department of Mineralogy and Petrology,*
*University of Cambridge, England*

## I. Introduction

Electron-probe analysis provides the means of determining the chemical composition of very small volumes at the surface of polished thin sections of rocks or mineral mounts. The name derives from the essential feature of a fine electron beam which is directed at the point to be analysed. The X-rays generated by the impact of the beam are characteristic of the elements present, and their intensity is an approximately linear function of concentration.

Although X-ray analysis by direct electron excitation of material coated on the anode of an X-ray tube was widely practised (von Hevesy, 1932) in the two decades following Moseley's fundamental discovery of the nature of X-ray spectra (Moseley, 1913, 1914), it was not until the early 1950's that a successful union of the established methods of X-ray spectroscopy and the more recently developed techniques of electron optics was achieved. Such a possibility had been indicated earlier in a patent specification filed by Hillier (1947), but the development of the first working instrument was due to Castaing and Guinier (1949). Much of the relevant theory of X-ray production and the design of an improved instrument were described by Castaing (1951) in his doctoral dissertation.

In its simplest form the electron-probe microanalyser consists of an electron-optical system which focuses an electron beam into an area about 1 $\mu$m diameter on the surface of the specimen, a stage on which the specimen and standards are mounted, a microscope which allows the area of interest to be selected and positioned in the electron beam, and one or more spectrometers which select and measure the intensity of the characteristic radiation of the elements to be determined. To the "static" probe developed by Castaing has been added the refinement of scanning, which allows the distribution of chosen elements on the surface of the specimen to be displayed as an enlarged image on the screen of a cathode-ray tube.

The characteristic X-ray spectra generated by the electron beam are identical to those produced in fluorescence excitation by X-ray irradiation. In addition, however, the slowing down of the electrons produces a continuous spectrum or "bremsstrahlung", which constitutes a background upon which the characteristic lines are superimposed. The basic measurement is a comparison of the net intensity of a particular X-ray line generated in the specimen with that generated in the standard by the same incident current. To a first approximation, concentrations and characteristic intensities are related according to:

$$C_{\text{spec}} = C_{\text{std}} \cdot \frac{I_{\text{spec}}}{I_{\text{std}}} \qquad (1)$$

Castaing showed that the correction factors for the effects of differences in absorption, secondary fluorescence and atomic number of the specimen and standard could be evaluated on a much more fundamental basis than in X-ray fluorescence analysis.

The method has developed with great rapidity during the succeeding period of 16 years, so that at the present time there are approximately twelve

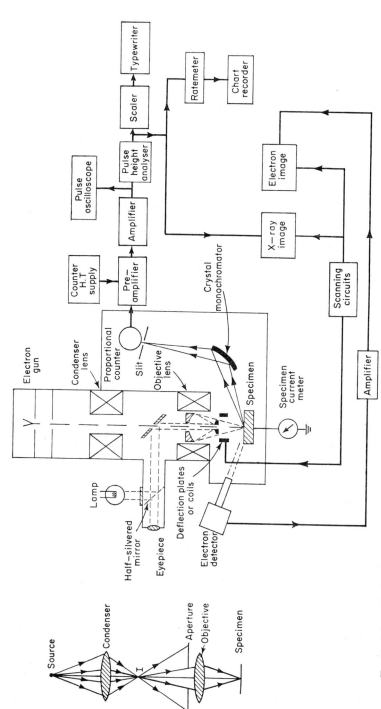

FIG. 1. Schematic diagram showing components of an electron-probe analyser and the ray paths in the electron-optical system.

different commercial instruments available and some four hundred instruments in use throughout the world. The major application of the technique has been in the field of metallurgy (for review papers and bibliography see Melford and Duncumb, 1958; Philibert, 1962; Heinrich, 1964, 1966a), but at least fifty instruments are now installed in laboratories concerned with the earth sciences.

## II. Instrumentation

The various components that make up an electron-probe analyser are shown diagrammatically in Fig. 1. The way in which these are arranged

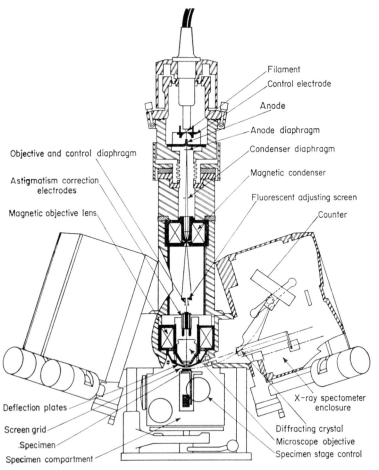

FIG. 2. Diagrammatic cross-section of the CAMECA MS 46 electron-probe analyser. (Reproduced with permission from the makers.)

Fig. 3. General view of the CAMECA MS 46 electron-probe analyser.

varies somewhat in different designs and is illustrated by reference to two recent commercial instruments; the French "MS 46" (Figs. 2 and 3), made by the CAMECA* Company and the British Geoscan (Figs. 4 and 5), made by the Cambridge Instrument Company.† The major fundamental difference between these two instruments lies in the angle of take-off of the X-rays (the angle between the specimen surface and the direction of the X-ray beam to the spectrometer). In the CAMECA design this is 20°, and in the Cambridge instrument 75°. The reasons for choosing a high or low take-off angle are discussed in later Sections.

### A. THE ELECTRON-OPTICAL SYSTEM

The electron beam originates in a triode electron gun, which, apart from its lower operating potential (5–50 kV), is essentially the same as those used in

* Compagnie D'applications Mécaniques à l'Électronique au Cinéma et à l'Atomistique, 103, Bd St Denis Paris Courbevoie (Seine).
† Cambridge Instrument Company Ltd., 13 Grosvenor Place, London W.C.1.

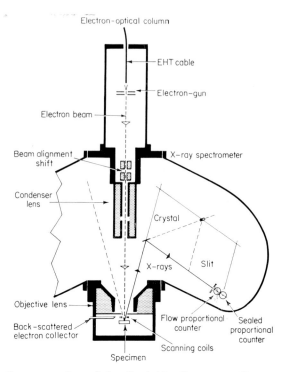

Fɪɢ. 4. Diagrammatic cross-section of the Cambridge Instrument Company Geoscan electron-probe analyser. (Reproduced with permission from the makers.)

electron microscopes. The effective electron source size is approximately 50 μm and the beam from the gun is slightly divergent. The subsequent magnetic lenses focus this beam in the same way that convex lenses focus a light beam, i.e., the condenser lens forms a reduced image, I (Fig. 1), of the source, which in turn acts as the object for the following lens, usually called the objective. The current flowing in the windings of the condenser lens determines the position and size of the intermediate image, I. As I is made smaller, the solid angle of the beam at I increases, and the fraction of the total beam current that is intercepted by the aperture of the final lens falls. A decrease in the final probe diameter produced by increasing the strength of the condenser lens is thus obtained at the expense of a smaller probe current.

The behaviour of systems for the formation of fine electron beams has been discussed by several authors, both with direct reference to the electron-probe analyser itself (Castaing, 1951; Mulvey, 1959; Duncumb, 1962; Fisher, 1964) and also with reference to the X-ray microscope, in which an almost exactly similar system is used (Cosslett and Nixon, 1960).

Fig. 5. General view of the Cambridge Instrument Company Geoscan.

## B. THE MECHANICAL STAGE AND THE LIGHT-OPTICAL SYSTEM

The specimen and appropriate standards are mounted on an insulated mechanical stage with fine micrometer drives that enable it to be moved in increments of the order of 1 μm. The stage of the Cameca instrument also incorporates facilities for electro-mechanical scanning (see p. 226). The Geoscan is arranged so that the stage may be operated by a servo system which allows push-button selection of the various standards and also permits traverses to be made on the specimen at controlled speed in a direction chosen by the operator.

The light-optical system affords the designer considerable scope for ingenuity. The basic problem is to provide a versatile microscope of good resolution and at the same time to preserve a short working distance in the final electron-optical lens and a free path for the emerging X-rays. One solution, used originally in the first Cameca instrument (Castaing, 1954) and retained in the MS 46, is to place a mirror objective lens within the bore of the magnetic objective (Fig. 6). The electron beam then passes through a small hole drilled axially through the mirrors. The advantages of this arrangement are that the light-optical and electron-optical axes are permanently fixed and coaxial and that the specimen may be viewed continuously during analysis. This system, however, does not lend itself readily to the use of

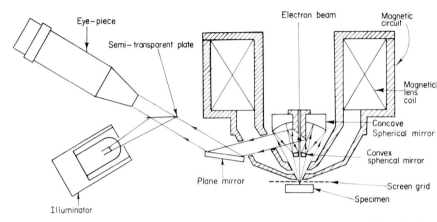

FIG. 6. Cross-section of magnetic objective lens showing coaxial-mirror light objective (Cameca MS 46). (Reproduced with permission from the makers.)

interchangeable objectives to provide alternative magnifications. The system used in the Geoscan, the principle of which is illustrated in Fig. 7, uses refracting objectives which are moved in front of the magnetic objective on an accurately made slide. A further objective is mounted behind the specimen plane so that transparent specimens may be viewed during electron bombardment. This lens also acts as the condenser when the objectives on the slide are used with transparent specimens. The disadvantage of this system is that it does not by itself allow luminescent effects to be observed in opaque specimens. It does, however, give the advantage of an alternative low-power objective, which facilitates the initial exploration of the specimen and which is usually more appropriate to the examination of sections of comparatively coarse-grained rocks in transmitted light.

### C. THE X-RAY DETECTION SYSTEM

The X-ray detection system consists of a crystal spectrometer, together with a proportional counter and its ancillary electronic equipment. For the analysis of elements below sodium in the periodic table, a non-dispersive arrangement has also been employed, in which the energy-resolving properties of the proportional counter only are used (Mulvey and Campbell, 1958; Dolby, 1963; Ranzetta and Scott, 1967). However, since the development of crystals such as potassium acid phthalate (KAP) (Henke, 1963), which has a $2d$ spacing of 26·6 Å, multilayer stearate crystals with spacings up to approximately 100 Å (Henke, 1964; Ong, 1964; Fischer and Baun, 1964; Henke,

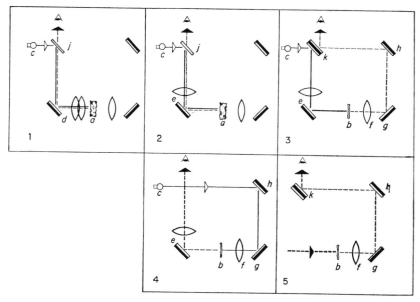

FIG. 7. The light microscope system of the Cambridge Geoscan, showing the alternative modes of operation. 1. High-power objective (*d*) and opaque specimen (*a*); reflected light. 2. Low-power objective (*e*) and opaque specimen; reflected light. 3. Low-power objective (*e*) and transparent thin section (*b*); transmitted light. 4. Low-power objective (*f*) and transparent thin section using objective behind specimen plane; transmitted light. 5. Analysis position; electron beam striking specimen: cathodoluminescence visible with transparent thin specimen.

1967), and efficient gratings (Franks, 1963; Wittry and Nicholson, 1964), the non-dispersive technique, at least with the detectors available at present, has assumed a rather less important position in this field. It remains useful on account of the high collection efficiency of a counter placed close to the specimen and is applicable to such problems as the identification of very small particles on extraction replicas (Cooke and Duncumb, 1967), where, with a dispersive system, the intensity at the detector would be very low.

The X-ray spectrometers used in microprobe analysis are invariably of the curved-crystal type and make use of the fact that the X-ray source may for practical purposes be regarded as a true point. There are essentially four different geometrical arrangements. The first CAMECA instrument and the design by Mulvey (1960) (subsequently the AEI* instrument) use the simple centre-bearing spectrometer shown in Fig. 8(a). Here the counter is mounted on the $2\theta$ arm which is geared at a ratio of 2:1 to the $\theta$ arm so that it maintains

* Associated Electrical Industries, Manchester, England.

the appropriate position relative to the crystal. The crystal may be of the Johan (bent only) or the Johansson (bent and ground) type. The merit of this design is its mechanical simplicity, but it suffers from the drawback that the angle of take-off of the X-rays (the angle between the surface of the specimen and the direction of the X-ray beam to the crystal) does not remain constant as the Bragg angle is changed.

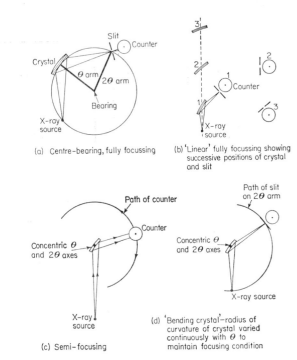

FIG. 8. Alternative arrangements for curved-crystal spectrometers for use with a "point" source of X-rays.

This latter defect is remedied in the "linear" spectrometer (Sandstrom, 1952) (Fig. 8b), which was first used in the electron-probe analyser by Fisher and Swartz (1956). It has also been adopted in the design by Wittry (1959) (the A.R.L. instrument*), in the Geoscan and in the MS 46. The basic Rowland circle geometry is maintained but now the centre of the circle is not fixed and the crystal is constrained to move in a straight line, so maintaining a constant angle of take-off. Several different ways of achieving the requisite

* Applied Research Laboratories, Glendale, California.

movement of the crystal and counter have been described and others no doubt remain to be explored.

The third type of spectrometer (Long and Cosslett, 1957), originally designed for X-ray absorption analysis with point-focus X-ray tubes, is usually described as "semi-focusing". The geometry (Fig. 8c) is the same as that of the flat-crystal spectrometers used in X-ray fluorescence analysis except that the crystal is curved and only very simple slits are used. The position of the source with respect to the crystal is equivalent to that in the Johan spectrometer at one Bragg angle for which $d = R \sin \theta$ (where $R = $ radius of curvature of the crystal). Owing to the fact that the crystals used are not perfect, the spectrometer works satisfactorily over a range of Bragg angles which extends on either side of the "design" angle, $\theta = \sin^{-1} d/R$. Three or four crystals of different radii will normally suffice to cover the angular range $\theta = 10°$–$70°$, the actual number depending on the resolution required and on the material of the crystal. With lithium fluoride, a useful performance may be obtained over the range $20°$–$60°$ with only one crystal bent for the angle $\theta = 30°$. Simple mechanics and compactness are the chief features of this design. For these advantages it sacrifices resolution (often no disadvantage since it simplifies setting of the spectrometer). More important are the comparatively poor peak/background ratios obtained, which seldom exceed 200–300 for pure elements. A considerable improvement is however obtained by inserting a two-dimensional "honeycomb" collimator in front of the counter.

The necessity for changing crystals to maintain the correct focusing conditions in the semi-focusing spectrometer has been overcome by Elion and Ogilvie (1962) who used the geometry shown in Fig. 8(d). This arrangement differs from the previous one, in that the curvature of the crystal is continuously varied by elastic bending in order to maintain the Johan geometry. Since the beam is now brought to an approximate focus on the Rowland circle for all values of $\theta$, it is possible to use a narrow slit in front of the detector, with consequent benefit to the ratio of peak to background.

The number of spectrometers varies; the MS 46 has four, the Geoscan two. An alternative to the use of multiple spectrometers is the addition of a servo system (Hall, 1964), which enables a number of Bragg angles, corresponding to the emission lines of different elements, to be pre-set, and subsequently selected at will. Such systems are incorporated in the A.E.I. instrument, in the Geoscan and in the A.R.L. AMX microanalyser.

The electronic equipment associated with the X-ray spectrometers is essentially identical with that used for fluorescence analysis; ratemeters and potentiometric recorders are used for plotting X-ray spectra and for following variations in concentration as a specimen is traversed with respect to the probe; digital recording, sometimes with automatic print-out facilities, is usually employed for quantitative analysis.

## D. THE SCANNING SYSTEM

The scanning system, developed in the first instance by Cosslett and Duncumb (1956; Duncumb and Cosslett, 1957) and described in more detail by Duncumb (1959), is now fitted to practically all instruments. A set of magnetic deflection coils or electrostatic deflection plates is arranged within the electron-optical column so that the beam may be traversed in a raster pattern over an area of the specimen surface about 500 $\mu$m square at such a speed that a complete "frame" occupies about 2–3 sec. The spot on a cathode-ray tube is scanned in synchronism with the electron probe but over a much larger area ($\sim$ 10 cm square), and its brightness is controlled by the electrical output of the X-ray spectrometer. The cathode-ray tube thus displays an enlarged image whose brightness at any point is determined by the concentration of the selected element at the corresponding point on the specimen. The image shows local fluctuations in brightness owing to statistical variations in the number of X-ray quanta contributing to the image at any point. This effect is reduced by the use of a long-persistence phosphor in the cathode-ray tube and still further by photographing the screen with an exposure time of several minutes (Fig. 9c, d) or by the use of a storage oscilloscope. The magnification of the image is readily controlled by varying the size of the scanned area on the specimen and in a typical instrument the useful range may be 150–3000 ×. In the CAMECA MS 46 instrument, a combination of mechanical scanning of the specimen and electronic scanning of the electron beam is used, an arrangement that avoids concentration distortion in the image due to movement of the X-ray source in a direction producing a change of Bragg angle at the X-ray spectrometer.

An alternative method of forming an image of the scanned area makes use of the signal produced by variations in the number of backscattered electrons from point to point on the surface. This variation occurs partly because the backscattering coefficient increases with increasing atomic number of the target element and partly because topographic features on the surface may act as partial Faraday cages and so increase the number of electrons retained within the specimen. The latter effect is enhanced if the electron detector is placed so as to view the specimen at a low angle, as in Fig. 1. In this case the "electron image" seen on the cathode-ray tube shows a shadowing rather similar to that which would be observed with the specimen viewed normally in a light microscope and illuminated by a light source placed in the position of the electron detector. An image with strong "atomic number" contrast is illustrated in Fig. 9(b).

A concentration profile may be obtained by traversing the electron probe slowly along a line on the specimen. The output of the ratemeter is then used to control the vertical deflection of the cathode-ray tube spot which is moved

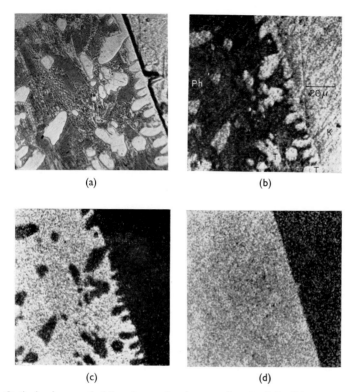

(a)                                (b)

(c)                                (d)

FIG. 9. Optical micrograph (a) and scanning images of part of the Edmonton meteorite
showing the reaction between a lamella of schreibersite [(NiFe)$_3$P] (Ph) and surrounding
kamacite (K) during atmospheric reheating. The electron picture (b) shows strong contrast between the metal and phosphide phases owing to the appreciable difference in
average atomic number. The X-ray pictures (c) and (d) show the distribution of
phosphorus and nickel, respectively.

horizontally in synchronism with the probe. This facility is particularly useful
for a rapid examination of zoning in a mineral grain. Melford (1962) and
Heinrich (1963) have developed "expanded contrast systems" whereby only
those parts of the specimen with a particular element present in concentrations
between selected limits appear bright in a scanning X-ray image. These
facilities, although valuable in certain cases, are limited by the normal laws of
statistics in the accuracy with which they delineate variations in concentration.

One of the most useful functions of the scanning system is as an aid to the
optical microscope in positioning the electron probe: when the scanning
circuits are switched off, the bright spot on the cathode-ray tube may be moved
by means of d.c. controls to any selected feature on the persistent X-ray or

electron image. Under these conditions, the deflection system of the cathode-ray tube and that of the column remain connected together so that the electron probe is thus directed at the corresponding point on the specimen.

## III. Specimen Preparation

Materials to be examined in the electron probe must have a well polished surface, free from relief, if accurate quantitative data are to be obtained. The required perfection of surface is greater for soft radiations such as $MgK_\alpha$ (9·89 Å) than for harder radiations in the 1–2 Å region. In general, a finish similar to that used in reflected-light microscopy of ore minerals is desirable. An appreciation of the importance of avoiding relief in polishing, which may be particularly severe near phase boundaries, can be gained by estimating the effect of an 0·5 $\mu$m deep scratch on the surface of a pyroxene, as illustrated in Fig. 10. The additional absorption introduced when the emergent X-rays pass

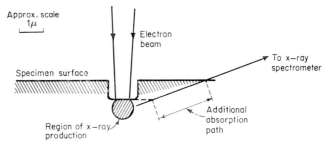

FIG. 10. Anomalous absorption of soft X-rays produced by a scratch in the surface of a polished specimen.

through the material at the edge of the scratch is such that the observed concentration of Mg in a pyroxene of composition $En_{50} Fs_{50}$ is reduced by approximately 10% for an accelerating voltage of 15 kV when the take-off angle is 20°. Such errors are clearly minimized by a higher take-off angle.

Flatness of the surface is also important with specimens of high absorption coefficient in instruments employing low take-off angles. For the $En_{50} Fs_{50}$ pyroxene, the error introduced in the determination of magnesium when the surface is inadvertently tilted at an angle of 2° to the plane normal to the beam is 8% for a nominal take-off angle of 20°. In contrast, the error for a take-off angle of 75° is only 0·5%.

Rock sections containing silicates and other non-conducting minerals require the addition of a thin conducting surface film, applied by vacuum evaporation. This film carries the current of the electron beam and so maintains the surface of the specimen very close to earth potential. A further requirement is that the materials used for mounting should have a low vapour pressure and be resistant to decomposition by the electron beam.

The techniques employed for specimen preparation vary from one laboratory to another. Those described briefly below have been used in the Department of Mineralogy and Petrology, Cambridge, for some years.

### A. POLISHED MOUNTS

Suitable mounting materials include Ceemar[1] resin and Araldite.[2] The sequence of polishing operations is as follows:

| | |
|---|---|
| Cast-iron lap | Norbide 600 abrasive[3] |
| Lead-alloy lap | Fast-cutting alumina[1] |
| Lead-alloy lap | Slow-cutting alumina[1] |
| Lead-alloy lap | Gamma polishing alumina[1] |

Abrasives are mixed with a light oil and applied as a paste to 8-in diameter laps running at approximately 100 rev/min with a load of up to 3 lb (1·4 kg) for a 3·7 cm diameter specimen. Short runs of about 15 min are used and between them the specimen is cleaned and inspected. A very short buffing with fine grade metallurgical polishing medium is sometimes used as a final stage to remove scratches.

### B. THIN SECTIONS

The procedure adopted consists of a combination of the normal technique for preparing covered thin sections and polishing methods similar to those described above. The choice of mounting material is important: Canada balsam is unsuitable owing to its high vapour pressure but Lakeside 70 cement[4] and epoxy resins have been found to be very satisfactory. It is generally advantageous to prepare thin sections as far as the hand-grinding stage (i.e. the stage where the cover-slip would normally be placed on the surface) with a thickness of about 60 $\mu$m. Such sections, particularly those of friable rocks, prove much easier to polish mechanically than those of the normal thickness of 30 $\mu$m. The individual glass slides are mounted in circular

[1] Griffin and George Ltd., 57 Uxbridge Road, London W.5.
[2] CIBA Ltd., Duxford, Cambs., England.
[3] Norton Grinding Co., Welwyn Garden City, Herts., England.
[4] Cutrock Engineering Co. Ltd., 35 Ballards Lane, London N.3.
[5] Engis Ltd., Park Road Trading Estate, Maidstone, Kent, England. (See page 230.)

Perspex (lucite) holders and inserted in the polishing machines in the same
way as the thick mounts. The polishing sequence is:

Hyprocell-paper lap[5]    6 $\mu$m Hypress diamond Compound[1]

Hyprocell-paper lap[5]    3 $\mu$m Hypress diamond Compound[1]

Hyprocell-paper lap[5]    1 $\mu$m Hypress diamond Compound[1]

Hyprocell-paper lap[5]    $\frac{1}{4}$ $\mu$m Hypress diamond Compound[1]

### C. STANDARDS

The method of mounting standards is often dictated by the specimen-
handling facilities provided by the electron-probe manufacturer. A technique
that has been found to be convenient is illustrated by Fig. 11. Individual

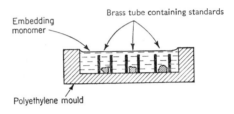

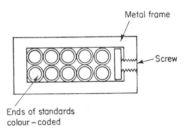

FIG. 11. Diagram showing one method of embedding standards for polishing. The plastic
disc is broken after polishing and the required combination of individual standards is
mounted in a metal frame.

standards are mounted in short brass tubes and embedded in plastic for
polishing, after which the tubes are broken out of the mounts and assembled
in the required combination in a metal frame.

### D. SURFACE COATING

The choice of surface coating lies between a metallic film of copper,
aluminium etc., evaporated *in vacuo* from a molybdenum boat or tungsten

wire, and a film of carbon obtained by vacuum evaporation from a carbon arc. Metallic films of the order of 50–100 Å in thickness are electrically satisfactory but leave only major differences in reflectivity of the specimen visible in the microscope. Carbon films, to be satisfactory, need to be somewhat thicker—200 Å—but even at this thickness allow much better visibility of the surface in reflected light. When using metallic films it is important to take account of variations in film thickness between specimen and standard. The techniques of vacuum evaporation are described fully in textbooks of vacuum technology and electron microscopy.

## IV. Evaluation of Experimental Measurements

### A. STANDARDS

Electron-probe analysis differs from X-ray fluorescence analysis in that the factors that relate concentration to the measured X-ray intensities from the standard and specimen may be calculated on a more fundamental basis. For many analyses it is thus possible to use pure elements as standards, or, alternatively, compounds of composition considerably different from the unknown: for example, wollastonite serves well as a standard for the determination of calcium in practically all silicates. It is important, however, to make sure that the emission wavelength of an element is the same in the standard and the unknown. Appreciable "chemical shifts" of the $K_\alpha$ wavelengths of the light elements and the $L$ wavelengths of heavier elements occur, and these cannot be ignored when high-resolution X-ray spectrometers are used. In the case of silicon, for example, the shift of the $K_{\alpha_1\alpha_2}$ wavelength between the pure element and a silicate is $0\cdot0022$ Å, and the error resulting from the use of pure silicon as a standard for silicate analysis can approach $50\%$ if the spectrometer is not reset (Koffman et al., 1967). However, the shift between different silicates is very small.

Some workers have used empirical calibration with chemically analysed standards of composition close to the unknowns. This procedure clearly minimizes the importance of the corrections, but is limited in its application, owing to the rarity of homogeneous natural minerals and the difficulty of preparing many compositions in the form of crystalline or glassy material, homogeneous on a sub-micron scale. The work of Smith and Stenstrom (1965) has shown, for example, the unreliability of some wet-chemical analyses of olivine, both on account of zoning in the mineral grains and errors in the analyses themselves.

Walter (1967) has recently described a method for preparing silicate standards by evaporating standardized solutions with silica gel and compressing the ignited powder at pressures of the order of 30 kbars. An inhomogeneity amounting to several per cent was, however, observed in such preparations, and it appears that this technique is not suitable for the most accurate analysis. Fredriksson (1967) has shown that chemical analysis of the standard may be avoided if homogeneous stoicheiometric near-end members of a solid solution series are available. For example, a nearly pure fayalite may be used to determine small quantities of iron in nearly pure forsterite. Conversely, the magnesium content of the fayalite may be measured with the forsterite as standard. The necessary corrections are thus applied only to the minor constituents, and do not need to be known with great accuracy. The content of trace elements, such as nickel, may similarly be determined by using pure metals as standards. With the additional knowledge that the compounds are stoicheiometric, the major-element concentrations may then be calculated with considerable accuracy.

### B. CALCULATION OF CORRECTION FACTORS

The method of deriving corrections was worked out initially by Castaing (1951). Castaing's theory has subsequently been modified and the present position is that the deficiencies in the correction procedure lie for the most part in the lack of sufficiently accurate fundamental constants, e.g. absorption coefficients, back-scattering data, etc., rather than in the theory. This is particularly true for measurements on the light elements below sodium in the periodic table. Comprehensive rigorous discussions of the theory of corrections are given by Castaing (1951, 1960), Duncumb and Shields (1963), Archard and Mulvey (1963), Philibert (1964) and Wittry (1964). It is convenient here to assume that the experimentally determined X-ray intensities have been corrected for counting losses (see Chapter 4, p. 180) and the intensity of the continuous background has been subtracted, giving directly the apparent concentration of each element determined according to equation (1). The apparent concentration is then subjected to three corrections, taking account of the effects of absorption, secondary fluorescence excitation and differences between the efficiency of X-ray production in the standard and specimen.

In order to calculate quantities such as absorption coefficients and mean atomic number of the specimen, which are required in the correction procedure, a provisional composition, based either on the measured concentrations or on *a priori* knowledge of the mineral, must be assumed. When the overall correction is large, it may be necessary to adopt an iterative method based on successive revisions of the assumed composition, in order to arrive at the true analysis; it is rarely necessary, however, to carry out the process more than two or at most three times. Moreover, it is often possible to estimate correction factors from past experience, which may be used in arriving at the assumed composition so that only one cycle of the correction procedure is required.

The nature and origin of the three corrections are as follows.


## C. THE ATOMIC-NUMBER CORRECTION

The assumption that the intensity of characteristic radiation generated within the specimen is proportional to concentration is only approximately correct. Divergence from strict proportionality is referred to as the "atomic-number effect", since its magnitude is dependent on the difference in the average atomic numbers of the specimen and the standard and also on the actual atomic number of the element determined. The mechanism may be understood by considering the fate of the beam incident on the specimen; the electrons suffer retardation as a result of inelastic scattering by the extra-nuclear electrons of the target atoms, a small fraction of the interactions resulting in the production of the spectra that are used for analysis. Since the ratio of the atomic number, $Z$ (i.e. the number of extra-nuclear electrons), to the atomic weight, $A$, increases with decreasing atomic number, it follows that the light elements in a compound target will absorb a disproportionately large amount of the energy of the electron beam. In other words, the "mass penetration" is less in elements of low $Z$ than in elements of high $Z$. Correspondingly, the "stopping power" of light elements for electrons is higher than that of heavy elements. The characteristic intensity produced per unit concentration of a heavy element is thus less when it is combined with a light element than in a standard of the pure element.

In opposition to the penetration effect is the effect of electron backscattering. The backscattering coefficient, $\eta$, increases with $Z$, as shown in Fig. 12. The fraction of the incident beam energy that is lost to the production of X-rays is thus greater in elements of high $Z$ than in light elements. In a compound target, therefore, the presence of light elements will reduce the backscattering loss, thus tending to make the characteristic intensity per unit concentration of a heavy element greater than in the pure standard.

The backscattering and retardation effects partially cancel one another. However, complete compensation seldom occurs, and departure from a linear concentration/intensity relationship results. A calculation of the atomic-number correction thus requires a knowledge of the magnitude of both the penetration factor and the backscattering factor. It is not possible here to give a complete discussion of this correction, but in illustration of the rigorous approach, it is useful to follow the treatment by Duncumb and Shields (1963) to the point where the salient features become clear.

K

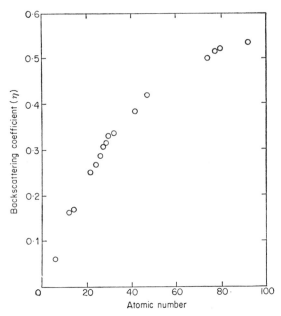

FIG. 12. Variation of electron backscattering coefficient ($\eta$) with atomic number. (After Bishop, 1967a.)

An electron incident on the surface of a target with energy $E_0$ may either remain within the target until it comes to rest, or it may be backscattered. If the energy of the electron at some distance $x$ along its path is $E$ then the number of $K$ ionizations of an element $A$ in the specimen which are produced in a path length $dx$ is:

$$dn_K = Q_K. \text{(no. of } A \text{ atoms per cm}^3). \ dx \qquad (2)$$

where $Q_K$ is the ionization cross-section of $A$ and a function of $E$ and $E_K$ only. ($E_K$ is the critical excitation potential of $A$.) Thus:

$$dn_K = Q_K \frac{N_0 \rho C_A}{A} \qquad (3)$$

where $N_0$ is Avogadro's number.

Once the energy of the electron falls below $E_K$ it no longer excites $K$ radiation, so that the total $K$ ionization is:

$$n_K = \frac{N_0 C_A}{A} \int_{E_0}^{E_K} \frac{\rho Q_K}{dE/dx} dE \qquad (4)$$

where $dE/dx$ is the loss of energy per unit path length. It is convenient to write $S = (-1/\rho)dE/dx$, where $S$ is the stopping power which determines the mass penetration into the target, measured along the path of the electron.

In order to allow for the loss of ionization by electrons that are back-scattered with energies $> E_K$, the factor $R$, which depends both on the backscattering coefficient, $\eta$, and on the energy distribution of the back-scattered electrons, is introduced. Thus we have:

$$n_K = \frac{N_0 C_A}{A} R \int_{E_K}^{E_0} \frac{Q_K}{S} dE \qquad (5)$$

Using the subscripts 0 and 1 to denote the standard and unknown, respectively, the corrected concentration becomes:

$$C_{\text{corrected}} = C_{\text{uncorrected}} \cdot \frac{R_0 \int_{E_K}^{E_0} Q_K/S_0 . dE}{R_1 \int_{E_K}^{E_0} Q_K/S_1 . dE} \qquad (6)$$

The most accurate expression for $S$ appears to be that due to Bethe (1930), which gives:

$$S = \text{const.} \; \overline{(Z/A)} \frac{1}{E} \ln(2E/11 \cdot 5 \, Z) \qquad (7)$$

where $E$ is in eV and $\overline{Z}$ and $\overline{(Z/A)}$ are given by:

$$\overline{Z} = \sum C_i Z_i \text{ and } \overline{(Z/A)} = \sum C_i (Z/A)_i.$$

Evaluation of the correction factor thus requires a numerical integration. Poole and Thomas (1962), however, have pointed out that the quantity $S_1/S_0$ is a very slowly varying function of $E$ and that it can therefore be taken outside the integrals. The latter then cancel leaving a penetration factor which is merely the ratio of stopping powers, evaluated at some mean potential between $E_0$ and $E_K$. The correction factor then becomes $R_0/R_1 . \bar{S}_1/\bar{S}_0$.

Reed (1964a) has used a similar approach and has derived the expression:

$$\bar{S} = \text{const.} \; \overline{(Z/A)} \frac{1}{E} (4 \cdot 54 + \ln \bar{E} - \ln \bar{Z}) \qquad (8)$$

from which the constant and the $1/E$ cancel when the ratio $\bar{S}_1/\bar{S}_0$ is taken; $\bar{E}$ is given by $\bar{E} = \frac{1}{2}(E_0 + E_K)$ and is in keV.

More recently da Casa and Duncumb (1967) have used the analyses of alloys compiled by Poole and Thomas to evaluate the mean ionization potential, $J$ (implicit in the Bethe equation), as a function of $Z$. The value of $S$ then becomes:

$$S = \text{const.} \, (Z/A) \frac{1}{E} \cdot \ln(1 \cdot 166 \, E/J) \tag{9}$$

where $E$ is in eV and equal to $\frac{1}{2}(E_0 + E_K)$. The value of $\bar{S}$ for a compound is obtained by calculating the mass concentration average of the $\bar{S}$ values for individual elements. Again the constant and $1/E$ cancel when the ratio $\bar{S}_1/\bar{S}_0$ is taken.

The calculation of $R$ is dependent on a knowledge of the energy distribution of the backscattered electrons. Duncumb and Shields (1963) use the data obtained by Kulenkampff and Spyra (1954), but the more recent measurements by Bishop (1967a) almost certainly lead to more accurate $R$ curves (Fig. 13). In order to obtain $R$, the mean $Z$ is first calculated, after which the value of $R$ may be read off directly from the curve appropriate to the overvoltage $U$ which has been used in the measurements.

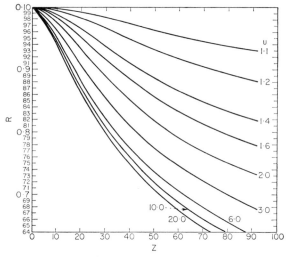

FIG. 13. Curves for the backscattering loss factor, $R$, as a function of mean atomic number ($\bar{Z}$) and overvoltage, ($U = E_0/E_K$), calculated from data by Bishop (1967a).

## D. THE ABSORPTION CORRECTION

Since X-rays are generated by the electron beam below the surface of the specimen, the emergent radiation suffers absorption to an extent that, in any material, depends on the absorption coefficient for the wavelength concerned and on the path length traversed in the specimen. As shown in Fig. 14,

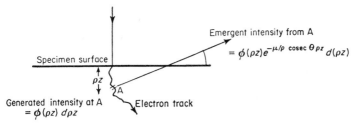

FIG. 14. Diagram showing dependence of absorption on the parameter $\chi$, where $\chi = (\mu/\rho)$ cosec $\theta$.

the path length is proportional to cosec $\theta$, where $\theta$ is the take-off angle and increases with increasing depth of production in the target. The absorption factor, by which the intensity generated in the specimen is attenuated, may be written as $f(\chi)$, where $\chi = (\mu/\rho)$cosec $\theta$. The magnitude of $f(\chi)$ was determined originally (Castaing, 1951) for a 50–50 Fe–Cr alloy by measuring the emergent intensity as a function of $\theta$. Later, Castaing and Descamps (1955) made experimental determinations of the distribution of X-ray production in depth, $[\phi(\rho z)]$, for gold, copper and aluminium. This data gives $f(\chi)$ directly from the ratio of the total emergent intensity to the total generated intensity:

$$f(\chi) = \frac{\text{emergent intensity}}{\text{generated intensity}} = \frac{\int\limits_0^\infty \phi(\rho z) \exp \chi(\rho z)\, \mathrm{d}(\rho z)}{\int\limits_0^\infty \phi(\rho z)\, \mathrm{d}(\rho z)} = \frac{F(\chi)}{F(0)} \qquad (10)$$

In order to extend the calculation of $f(\chi)$ to all elements and to compound targets, Philibert (1963) derived a formula based on theoretical argument but fitted to the $\phi(\rho z)$ data of Castaing and Descamps. In its original form this gives:

$$F(\chi) = \frac{1}{\left(1 + \dfrac{\chi}{\sigma}\right)\left(1 + h\left[1 + \dfrac{\chi}{\sigma}\right]\right)} \qquad (11)$$

where $h = 1 \cdot 2A/Z^2$ and $\sigma$ is an adjusted Lenard coefficient. Since $F(\chi)$ is proportional to the emergent intensity rather than the ratio of emergent-to-generated intensity, the expression must also contain the atomic-number factor $F(0) = 1/(1 + h)$ obtained by setting $\chi = 0$. There is little theoretical justification for this form of the atomic-number correction and it seems better to use the Philibert formula for absorption alone and to apply an atomic number factor as described above. The absorption factor alone is then:

$$f(\chi) = \frac{F(\chi)}{F(0)} = \frac{1+h}{\left(1+\dfrac{\chi}{\sigma}\right)\left(1 + h\left[1 +\dfrac{\chi}{\sigma}\right]\right)} \tag{12}$$

A deficiency in this formula has become apparent, particularly in its application to the light elements, as a result of the work of Green (1962). It stems from the fact that no account is taken of the excitation potential of the element for whose radiation the correction is made. A dependence on $E_c$, the critical excitation potential, has been introduced by Duncumb and Shields (1966) who use the formula as in equation (12), but with $\sigma$ now replaced by $\sigma_c$, where:

$$\sigma_c = \frac{2 \cdot 39 \times 10^5}{E_0^{1 \cdot 5} - E_c^{1 \cdot 5}} \tag{13}$$

$E_0$ being the beam accelerating potential (both $E_0$ and $E_c$ are measured in kV).

The voltage-modified form of the Philibert formula probably represents the best means of applying the absorption correction available at the present time, at least for radiations up to about 10 Å in wavelength.

One of the main sources of uncertainty has been in the values of the mass-absorption coefficients themselves. There is considerable disagreement between the experimental values quoted in the literature, particularly with regard to the rate of variation of $\mu/\rho$ with wavelength. Heinrich (1966b) has computed a comprehensive table by interpolation of the most reliable experimental data. Some of the data in this table has been confirmed experimentally by Hughes and Woodhouse (1967) and it undoubtedly represents the most reliable compilation at present available. In particular, the values given for the coefficients of long-wavelength radiation appear to be considerably higher than previously supposed.

### E. ELECTRON-TRAJECTORY MODELS AND MONTE CARLO CALCULATIONS

The methods described above for the determination of the absorption and atomic-number corrections are based on expressions and data that describe the average behaviour of a large number of electrons in the target. A completely different approach is to sum the effect of a large number of individual electrons. A high-speed digital computer is used to calculate the electron trajectory and the X-ray intensity generated at each section of the path until the point where the energy falls below the critical excitation potential or the electron is lost by backscattering. Such calculations have been performed by

Archard and Mulvey (1963) by using a model of the scattering process and by Green (1963b) and Bishop (1967b) using Monte Carlo techniques. The computations give directly the depth distribution of primary characteristic X-ray production $[\phi(\rho z)]$ and hence $f(\chi)$. The atomic-number correction may also be obtained if the computation is made for a compound or an alloy system.

At present the results may be said to be in encouraging agreement with the experimentally determined $f(\chi)$ values, one limitation being the accuracy of the available electron-scattering cross-sections at the low energies of interest in X-ray microanalysis. In particular, the data have been used by Duncumb and Melford (1967) for the correction of experimental measurements of the carbon content of different compounds. In this long-wavelength region the method appears to be more satisfactory than the Philibert model.

## F. FLUORESCENCE CORRECTION

Part of the measured characteristic intensity from the specimen may arise from fluorescence excited by other characteristic lines produced by the electron beam. Such secondary fluorescence can only occur when the spectrum contains lines with quantum energies greater than the critical excitation potential of the analysed element. (Fluorescence may also be excited by continuous radiation, but the magnitude of this effect in homogeneous specimens is generally small [Henoc, 1962].) Fluorescence excitation by characteristic lines will thus occur in specimens such as $CuFeS_2$, where $CuK_\alpha$ and $CuK_\beta$ excite $FeK_\alpha$ and in $FeTiO_3$ where $FeK_\alpha$ and $FeK_\beta$ excite $TiK_\alpha$. However, the effect is generally negligible for the light elements, e.g. for excitation of $MgK_\alpha$ by $SiK_\alpha$ in a silicate, owing to the rapid decrease of fluorescence yield with decrease in atomic number.

If the ratio of the primary to the total (primary + fluorescence) intensity is denoted as $\gamma$, the correction factor to be applied to the measured concentration will be $1/(1 + \gamma)$. The original formula (Castaing, 1951) for the calculation of $\gamma$ has been modified by Wittry (1962) and by Reed and Long (1963) to include a dependence on the beam-accelerating voltage. The most comprehensive treatment of the fluorescence correction presently available is that of Reed (1965a), who gives a general formula, based on the voltage-dependent form of Castaing's expression, which is applicable to the excitation of one $K$ line by another ($K-K$ fluorescence), and also to $K-L$, $L-K$ and $L-L$ fluorescence. In addition, the equation for $\gamma$ has been arranged to facilitate computation, and all necessary data, apart from absorption coefficients, are included in the paper in tabular or graphical form. The value of $\gamma$ is rarely larger than 0·2 and often very much smaller, so that comparatively large errors in its calculation—for example due to the use of an assumed rather

TABLE 1

*Summary of correction procedure*

| 1. Intensity measurements. | Standard | Specimen | | | |
|---|---|---|---|---|---|
| | peak $N^{st}$ | background $N_b^{st}$ | peak $N^{sp}$ | background $N_b^{sp}$ | |

| 2. Resolving-time correction for peak intensities. | true intensity $= N_t$ $$= \frac{N(\text{measured})}{1 - N(\text{measured})\,\tau}$$ $N$ in counts per sec, $\tau$ in sec) | Corrects for loss of pulses in resolving time of counting circuits. Necessary only for high counting rates. |
|---|---|---|
| 3. Subtract background. | $N_t^{st} - N_b^{st} = N_0$    $N_t^{sp} - N_b^{sp} = N_1$ | |
| 4. Calculate apparent concentration. | apparent conc. (wt. %) $C = N_1/N_0 \times C_0$ ($C_0 =$ wt. % of element in standard) | Based on assumption that intensity is proportional to concentration. |
| 5. Correct for absorption. | $C' = C \times f(\chi_0)/f(\chi_1)$ ($\chi = (\mu/\rho)$ cosec $\theta$ and is calculated either from the assumed or the approximate measured composition). | X-rays suffer different absorption in specimen and standard. |
| 6. Correct for fluorescence. | $C'' = C' \times 1/1 + \gamma$ Only required when energy criterion for excitation is satisfied. | Primary radiation from one element in the specimen may excite additional characteristic radiation from the measured element. |
| 7. Correct for atomic number effect. | $C''' = C'' \times \dfrac{\bar{S}_1 \, R_0}{\bar{S}_0 \, R_1}$ ($\bar{S} =$ retardation factor, $R =$ backscattering factor) | Variation of retardation and backscattering with atomic number do not quite balance, so intensity is not exactly proportional to concentration. |
| 8. True concentration. | $C'''$ is the "true" concentration, but is still subject to experimental error. | |

than the true composition, have only a small effect on the accuracy of the analysis. (In the case of Fe TiO$_3$, the correction factor $1/(1 + \gamma)$, to be applied to the observed concentration for fluorescence enhancement at 20 kV in an instrument working at 40° angle of take-off is 0·97.) The correction factor is greatest when the energy of the exciting radiation is just above the critical excitation potential of the analysed element and when the exciting element is present in high concentration.

The complete correction procedure, including the initial corrections for background and resolving time, is summarized in Table 1.

Iteration, necessary if the overall correction is large, must of course be carried out after all three separate corrections have been applied.

## V. Fundamental Limitations

### A. RESOLUTION AND SENSITIVITY

The spatial resolution and the minimum detectable concentration for a given resolution together provide a good criterion for assessing the performance of an electron-probe analyser. The resolution must clearly be limited at some point by the diameter of the electron beam, but may be considerably worse than the beam diameter as a result of electron scattering within the target, or of fluorescence excitation of the measured element in the region surrounding that excited by the electron beam. The concentration sensitivity is determined by the minimum characteristic intensity that can be detected above the fluctuations in the background, and is thus dependent on the peak intensity per unit concentration, the background intensity and the stability of the probe current. The peak intensity is directly proportional to the current delivered by the electron-optical system and to the efficiency of the X-ray spectrometer. The background is determined by the efficiency with which the spectrometer excludes unwanted radiation. This factor is most conveniently expressed in terms of the peak/background ratio for a pure element. Resolution and sensitivity are related because of the strong dependence of probe current on probe diameter.

The maximum current obtainable in a beam of given diameter is fixed by the brightness of the electron source and the severe spherical aberration of the electron lenses, which necessitates the use of very small apertures. The diameter, $d_s$, of the spherical-aberration disc of least confusion for a given semi-aperture $\alpha$ is given by:

$$d_s = \tfrac{1}{2} C_s \alpha^3 \tag{14}$$

where $C_s$ is the spherical aberration coefficient, which for a typical objective lens will be of the order of 4 cm. If the enlargement of the focused beam diameter by spherical aberration is limited to 10%, it may be shown that the maximum current obtainable in a probe of diameter $d_0$ is given by:

$$i = j \frac{V_0}{V_e} \frac{\pi}{4} \frac{d_0^{\frac{8}{3}}}{C_s^{\frac{2}{3}}} \quad (15)$$

where $V_0$ is the accelerating voltage, $V_e$ is the mean energy of emission of electrons from the filament ($\sim 0.25$ eV) and $j$ is the current density at the filament (typically $2A/cm^2$). This equation shows that the current obtainable, and hence the X-ray intensity, falls off very rapidly as the probe is made smaller. Compensation for this decrease is difficult to obtain; $j$ cannot be raised appreciably without drastic reduction of filament life (Bloomer, 1957), and $C_s$ is fixed by the fundamental properties of the electron lenses (Liebmann, 1955) for which no satisfactory means of correcting spherical aberration has yet been devised. $V_0$ is normally restricted by the need to control the penetration and diffusion of electrons in the specimen.

A typical instrument will be capable of producing a probe of diameter 1 $\mu$m carrying a current of 0.5 $\mu$A for an accelerating voltage of 20 kV. The characteristic X-ray intensity produced by this current at the detector varies according to the position of the target element in the periodic table and on the efficiency of the spectrometer. Typical figures for peak intensity and peak-to-background ratio are shown in Table 2.

TABLE 2

| Line | Crystal | Slit | Counts/sec/$\mu$A | Peak/background |
|------|---------|------|-------------------|-----------------|
| Ni$K\alpha$, 1st order (pure Ni, 30 kV) | LiF | 0.010″ | $517 \times 10^3$ | 627 |
| Cr$K\alpha_{1,2}$ 1st order (pure Cr 30 kV) | Quartz | 0.010″ | $192 \times 10^3$ | 677 |
| Na$K\alpha$ 1st order (NaCl, 15 kV) | Mica | 0.012″ | $1.3 \times 10^3$ | 570 (with pulse-height analysis) |

Data obtained with Cambridge Instrument Company Geoscan.

The minimum detectable concentration (defined as that which produces a peak equal to three times the standard deviation of the background) may be obtained from such data by means of the nomogram in Fig. 15 (J. V. P. Long, unpublished). $Z$ is the mean atomic number of the specimen and $Z_A$ the atomic number of the pure standard. $N_0$ and $R_A$ are the counting rate and the peak-to-background ratio, respectively, for the pure standard. $T$ is the measuring time (equal for peak and background) and $k$ is the stability of the probe current during the period of the determination.

The nomogram is used by placing a rule through the $P/B$ figure for the specimen ($R.Z_A/\bar{Z}$) and $N_0$ so as to intersect the reference line 1. A point on reference line 2 is then found by projecting horizontally the point of intersection of the counting time and the appropriate stability curve. A straight edge placed on the two points obtained on the reference lines will then enable the limit of detection to be read off from the extreme right-hand scale.

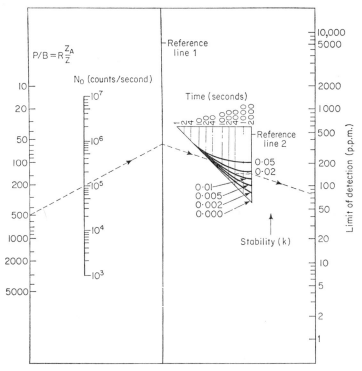

FIG. 15. Nomogram for the calculation of the statistical limit of detection.

For the determination of nickel in an olivine, typical values might be $R = 1,000$, $Z_A/\bar{Z} = 2\cdot3$, $T = 400$ sec, $N_0 = 10^5$ counts/sec and $k = 0\cdot01$ (1%), giving a limit of detection of approximately 15 p.p.m.

The problem of optimizing intensity and resolution has been considered by Duncumb (1960) and more recently by Reed (1967). Penetration and diffusion set a minimum size to the volume excited by the electron beam, no matter how small its diameter. A very approximate estimate of the diameter

of the excited volume may be obtained by equating it to the depth of penetration, $p$, as calculated from the Thomson–Whiddington law, which with the
excitation potential taken into account may be written:

$$p = \frac{K(E_0^2 - E_K^2)}{\rho} \qquad (16)$$

where $K$ is approximately 0·025 for $E$ in keV and $p$ in $\mu$m. Thus for a silicate
of density 3, the radius of the volume excited at 20 kV when $FeK_\alpha$ is measured,
comes out at approximately 2·8 $\mu$m. A more accurate estimate of the shape
and size of the excited volume is obtained as a result of computer calculations
of the depth and lateral distributions of ionization in the target (Bishop
1965). These lead to a picture of the ionized volume such as that shown in
Fig. 16.

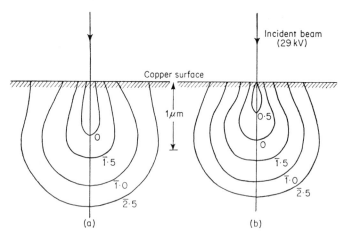

FIG. 16. Contours of equal energy loss and ionization concentrations. Units: fraction of
total energy loss or ionization/$\mu$m³. (After Bishop, 1965.)

The penetration may be reduced by reducing $E_0$, but difficulties arise when
the accelerating voltage approaches the excitation potential, both in the
application of atomic-number corrections and as a result of the significant
loss of electron energy in the conducting film on the surface of insulating
specimens. An alternative course is to use a longer-wavelength radiation,
e.g. $L$ or $M$ instead of $K$. This allows a reduction of $E_0$ while still maintaining
a reasonable value of the overvoltage. However, the use of long wavelengths
introduces uncertainties in the application of the absorption correction,
particularly owing to the rather unreliable absorption-coefficient data in this
region.

## B. FLUORESCENCE UNCERTAINTY

Whereas electron-penetration effects become important only for resolutions below about 3 $\mu$m (depending on $E_0$, $E_K$ and $\rho$) secondary fluorescence excitation is effective over a much larger volume. This effect is important in mineralogy in measurements on fine exsolution lamellae or near phase boundaries. Since excitation may occur in phases of unknown composition below the visible surface, this type of fluorescence is not, except in certain ideal cases, amenable to correction and has come to be known as the "fluorescence uncertainty".

The case of a boundary normal to the surface has been examined theoretically and experimentally by Reed and Long (1963). As shown in Fig. 17 an

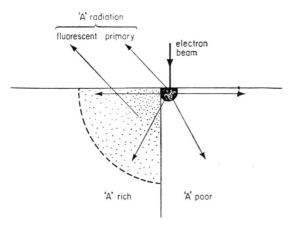

FIG. 17. Fluorescence excitation across a phase boundary.

analysis for the element A in the A-poor phase will give high results owing to the secondary excitation of A characteristic radiation in the neighbouring A-rich phase. The effect is particularly bad when the A-poor phase contains a high proportion of an element whose characteristic radiation is just shorter than the absorption edge of A. Fig. 18 shows the experimental data obtained on an artificial Fe–Ni boundary. With the electron probe entirely absorbed within the nickel, the observed iron content of the pure nickel near the boundary was 5·4%. Similar experiments have been carried out by Dils et al. (1963) with respect to excitation by the continuous spectrum, and more recently by Maurice et al. (1967).

Since the mean depth of production of the fluorescence radiation is greater than that of the primary radiation, it follows that it will be attenuated to a greater extent by absorption, particularly at low take-off angles. In the analysis of small inclusions in metals the reduction in fluorescence uncertainty obtained by using a low take-off angle is undoubtedly valuable. In mineralogical systems, however, where fluorescence effects are infrequent and, owing

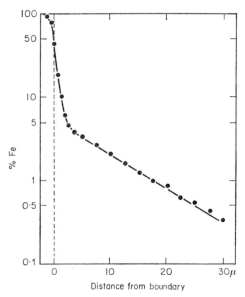

Fig. 18. Apparent iron content of pure nickel as a function of distance from boundary with pure iron. (After Reed and Long, 1963.)

to the diluent effect of matrix elements such as oxygen and silicon, generally much smaller, the advantage to be gained from a low take-off angle is generally outweighed by the much greater absorption correction and associated "absorption uncertainty" that is entailed in the measurement of the common elements sodium, magnesium, aluminium and silicon.

### C. THERMAL EFFECTS IN THE SPECIMEN

The sensitivity of some minerals to thermal decomposition imposes a limit both to concentration sensitivity and resolution, since in order to restrict the temperature rise of the specimen, the current density in the electron probe must be reduced. The thermal conductivity is often less than 0·01 (cf. 0·9 for copper) and the temperature rise at the point of impact of the probe may amount to several hundred degrees. Loss of water thus commonly occurs; more unexpectedly, the alkali metals are found to evaporate from certain silicates such as the feldspars and biotite. As shown by Fig. 19 the rate of loss of potassium from a feldspar may be sufficiently rapid to interfere seriously with quantitative analysis. In such cases, there is little alternative but to sacrifice resolution by enlarging the diameter of the beam in order to reduce the target temperature.

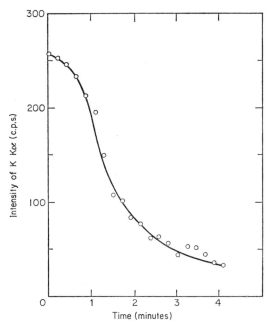

FIG. 19. Variation with time of potassium $K_\alpha$ intensity from orthoclase for an electron beam approx. $1\mu$m diameter and a beam current of $2\cdot0 \times 10^{-7}$A at 15 kV.

## VI. Some Applications of the Method in Mineralogy

An analytical technique with a resolution of $1\mu$m, an accuracy of the order of 1% and a lower limit of detection in the region of 10–100 p.p.m. must inevitably be applicable to many problems of mineral analysis for which conventional techniques involving separation were previously used; it also makes possible many analyses that could not otherwise be attempted. It is not surprising therefore that the number of papers including electron-probe data on minerals is already large; as a consequence the examples quoted below constitute a cross-section, but by no means a complete bibliography. Other references are to be found in review papers (Guillemin and Capitant, 1960; Capitant and Weinryb, 1960; Long, 1963a) and in the Heinrich bibliography (1966a).

Probably the simplest application of the technique is in the identification of minerals in polished mounts or thin sections in cases where the number of possibilities is limited. A qualitative estimation of the major constituents without detailed correction of the results is then often all that is needed. Keil (1965) has recently described an extension of this procedure in which the electron probe is used for modal analysis.

The examination of opaque minerals (Bizouard and Roering, 1958; Birks et al., 1959; Andronopoulos, 1961; Fauquier, 1961; Bahezre et al., 1961; Kuovo et al., 1963; Takeno, 1963; Fredriksson and Andersen, 1964; Chamberlain and Delabio, 1965; Stumpfl and Clark, 1965; Springer et al., 1964) has proved a particularly fruitful field on account of the limited information on variations of composition that is obtainable by other methods. Several new mineral phases have been found (Stumpfl, 1961; Long et al., 1963; Stumpfl and Clark, 1964, 1966), among them the metallic CoFe in association with the previously known mineral awaruite, $(Ni_3Fe)$, in the Red Hills serpentinites, New Zealand (Challis and Long, 1964) and in the Muskox intrusion, Canada (Chamberlain et al., 1965). This latter example illustrates the power of the technique in obtaining good chemical data from minute grains in thin sections. The photographs in Fig. 20 of this mineral also show the use of the scanning technique in demonstrating the duplex nature of the largest grain found in the New Zealand material. One new mineral, castaingite (Schüller and Otteman, 1963), has been named after the originator of the electron-probe technique.

Meteorites have received much attention. Early work (Yavnel et al., 1958; Agrell and Long, 1960; Feller-Kniepmeier and Uhlig, 1961; Agrell et al., 1963) confirmed and established the extent of the already known zoning of the taenite lamellae of irons and revealed concentration gradients in the kamacite near the boundary with taenite.

Later work on the irons has been concerned with detailed studies of phase compositions and with a comparison of the observed concentration gradients with those calculated on the basis of various models of the cooling of the parent bodies (Goldstein and Ogilvie, 1963; Lovering, 1964; Wood, 1964; Goldstein, 1965; Goldstein and Ogilvie, 1965; Reed, 1965b,c).

Similarly, much new data on the composition of minerals in stony meteorites has been obtained (Fredriksson and Keil, 1964; Reed, 1964b; Keil et al., 1964; Keil and Andersen, 1965a; Fredriksson and Reid, 1965; Marshall and Keil, 1965). Keil and Fredriksson (1964) who examined the coexisting olivines and orthorhombic pyroxenes in 85 chondrites have demonstrated a close correlation between the $Fe/(Fe + Mg)$ ratio in these phases and the Urey–Craig groupings (Fig. 21). The occurrence of the new mineral sinoite $(Si_2N_2O)$ in meteorites has also been established (Andersen et al., 1964; Keil and Andersen, 1965b) by electron-probe analysis.

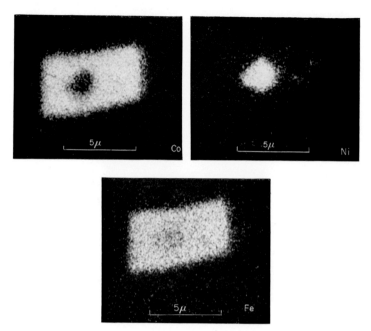

FIG. 20. X-ray scanning images showing the distribution of Co, Ni and Fe in a metal grain having a core of awaruite ($Ni_3Fe$) and a rim of wairauite (CoFe). (After Challis and Long, 1964.)

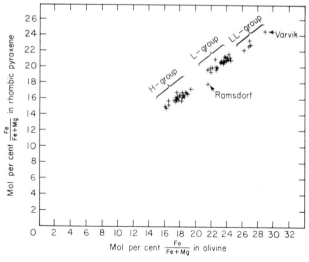

FIG. 21. Ratios Fe/(Fe + Mg) in olivine plotted against ratios Fe/(Fe + Mg) in orthorhombic pyroxene for 85 chondrites. (After Keil and Fredriksson, 1964.)

Clearly the technique is applicable to the determination of the composition of coexisting phases and to the study of zoning and exsolution in individual minerals and has been widely used in this field (Chinner, 1962; Binns *et al.*, 1963; Green, 1963a; de Bethune *et al.*, 1965; Muir and Long, 1965). Recent papers by Smith (1965, 1966) demonstrate well the high accuracy attainable in the analysis of rock-forming minerals. Some measurements have been made on synthetic minerals, and recently Terrier and Capitant (1967) have used the technique in a study of the hydration of Portland cement.

## VII. Extensions of the Basic Technique

The basic method of analysis by the examination of X-ray emission spectra has been extended in a number of ways. Castaing (1951) showed that the very small and intense source of X-rays generated by the electron probe was particularly suitable for the formation of divergent beam X-ray diffraction patterns. This technique, which is capable of giving very accurate cell dimensions has, however, so far found more application in the study of metals than minerals (Sawkill and Schwartzenberger, 1960; Hanneman *et al.*, 1962).

### A. CATHODOLUMINESCENCE

From the point of view of the mineralogist, probably the most useful auxiliary technique that has resulted from the use of the electron-probe is the simple one of visual observation of the luminescence generated by the electron beam. Cathodoluminescence in thin sections or polished mounts of minerals and rocks may be observed in the electron probe, usually with the objective lens defocused so as to illuminate an area of the specimen comparable with that seen in the light microscope. Alternatively, a simple arrangement for generating a roughly focused electron beam may be attached to a normal microscope. Such a device may either use a hot filament (Long, 1960; Le-Poole *et al.*, 1965) or consist of a cold cathode-discharge tube operating under poor vacuum conditions (Sippel, 1965). Cathodoluminescence in the visible part of the spectrum is sometimes an intrinsic property of a mineral, as for example in the case of scheelite, though more frequently it is the result of the presence of trace impurities—often manganese or rare earths—in an otherwise non-luminescent host. Figure 22 shows alternate luminescent and dark zones in an apparently homogeneous calcite crystal. In this case the variations in colour may be shown to be closely correlated with variations in the Mn/Fe ratio in the various zones (Long, 1963b; Long and Agrell, 1965). The dependence of fluorescence colour on trace element content may in some cases be used to determine the provenance of minerals in sedimentary deposits

(Stenstrom and Smith, 1964, 1965). Where a mineral occurs only as a minor constituent, a characteristic luminescence can be an aid to the identification of small grains in the matrix; for example, the new mineral sinoite (Andersen *et al.*, 1964) is readily identified in meteorites by its green cathodoluminescence. Although the observation of cathodoluminescence of minerals in thin section and polished mounts has only recently become prominent as a mineralogical technique, it is perhaps worth noting that Crookes (1879) made a careful study of the cathodoluminescence of several minerals and recorded the fact that in some cases the emitted light was polarized.

Fig. 22. Cathodoluminescence of a single crystal of calcite showing rhythmic variations in colour and intensity due to fluctuations in the Mn/Fe ratio. (After Long and Agrell, 1965.)

## B. SHIFTS IN WAVELENGTH OF CHARACTERISTIC EMISSION LINES

Small but detectable differences between the wavelengths of the characteristic lines of elements in the pure state and in compounds were recorded in the 1920's; a summary of early work is provided by Siegbahn (1925). Since the shifts in wavelength are produced by differences in the electrostatic field surrounding the emitting atoms in different structures, they are greater when a line arises from a transition involving an outer level, e.g. $CuK_{\beta_5}$ ($KM_{IV,V}$) than when the electron starts in a comparatively shielded inner level, e.g. $CuK_{\alpha_1}$ ($KL_{III}$). Further, since the degree of perturbation of the levels in a given structure is constant, it follows that the relative shift will be greater in the long wavelength spectra: thus the relative shift for $CuL_{\alpha_{1,2}}$ ($L_{III}M_{IV,V}$) is greater than that of $CuK_{\beta_5}$ although they both involve the same outer level.

A measurement of the shift in a particular material thus, in principle, provides the possibility of obtaining information about the chemical state of an atom in the structure. While many detailed measurements have appeared in recent years [e.g. the work of Fischer (1964) and Fischer and Baun (1967)], the method has so far been little used in mineralogy. One paper of note is that of Day (1963) who used X-ray excitation in a fluorescence spectrometer in order to study the shift of $AlK_{\alpha_{1,2}}$ as a function of co-ordination number in some aluminium minerals (Fig. 23). Figure 24 shows the line profile and peak position of the $CuL_{\alpha_{1,2}}$ line from a number of copper minerals, measured in the electron probe (Long, 1964). It is clear from these results that both co-ordination and valency must be taken into account in attempting to use

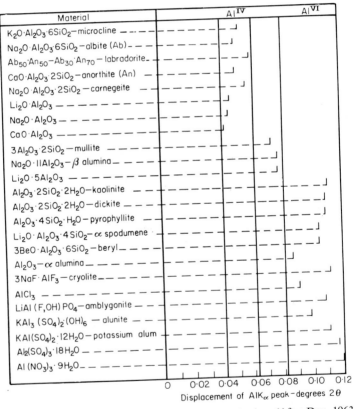

FIG. 23. Shift in the $AlK_\alpha$ line using secondary excitation. (After Day, 1963.)

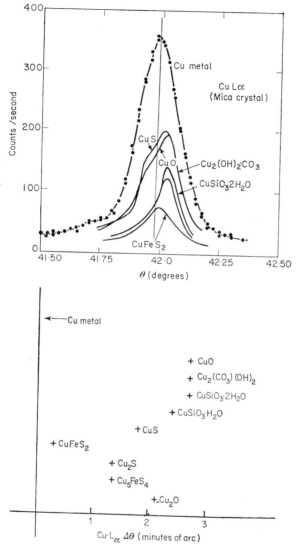

FIG. 24. Shifts of $CuL_\alpha$ line in different minerals measured in the electron-probe analyser. (After Long, 1964.)

wavelength-shift measurements as an aid to structure determination. Conversely, such measurements provide a potential method of estimating relative amounts of ions in different oxidation states (e.g. $Ti^{3+}/Ti^{4+}$, $Fe^{2+}/Fe^{3+}$, $Mn^{2+}/Mn^{3+}$ in ferromagnesian silicates) and different co-ordination symmetries (e.g. $Al^{3+}$ in 6-fold and 4-fold co-ordination in pyroxenes).

L*

## C. THE COMBINATION OF ELECTRON-PROBE ANALYSIS
## WITH ELECTRON MICROSCOPY AND DIFFRACTION

The possibility of performing electron-probe analysis and obtaining electron micrographs and diffraction patterns in the same instrument has been explored (Nixon, 1963; Duncumb, 1966), although no use of the technique in mineralogy has so far been reported. The main application has so far been in the identification of particles down to about 0·1 μm in diameter in extraction replicas of metals.

### REFERENCES

Agrell, S. O. and Long, J. V. P. (1960). The application of the scanning X-ray Microanalyser to mineralogy. *In* "X-ray Microscopy and X-ray Microanalysis." Ed. A. Engström, V. E. Cosslett and H. H. Pattee, p. 391. Elsevier, Amsterdam.

Agrell, S. O., Long, J. V. P. and Ogilvie, R. E. (1963). Nickel content of kamacite near the interface with taenite in iron meteorites. *Nature, Lond.* **198**, 749.

Andersen, C. A., Keil, K. and Mason, B. (1964). Silicon Oxynitride: A meteoritic mineral. *Science, N.Y.* **146**, 256.

Andronopoulos, B. (1961). Association de magnétite-chromite-pentlandite dans quelques gites de fer en Grèce. *Bull. Soc. fr. Minér. Cristallogr.* **84**, 345.

Archard, G. D. and Mulvey, T. (1963). The present state of quantitative X-ray microanalysis, Part 2: computational methods. *Br. J. appl. Phys.* **14**, 626.

Bahezre, C., Capitant, M. and Phan, K. D. (1961). Analyse ponctuelle d'un cassiterite zonée. *Bull. Soc. fr. Minér. Cristallogr.* **84**, 321.

Bethe, H. A. (1930). Zur Theorie des Durchgangs schnellen Korpuscularstrahlen durch Materie. *Ann. Phys.* **5**, 325.

de Bethune, P., Goossens, P. and Berger, P. (1965). Emploi des grenats zonaires comme indicateurs du degré de métamorphisme. *C. r. hebd. Séanc. Acad. Sci., Paris* **260**, 6946.

Binns, R. A., Long, J. V. P. and Reed, S. J. B. (1963). Some naturally occurring members of the clinoenstatite-clinoferrosilite mineral series. *Nature, Lond.* **198**, 777.

Birks, L. S., Brooks, E. J., Adler, I. and Milton, C. (1959). Electron-probe analysis of minute inclusions of a copper-iron mineral. *Am. Miner.* **44**, 974.

Bishop, H. E. (1965). A Monte-Carlo calculation of the scattering of electrons in copper. *Proc. phys. Soc.* **85**, 855.

Bishop, H. E. (1967a). Some electron backscattering measurements for solid targets. *In* "Optique des Rayons X et Microanalyse." University of Paris. (In press.)

Bishop, H. E. (1967b). Calculations of electron penetration and X-ray production in a solid target. *In* "Optique des Rayons X et Microanalyse." University of Paris. (In press.)

Bizouard, H. and Roering, C. (1958). An investigation of sphalerite. *Geol. För. Stockh. Förh.* **80**, 309.

Bloomer, R. N. (1957). The lives of electron microscope filaments. *Brit. J. appl. Phys.* **8**, 83.

Capitant, M. and Guillemin, C. (1960). Utilisation de la microsonde électronique de Castaing pour les études minéralogiques. International Geological Congress, 21st Session, Norden, part 21, P.L. 201, Det Berlingske, Copenhagen.

Capitant, M. and Weinryb, E. (1960). Études minéralogiques realisées avec la sonde électronique de Castaing. *Bull. Soc. fr. Minér. Cristallogr.* **83**, XXXVIII.

da Casa, C. and Duncumb, P. (1967). "Choice of mean ionization potential, *J*, in atomic number corrections." Presented at the Conference on Recent Developments in Theory, Technique and Application of Electron-probe microanalysis. (London.) (To appear under Duncumb, P., Reed, S. J. B., Mason, P. K. and da Casa, C. (1967) *Br. J. appl. Phys.* Submitted for publication.)

Castaing, R. (1951). Application des sondes électroniques a une methode d'analyse ponctuelle chimique et cristallographique. Thesis, University of Paris. O.N.E.R.A. Publication No. 55.

Castaing, R. (1954). État actuel du microanalyseur à sonde électronique. Proceedings of the International Conference on Electron Microscopy. London, pp. 300–304.

Castaing, R. (1960). Electron probe microanalysis. *In* "Advances in Electronics (and Electron Physics)." Ed. L. Marton, Vol. 13, p. 317. Academic Press, New York.

Castaing, R. and Descamps, J. (1955). Sur les bases physiques de l'analyse ponctuelle par spectrographie X. *J. Phys. Radium* **16**, 304.

Castaing, R. and Guinier, R. (1949). Application des sondes électroniques a l'analyse metallographique. Proceedings of the Conference on Electron Microscopy, Delft **60**. Martinus Nijhoff, The Hague.

Challis, G. A. and Long, J. V. P. (1964). Wairauite – a new cobalt-iron mineral. *Mineralog. Mag.* **33**, 942.

Chamberlain, J. A. and Delabio, R. N. (1965). Mackinawite and valleriite in the Muskox intrusion. *Am. Miner.* **50**, 682.

Chamberlain, J. A., McLeod, C. R., Traill, R. J. and Lachance, G. R. (1965). Native metals in the Muskox intrusion. *Can. J. Earth Sci.* **2**, 188.

Chinner, G. A. (1962). Almandine in thermal aureoles. *J. Petrology* **3**, 316.

Cooke, C. J. and Duncumb, P. (1967). Comparison of a non-dispersive method of microanalysis with conventional crystal spectrometry. *In* "Optique des Rayons X et Microanalyse." University of Paris. (In press.)

Cosslett, V. E. and Duncumb, P. (1956). Microanalysis by a flying-spot X-ray method. *Nature, Lond.* **177**, 1172.

Cosslett, V. E. and Nixon, W. C. (1960). "X-ray Microscopy." Cambridge University Press.

Crookes, W. (1879). Contributions to molecular physics in high vacua. *Phil. Trans. R. Soc.* **170**, 641.

Day, D. E. (1963). Determining the coordination number of aluminium ions by X-ray spectroscopy. *Nature, Lond.* **200**, 649.

Dils, R. R., Zeitz, L. and Huggins, R. A. (1963). A suggested secondary fluorescence correction technique for electron-probe analyses in the vicinity of a steep concentration gradient. *In* "X-ray Optics and X-ray Microanalysis." Ed. H. E. Pattee, V. E. Cosslett and A. Engström, p. 341. Academic Press, New York.

Dolby, R. M. (1963). An X-ray microanalyser for elements of low atomic number. *In* "X-ray Optics and X-ray Microanalysis." Ed. H. E. Pattee, V. E. Cosslett and A. Engström, p. 483. Academic Press, New York.

Duncumb, P. (1959). The X-ray scanning microanalyser. *Br. J. appl. Phys.* **10**, 420.

Duncumb, P. (1960). Improved resolution with the X-ray scanning microanalyser. *In* "X-ray Microscopy and X-ray Microanalysis." Ed. A. Engström, V. E. Cosslett and H. E. Pattee, pp. 365–371. Elsevier, Amsterdam.

Duncumb, P. (1962). The design of electron-probe microanalysers. *J. Inst. Metals* **90**, 154.

Duncumb, P. (1966). Precipitation studies with EMMA a combined electron micro-scope and microanalyser. *In* "The Electron Microprobe," p. 490. Wiley, New York.

Duncumb, P. and Cosslett, V. E. (1957). A scanning microscope for X-ray emission pictures. *In* "X-ray Microscopy and Microradiography." Ed. V. E. Cosslett, A. Engström and H. H. Pattee, Jr., p. 374. Academic Press, New York.

Duncumb, P. and Melford, D. A. (1967). Quantitative applications of ultra-soft X-ray microanalysis in metallurgical problems. *In* "Optiques des Rayons X et Microanalyse." University of Paris. (In press.)

Duncumb, P. and Shields, P. K. (1963). The present state of quantitative X-ray microanalysis, Part 1: Physical basis. *Br. J. appl. Phys.* **14**, 617.

Duncumb, P. and Shields, P. K. (1966). The effect of excitation potential on the absorption correction. *In* "The Electron Microprobe," p. 284. Wiley, New York.

Elion, H. A. and Ogilvie, R. E. (1962). Reflecting variable bent crystal spectro-meter. *Rev. scient. Instrum.* **33**, 753.

Fauquier, D. (1961). Etude de la répartition des éléments dans les niobiotantalates metamictes, à l'aide d'un procédé d'analyse ponctuelle basé sur l'emploi des sondes électroniques. *C. r. hebd. Séanc. Acad. Sci. Paris* **252**, 3283.

Feller-Kniepmeier, M. and Uhlig, H. H. (1961). Nickel analyses of metallic meteo-rites by the electron-probe microanalyser. *Geochim. cosmochim. Acta* **21**, 257.

Fischer, D. W. (1964). Changes in the soft X-ray L-emission spectra with oxidation of the first series transition metals. *J. appl. Phys.* **36**, 2048.

Fischer, D. W. and Baun, W. L. (1964). Experimental dispersing devices and detec-tion systems for soft X-rays. *In* "Advances in X-ray Analysis," Vol. 7, pp. 489–496. Plenum Press, New York.

Fischer, D. W. and Baun, W. L. (1967). The effect of chemical combination on some soft K- and L-emission spectra. *In* "Advances in X-ray Analysis," Vol. 9. (Proc. 1965 Conf.). Plenum Press, New York. (In press.)

Fisher, R. M. (1964). "Electron-optical design of electron probes. Symposium on X-ray and electron-probe analysis." A.S.T.M. Special Technical Publication No. 349.

Fisher, R. M. and Swartz, J. C. (1956). Proceedings, 5th Conference on Industrial Applications of X-ray Analysis.

Franks, A. (1963). Optical techniques applied to X-ray wavelengths. "X-ray Optics and X-ray Microanalysis." Ed. H. E. Pattee, V. E. Cosslett and A. Engström, p. 199. Academic Press, New York.

Fredriksson, K. (1967). Standards and correction procedures for microprobe analysis of minerals. *In* "Optique des Rayons X et Microanalyse." University of Paris. (In press.)

Fredriksson, K. and Andersen, C. A. (1964). Electron-probe analyses of copper in meneghinite. *Am. Miner.* **49**, 1467.

Fredriksson, K. and Keil, K. (1964). The magnesium, calcium and nickel distribu-tion in the Murray carbonaceous chondrite. *Meteoritics* **2**, 201.

Fredriksson, K. and Reid, A. M. (1965). A chondrule from the Chainpur meteorite. *Science, N.Y.* **149**, 856.

Goldstein, J. I. (1965). The formation of the kamacite phase in metallic meteorites. *J. geophys. Res.* **70**, 6223.

Goldstein, J. I. and Ogilvie, R. E. (1963). Electron microanalysis of metallic meteorites, Part 1. Phosphides and sulfides. *Geochim. cosmochim. Acta* **27**, 623.

Goldstein, J. I. and Ogilvie, R. E. (1965). The growth of the widmanstatten pattern in metallic meteorites. *Geochim. cosmochim. Acta* **29**, 893.

Green, D. H. (1963). Alumina content of enstatite in a Venezuelan high-temperature peridotite. *Bull. geol. Soc. Am.* **74**, 1397.

Green, M. (1962). The target absorption correction in X-ray microanalysis. *In* "X-ray Optics and X-ray Microanalysis." Ed. H. E. Pattee, V. E. Cosslett and A. Engström, p. 361. Academic Press, New York.

Green, M. (1963). A Monte Carlo calculation of the spatial distribution of characteristic X-ray production in a solid target. *Proc. phys. Soc.* **82**, 204.

Hall, E. T. (1964). Some modifications to the A.E.I. microanalyser. Proceedings of the Conference on Electron-probe Microanalysis, Reading (1963). See Mulvey, T., *J. scient. Instrum.* **41**, 61.

Hanneman, R. E., Ogilvie, R. E. and Modzrejewski, A. (1962). Kossel line studies of irradiated nickel crystals. *J. appl. Phys.* **33**, 1429.

Heinrich, K. F. J. (1963). Oscilloscope readout of electron microprobe data. *In* "Advances in X-ray Analysis," Vol. 6, p. 291. Plenum Press, New York.

Heinrich, K. F. J. (1964). "Electron-probe analysis in metallurgical research." A.S.T.M. Publication STP 349.

Heinrich, K. F. J. (1966a). Bibliography on electron-probe microanalysis and related subjects. (Second Revision) *In* "The Electron Microprobe". Wiley, New York.

Heinrich, K. F. J. (1966b). X-ray absorption uncertainty. *In* "The Electron Microprobe," p. 296. Wiley, New York.

Henke, B. (1963). Sodium and magnesium fluorescence analysis. Part I: method. *In* "Advances in X-ray Analysis," Vol. 6, pp. 361–376. Plenum Press, New York.

Henke, B. (1964). X-ray fluorescence analysis for sodium, fluorine, oxygen, nitrogen, carbon and boron. *In* "Advances in X-ray Analysis," Vol. 7, pp. 460–488.

Henke, B. (1967). Spectroscopy in the 10–100 Ångström region. *In* "Optique des Rayons X et Microanalyse." University of Paris. (In press.)

Henoc, J. (1962). Contribution à la microanalyse par sonde électronique Part 3. Calcul de l'émission secondaire due au fond continu. Centre National d'Études des Telecommunications, Department Physique, Chimie, Métallurgie. Etude No. 655 P.C.M.

von Hevesy, G. (1932). "Chemical Analysis by X-rays and its Applications." McGraw-Hill, New York.

Hillier, J. (1947). Electron probe analysis employing X-ray spectrography. U.S. Patent 2,418,029.

Hughes, G. D. and Woodhouse, J. B. (1967). Some studies on the mass absorption coefficients and fluorescent yields of a number of elements. *In* "Optique des Rayons X et Microanalyse." University of Paris. (In press.)

Keil, K. (1965). Mineralogical modal analysis with the electron microprobe X-ray analyser. *Am. Miner.* **50**, 2089.

Keil, K. and Andersen, C. A. (1965a). Electron microprobe study of the Jajh Deh Kot Lalu enstatite chondrite. *Geochim. cosmochim. Acta* **29**, 621.

Keil, K. and Andersen, C. A. (1965b). Occurrences of sinoite, $Si_2N_2O$, in meteorites. *Nature, Lond.* **207**, 745.

Keil, K. and Fredriksson, K. (1964). The iron, magnesium and calcium distribution in coexisting olivines and rhombic pyroxenes of chondrites. *J. geophys. Res.* **69**, 3487.

Keil, K., Mason, B., Wilk, H. B. and Fredriksson, K. (1964). The Chainpur meteorite. *Am. Mus. Novit.* No. **2173**.

Koffman, D. M., Moll, S. H. and Norton, J. T. (1967). Influence of wavelength shift due to state of chemical combination on quantitative microbeam analysis. *In* "Optique des Rayons X et Microanalyse." University of Paris. (In press.)

Kouvo, O., Vuorelainen, Y. and Long, J. V. P. (1963). A tetragonal iron sulfide. *Am. Miner.* **48**, 511.

Kulenkampff, H. and Spyra, W. (1954). Energieverteilung rückdiffundierten Electronen. *Z. Phys.* **137**, 416.

LePoole, J. B., Bok, A. B. and Boogerd, W. J. (1965). Een electronenluminescentie-apparaat. *T.N.O.-Nieuws* **20**, 917.

Liebmann, G. (1955). A unified representation of magnetic lens properties. *Proc. phys. Soc. Lond.* **68B**, 737.

Long, J. V. P. (1960). Some observations of the electron-induced luminescence in rock-forming minerals. (Exhibit) *Mineralog. Mag.* **32**. lxxx.

Long, J. V. P. (1963a). The application of the electron-probe microanalyser to metallurgy and mineralogy. *In* "X-ray Optics and X-ray microanalysis." Ed. H. E. Pattee, V. E. Cosslett and A. Engström, Vol. 3, p. 279. Academic Press, New York.

Long, J. V. P. (1963b). Recent advances in electron-probe analysis. *In* "Advances in X-ray Analysis." Vol. 6, p. 276. Plenum Press, New York.

Long, J. V. P. (1964). Electron-probe analysis in mineralogy – a review. Conference on Electron-Probe Analysis, Delft.

Long, J. V. P. and Agrell, S. O. (1965). The cathode-luminescence of minerals in thin section. *Mineralog. Mag.* **34**, 318.

Long, J. V. P. and Cosslett, V. E. (1957). Some methods of X-ray microchemical analysis. *In* "X-ray Microscopy and Microradiography." Ed. V. E. Cosslett, A. Engström and H. H. Pattee, Jr., Vol. 1, p. 435. Academic Press, New York.

Long, J. V. P., Vuorelainen, Y. and Kouvo, O. (1963). Karelainite, a new vanadium mineral. *Am. Miner.* **48**, 33.

Lovering, J. F. (1964). Electron microprobe analysis of terrestrial and meteoritic cohenite. *Geochim. cosmochim. Acta* **28**, 1745.

Marshall, R. R. and Keil, K. (1965). Polymineralic inclusions in the Odessa iron meteorite. *Icarus* **4**, 461.

Maurice, F., Seguin, R. and Henoc, J. (1967). Phenomène de fluorèscence dans les couples de diffusion. (Fluorescence phenomena in diffusion couples). *In* "Optique des Rayons X et Microanalyse." University of Paris. (In press.)

Melford, D. A. (1962). The use of electron-probe microanalysis in physical metallurgy. *J. Inst. Metals* **90**, 217.

Melford, D. A. and Duncumb, P. (1958). The metallographic application of X-ray scanning microanalysis. *Metallurgia* **57**, 159.

Moseley, H. G. J. (1913). The high frequency spectra of the elements (I). *Phil. Mag.* **26**, 1024.

Moseley, H. G. J. (1914). The high frequency spectra of the elements (II). *Phil. Mag.* **27**, 703.

Muir, I. D. and Long, J. V. P. (1965). Pyroxene relations in two Hawaiian hypersthene-bearing basalts. *Mineralog. Mag.* **34**, 358.

Mulvey, T. (1959). Electron-optical design of an X-ray microanalyser. *J. scient. Instrum.* **36**, 350.

Mulvey, T. (1960). A new X-ray microanalyser. *In* "X-ray Microscopy and X-ray Microanalysis." Ed. H. E. Pattee, V. E. Cosslett and A. Engström, Vol. 2, pp. 372–378. Academic Press, New York.

Mulvey, T. and Campbell, A. J. (1958). Proportional counters in X-ray spectrochemical analysis. *Br. J. appl. Phys.* **9**, 406.

Nixon, W. C. (1963). An experimental optical bench for electron microscopy and X-ray microanalysis. In "X-ray Optics and X-ray Microanalysis." Ed. H. E. Pattee, V. E. Cosslett and A. Engström, p. 441. Academic Press, New York.

Ong, P. S. (1964). Microprobe analysis of the elements fluorine through boron. In "The Electron Microprobe," p. 43. Wiley, New York.

Philibert, J. (1962). The Castaing "Microsonde" in metallurgical and mineralogical research. J. Inst. Metals 90, 241.

Philibert, J. (1963). A method for calculating the absorption correction in electron-probe microanalysis. In "X-ray Optics and X-ray Microanalysis." Ed. H. E. Pattee, V. E. Cosslett and A. Engström, Vol. 3, p. 379. Academic Press, New York.

Philibert, J. (1964). L'analyse quantitative en microanalyse par sonde électronique. Métaux. Corros. Inds. 40, 157, 216, 325.

Poole, D. M. and Thomas, P. M. (1962). Quantitative electron-probe microanalysis. J. Inst. Met. 90, 228.

Ranzetta, G. V. T. and Scott, V. D. (1967). Light element microanalysis of oxides and carbides. In "Optique des Rayons X et Microanalyse." University of Paris. (In press.)

Reed, S. J. B. (1964a). Some aspects of X-ray microanalysis in mineralogy. Thesis, University of Cambridge.

Reed, S. J. B. (1964b). Electron microprobe analysis of the metallic phase in basic achondrites. Nature, Lond. 203, 70.

Reed, S. J. B. (1965a). Characteristic fluorescence corrections in electron-probe microanalysis. Br. J. appl. Phys. 16, 913–926.

Reed, S. J. B. (1965b). Electron-probe microanalysis of schreibersite and rhabdite in iron meteorites. Geochim. cosmochim. Acta 29, 513.

Reed, S. J. B. (1965c). Electron-probe microanalysis of the metallic phases in iron meteorites. Geochim. cosmochim. Acta 29, 535.

Reed, S. J. B. (1967). Spatial resolution in electron-probe microanalysis. In "Optique des Rayons X et Microanalyse." University of Paris. (In press.)

Reed, S. J. B. and Long, J. V. P. (1963). Electron-probe measurements near phase boundaries. In "X-ray Optics and X-ray Microanalysis." Ed. H. E. Pattee, V. E. Cosslett and A. Engström, Vol. 3, p. 317. Academic Press, New York.

Sandstrom, A. E. (1952). A large bent crystal vacuum spectrograph. Ark. Fys. 4, 517.

Sawkill, J. and Schwartzenberger, D. R. (1960). X-ray microscopy of beryllium. Br. J. appl. Phys. 11, 498.

Schüller, A. and Otteman, J. (1963). Castaingit, ein neues mit hilfe den electronen Microsonde bestimmtes Mineral aus den Mansfelden "Rucken". Neues Jb. Miner. Abh. 100, 317.

Siegbahn, M. (1925). "The Spectroscopy of X-rays" (Translated by G. A. Lindsay). Oxford University Press, London.

Sippel, R. F. (1965). Simple device for luminescence petrography. Rev. scient. Instrum. 36, 1556.

Smith, J. V. (1965). X-ray emission microanalysis of rock-forming minerals, I: experimental techniques. J. Geol. 73, 830.

Smith, J. V. (1966). X-ray emission analysis of rock-forming minerals, II: olivines. J. Geol. 74, 1.

Smith, J. V. and Stenstrom, R. C. (1965). Chemical analysis of olivines by the electron microprobe. Mineralog. Mag. 34, 436.

Springer, G., Schachner-Korn, D. and Long, J. V. P. (1964). Metastable solid solution relations in the system $FeS_2$–$CoS_2$–$NiS_2$. Econ. Geol. 59, 475.

L.**

Stenstrom, R. C. and Smith, J. V. (1964). Electron-excited luminescence as a petrologic tool. *Bull. Geol. soc. Am.* (Special Paper) No. 76, 158.

Stenstrom, R. C. and Smith, J. V. (1965). Electron-excited luminescence as a petrologic tool. *J. Geol.* **73**, 627.

Stumpfl, E. F. (1961). Some new platinoid-rich minerals identified with the electron microanalyser. *Mineralog. Mag.* **32**, 833.

Stumpfl, E. F. and Clark, A. M. (1964). A natural occurrence of $Co_9S_8$ identified by X-ray microanalysis. *Neues Jb. Miner. Mh.* **8**, 240.

Stumpfl, E. F. and Clark, A. M. (1965). Electron-probe microanalysis of gold-platinoid concentrates from south-east Borneo. *Trans. Instn Min. Metall.* **74**, 933.

Stumpfl, E. F. and Clark, A. M. (1966). Hollingworthite, a new rhodium mineral, identified by electron-probe microanalysis. *Am. Miner.* **50**, 1068.

Takeno, S. (1963). On the ore minerals of the lower ore deposit of the kawayama mine, Yamaguchi prefecture. *Geol. Rep. Hiroshima Univ.* No. **12**, 343.

Terrier, P. and Capitant, M. (1967). Étude de l'hydration du ciment Portland par microanalyse (emission X). (A study of the hydration of Portland cement using X-ray microanalysis.) *In* "Optique des Rayons X et Microanalysis." University of Paris. (In press.)

Walter, L. S. (1967). A convenient method for the preparation of silicate standards for electron microprobe analyses. *In* "Optique des Rayons X et Microanalyse." University of Paris. (In press.)

Wittry, D. B. (1959). Instrumentation for electron-probe microanalysis. *In* "Advances in X-ray Analysis," Vol. 3, p. 185. Plenum Press, New York.

Wittry, D. B. (1962). Fluorescence by characteristic radiation in electron-probe microanalysis. *U.S.C.E.C.* Report 84.

Wittry, D. B. (1964). Methods of quantitative electron-probe analysis. *In* "Advances in X-ray Analysis," Vol. 7, p. 395. Plenum Press, New York.

Wittry, D. B. and Nicholson, J. B. (1964). A comparison of the performance of gratings and crystals in the 20–115 Ångstrom region. *In* "Advances in X-ray Analysis," Vol. 7, pp. 497–511. Plenum Press, New York.

Wood, A. J. (1964). The cooling rates and parent planets of several iron meteorites. *Icarus* **3**, 429.

Yavnel, A. A., Borovskii, I. V., Il'in, N. P. and Marchukova, I. D. (1958). A study of the phase composition of meteoritic iron by X-ray microanalysis. *Doklady Akad. Nauk. SSSR.* **123**, 2, 256. (A.E.I. translation T/1408.)

# CHAPTER 6

# X-ray Diffraction

## J. ZUSSMAN

*Department of Geology and Mineralogy, University of Oxford, England*

## I. Introduction

There are now available many textbooks and reference books on X-ray diffraction; most of these cover the subject in considerable detail, thus giving a foundation for a wide diversity of applications. The present Chapter is intended to outline primarily those aspects of X-ray diffraction that are most useful in determinative mineralogy. It is assumed that the reader is already familiar with the fundamentals of crystallography and also has some general knowledge of the X-ray diffraction method. Theoretical treatments are few and brief, and the emphasis is on procedures, possibilities and limitations in respect of the more routine methods. For special problems, it will be necessary to refer to a more detailed work, and to this end a bibliography is given as well as a list of references.

The materials used for testing the first theories about X-ray diffraction were minerals, so it is perhaps apt that X-ray diffraction should now play such an important part in mineralogical work. In a very high proportion of the studies reported in mineralogical journals in the past decade, X-ray methods were used in one way or another. There is no sharp dividing line between determinative and other uses of X-ray diffraction, since a mineral may be "determined" or "characterized" in greater or less detail. X-ray studies of textures, phase transformations, or topotactic relationships may, for example, each in some circumstances provide useful clues to the exact nature of a mineral, and it may be that the distinction of one variety of a mineral from another may require nothing short of a full crystal-structure determination. Such applications are, however, beyond the scope of a chapter in a book on physical methods in general, and attention has therefore been confined almost entirely to the more obvious problems of identifying a mineral and determining its chemical composition.

In general, the most useful and most easily executed X-ray work in determinative mineralogy is that which uses the powder pattern, and so the bulk of this Chapter is concerned with the latter subject, covering both photographic and counter-recording methods. Although single-crystal methods are not so widely used in the present context, there are occasions when they are advisable or even essential, and so an introduction to these methods is given here. Use is made of the reciprocal lattice, and the reader who is less mathematically inclined may prefer the geometric treatment given here although it is less general and less elegant than that using vector algebra.

Some notes on published work in which X-ray methods have been used in determinative mineralogy are given at the end of this Chapter.

## II. The X-ray Beam

The nature of X-rays and their production are discussed in Chapter 4. It will be sufficient here, therefore, to repeat that the target of an X-ray diffraction tube emits under electron bombardment a spectrum of X-rays, which includes the characteristic emission lines of the target material and a continuum of "white" radiation.

## A. FILTERS

In most X-ray diffraction experiments it is desirable to use a very narrow range of wavelengths, usually the $K_\alpha$ emission lines of the target, and to have as little intensity as possible at other wavelengths. This is achieved to a satisfactory degree by use of a filter. For a copper target, for example, a nickel foil is placed in the primary X-ray beam; its effect is to reduce considerably the intensity of white radiation and the $K_\beta$ emission lines of the target. Table 1 lists suitable thicknesses of the various metal foils that may

TABLE 1

X-ray wavelengths and suitable $\beta$ filters to give $\dfrac{K_{\beta_1}}{K_{\alpha_1}} = \dfrac{1}{100}$

| Target element | Wavelength (Å) | | | $\beta$ filter | Thickness (mm) |
|---|---|---|---|---|---|
| | $K_{\alpha_1}$ | $K_{\alpha_2}$ | $K_\beta$ | | |
| Mo | 0·7093 | 0·7135 | 0·6323 | Zr | 0·08 |
| Cu | 1·5405 | 1·5443 | 1·3922 | Ni | 0·015 |
| Co | 1·7889 | 1·7928 | 1·6207 | Fe | 0·012 |
| Fe | 1·9360 | 1·9399 | 1·7565 | Mn | 0·011 |

be used as $\beta$ filters for common radiations. The intensity of the $\beta$ line and white radiation remaining after filtering is not altogether negligible, and care must be taken in interpreting powder patterns, since very strong X-ray reflections may be registered as weak $\beta$ peaks additional to the $\alpha$ pattern. Even the white radiation that remains after filtering may give rise to a spurious broad diffraction peak, since it has a maximum at about $\lambda = 0·5$ Å. Thus, for example, a strong reflection at $d = 3·5$ Å will occur at $2\theta = 25·45°$ with $CuK_\alpha$ radiation, but the $\beta$ line may give a peak at $22·94°$ and the white radiation at about $8°$ $2\theta$ (see Bragg Law, p. 265).

It is useful to note that the intensities of the $K_\beta$ lines of unfiltered radiation are approximately one-sixth of the intensities of the $K_\alpha$ lines, and that $K_{\alpha_2}$ has about half the intensity of $K_{\alpha_1}$. The $K_\alpha$ doublet cannot be separated by filters, and whether or not it is resolved at a given angle in a diffraction experiment depends upon the resolving power of the instrument. The $K_\beta$ line occurs at lower angle than $K_\alpha$ and the $K_{\alpha_1}$ at lower angle than $K_{\alpha_2}$.

For special purposes it may be desirable to have a more strictly monochromatic X-ray beam than that obtained by a $\beta$ filter, and for this purpose a crystal monochromator can be used (see Lonsdale, 1962).

In X-ray diffractometry, pulse-height discriminator circuits can be used with a proportional counter, and combined with a $\beta$ filter: this effectively suppresses nearly all of the white-radiation contribution to the diffraction pattern.

## B. CHOICE OF TARGET

If an element is present in the specimen under investigation that is itself excited by the primary X-ray beam, it emits fluorescent X-radiation, which results in a higher background and therefore poorer clarity of the diffraction pattern. An element in the specimen will fluoresce appreciably if its $K$ absorption edge is close to and on the long-wavelength side of the target-emission line. Unwanted fluorescence effects are avoided by choosing a suitable target. For example, if the specimen has a high iron content, a copper target would not be suitable, and cobalt or iron radiation should be used.

The avoidance of a copper target for investigating iron-containing minerals is, however, often over-emphasized. If the iron content is only moderate, the disadvantages of an alternative radiation (lower intensity, less suitable Bragg angles) may in some cases be more important than the disadvantage of a rather high background with copper radiation. The background on photographic film due to fluorescence radiation is mostly on the side of the film nearest the specimen, and the removal of the emulsion, after processing, from this side of the commonly used double-coated film is one way of reducing fluorescent background. The removal of photographic emulsion can be effected with a hypochlorite solution. An alternative method of gaining an improved result is to have the filter foil wrapped on the inside of the film instead of having it placed in the X-ray beam before it hits the specimen. When diffractometer methods are being used, a proportional counter and pulse-height-analysis circuits can be used to discriminate against the unwanted fluorescent radiation.

The use of a short X-ray wavelength may sometimes be desirable in order to record a greater number of X-ray reflections. With $MoK_\alpha$ radiation ($\lambda = 0.71$ Å), however, although many more reflections could theoretically be recorded in accordance with the Bragg law, in fact many are lost because of a more rapid fall off of diffracted intensity with increasing Bragg angle. Also, two similar $d$ values are more difficult to resolve when shorter wavelengths are used. The use of molybdenum radiation, however, for single-crystal studies of the structures of minerals is quite common since this reduces errors due to absorption.

Longer wavelengths, e.g. from cobalt and iron targets, may be used in order to increase resolution between close reflections, and they also have the sometimes desirable effect of putting a given reflection at a high Bragg angle. The power rating of Co and Fe X-ray tubes is usually about one-third that of a similar Cu tube, so that there is the disadvantage of lower intensities and longer exposure times.

The target which is most generally useful for powder-diffraction work is undoubtedly one of copper.

## C. SAFETY PRECAUTIONS

The danger of exposure to X-radiation is well known, and many pieces of standard equipment are designed by the manufacturers to be as safe as possible in use. The user, however, must always ensure that his experimental arrangement is safe by using suitable shielding against scattered radiation (checked by a radiation monitor) and preventing, by use of interlock devices, access to the direct X-ray beam. Guidance on these matters is given in the report of the International Union of Crystallography (*Acta cryst.* 1963, **16**, 324).

## III. The Powder Method

Although the fundamental process that occurs when X-rays strike a crystal is one of scattering (or diffraction) we can regard it conveniently as one of reflection. The reflection is similar to that of light in a mirror in that $i = r$, but there is an important difference in that the X-rays penetrate below the surface of the crystal and rays reflected from successive atomic layers may or may not be in phase. The condition for a maximum of reflected intensity is that the contribution from successive planes should be in phase. If the interplanar spacing is $d$, this condition is expressed by $n\lambda = 2d \sin \theta$ (see Fig. 1).

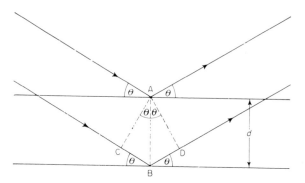

Path difference = $CB + BD = 2d \sin \theta = n\lambda$

Fig. 1. Diagram illustrating the Bragg Law. $d =$ interplanar spacing, $\theta =$ Bragg angle, $\lambda =$ X-ray wavelength, $n =$ integer.

This is the well known Bragg Law, which is usually given as $\lambda = 2d \sin \theta$, dropping the $n$. The justification of this is that for a given set of planes with spacing $d$ we can regard reflection as occurring with phase difference $n\lambda$, or with phase difference of $\lambda$ from planes with one $n$th of the spacing, i.e.

$\lambda = 2 \, (d/n) \sin \theta$. Thus, for example, the second-order reflection from (111) planes can be regarded as the first order from (222) planes (which have half the (111) spacing).

## A. POWDER PHOTOGRAPHY

### 1. Details of Camera

A typical powder camera is illustrated in Fig. 2. The essential features of the instrument are as follows. A cylindrical casing, which forms a light-tight film cassette. A specimen holder with centring device, which is rotatable about the camera axis. An entrance collimator, which in some cameras defines a narrow slit parallel to the camera axis and in some has a small circular opening. An appropriate beam trap.

The entrance collimator and beam trap are designed and arranged to give the maximum X-ray intensity falling on the specimen and the minimum falling on the film, apart from that diffracted by the specimen. A beam stop is provided to prevent emergence of the X-ray beam from the cameras.

FIG. 2. An X-ray powder camera in which the Straumanis film position is used.

The powdered specimen is usually made into the shape of a cylindrical rod and is mounted on the axis of the cylindrical camera. The film is wrapped around the inside of the camera and there are gaps or holes in the film for the entry and exit of the primary beam, which is usually filtered to give predominantly $K_\alpha$ radiation.

Crystal planes with a given $d$ value, which in a finely powdered specimen occur in all possible orientations, will give reflections along the surface of a cone with semi-angle $2\theta$ (Fig. 3). Such a cone will blacken the film in a pair of

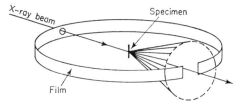

FIG. 3. Formation of a pair of powder arcs by the intersection of the film and a cone of reflections from planes of spacing $d$ in many crystallites.

arcs; reflections from planes with other $d$ values produce other pairs of arcs. When the film is processed the distances, $s$, between pairs of arcs can be measured and from these the corresponding Bragg angles can be determined and $d$ values obtained from the Bragg law. The relation between $s$ and $\theta$ would be simple $\left(\dfrac{s}{2\pi r} = \dfrac{360 - 4\theta}{360}\right)$ if the radius of the camera were accurately known, but usually it is not, and there are a variety of ways of determining $\theta$ from $s$ values. The methods described below also overcome the problem of distortion of the film on processing, providing that this takes place uniformly over its length.

*Knife edges.* Built into the camera there can be sharp-edged metal pieces that are in contact with the film, and which, by obstructing the generally scattered radiation, produce a sharp demarcation between blackened and unblackened film. The angular separation of the knife edges is a known or measurable constant of the camera. Figure 4 illustrates the van Arkel mounting commonly used with this method.

*Use of standard.* A substance with known $d$ spacings can be used to give a calibration curve of $s$ against $\theta$. Alternatively, approximate values of $\theta$ for the specimen may be determined by a simple formula as above and the corrections necessary at various $\theta$ values may be determined by use of the standard (see p. 276). In order to make the conditions for the standard and specimen as similar as possible, the standard can be used internally, i.e. mixed in suitable

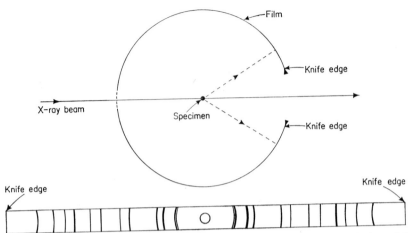

FIG. 4. The van Arkel mounting for X-ray powder photography. Note the resolution of the $K_\alpha$ doublet at high Bragg angle.

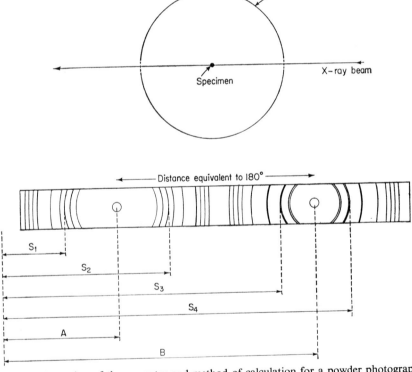

FIG. 5. Illustration of the geometry and method of calculation for a powder photograph in the Straumanis setting.

proportion with the specimen powder. For this method a standard must be chosen that does not have powder arcs overlapping reflections of interest arising from the specimen.

*Straumanis method.* The film is mounted in such a way that the position of the entrance and exit of the primary beam can readily be measured (Fig. 5). In this way the radius of the camera is implicitly determined, and moreover a correction is automatically made for any uniform shrinkage or expansion of the film. A useful way of tabulating results obtained with the Straumanis method is illustrated below.

As shown in Fig. 5, $s_1$, $s_2$, $s_3$, $s_4$, are successive readings on a millimetre scale for a pair of low-angle and a pair of high-angle reflections.

| | Line No. | I | $s_1$ | $s_2$ | $s_2 - s_1$ | $s_2 + s_1$ | | $\theta$ | $d$ |
|---|---|---|---|---|---|---|---|---|---|
| | 1 | — | — | — | — | — | | — | — |
| | 2 | — | — | — | — | — | | — | — |
| $2\theta < 90°$ | 3 | — | — | — | — | — | | — | — |
| | . | | | | | | | | |
| | . | | | | | | | | |
| | . | | | | | | | | |

| | | | $s_3$ | $s_4$ | $s_4 - s_3$ | $s_4 + s_3$ | $90 - \theta$ | $\theta$ | $d$ |
|---|---|---|---|---|---|---|---|---|---|
| | . | — | — | — | — | — | — | — | — |
| $2\theta > 90°$ | . | — | — | — | — | — | — | — | — |
| | . | | | | | | | | |
| | . | | | | | | | | |

$$s_2 + s_1 = 2A \qquad\qquad s_4 + s_3 = 2B$$

Starting from the left all $s_1$'s are recorded, then $s_2$'s, pairing off with $s_1$'s, then $s_3$ and $s_4$'s in a similar manner. All sums $s_1 + s_2$ should be equal, and similarly for $s_3 + s_4$. Any large deviation from a constant can lead to the detection of an error in measurement. For a pair of powder arcs at $2\theta < 90°$:

$$\frac{s_2 - s_1}{2B - 2A} = \frac{4\theta}{360} \text{ and } \theta = (s_2 - s_1) \cdot \frac{90}{2B - 2A}$$

and for reflections with $2\theta > 90°$:

$$90 - \theta = (s_4 - s_3) \cdot \frac{90}{2B - 2A}$$

A convenient radius of camera is 57·3 mm, since in this case 180 mm of film is subtended by an angle of 180°. Direct measurement with a mm rule of the distance between a pair of powder arcs then gives a close approximation to $4\theta$ (or $360 - 4\theta$) in degrees.

One difficulty with the normal Straumanis mounting that is sometimes encountered is that the fall off of intensity with angle means that no pairs of lines can be observed that are centred on the high-angle hole. A modified technique (Wilson, 1949) overcomes this by shifting the film round in relation to the X-ray beam; the holes in the film need to be punched differently (Fig. 6b).

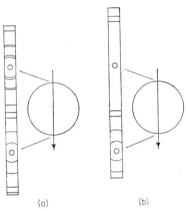

(a)                              (b)

Fig. 6. (a) Normal Straumanis setting. (b) Modified Straumanis setting, which is useful if high-angle reflections are very weak or absent. (From Wilson, 1949.)

## 2. Specimen Preparation

Whatever method of specimen preparation is used, an important prerequisite is that the grain size should be made small. The larger the grain size, the fewer the number of grains in a given specimen, and so all grain orientations are less likely to be represented. The result is spotty instead of evenly blackened powder arcs, and greater difficulty of positional measurement and of intensity estimation is encountered. Adequate grinding is also important in order to ensure homogeneity of the specimen. It is, however, possible to be over-zealous in grinding; excessive grinding can damage the crystal structure, and in consequence powder lines may be broadened and weakened (see p. 272). For most specimens a grain size of about 300 mesh is suitable. In order to help ensure that a sufficient number of grains are present in the specimen in all orientations, it is usual practice to have the specimen rotating about the camera axis, by means of a motor drive.

A convenient shape of specimen is a thin cylinder, and this can be produced by the following methods.

(a) Mixing the specimen with a gum (e.g. moistened gum tragacanth) that gives very little X-ray scattering itself, and rolling a ball of the moistened mixture first with the fingers and then between glass plates (microscope slides). The ball elongates and finally should set hard in a thin straight cylinder (0·1–0·3 mm diameter). The proportion of specimen to gum desirable is roughly 4:1.

(b) Extrusion of a gummy mixture (fish glue or collodion plus specimen) with a hypodermic needle (Brown *et al.*, 1956).

(c) A thin-walled cylindrical container can be filled with the powdered specimen. A well known method of producing such containers is by coating a copper wire with a suitable solution, and after setting, stretching the wire in order to remove the tube so formed (Williams, 1963). An alternative tube is one which is commercially available and is made of special thin-walled low-absorbency glass. The use of any container for the specimen does, however, result in a heavier background blackening of the film. Methods in which an enclosing tube is used are essential when dealing with substances that are affected by solvents, and a sealed tube is necessary for specimens that are unstable in air.

(d) A thin lead-free glass fibre can be coated on its outside with shellac and then with powdered specimen which adheres to it.

An alternative shape for the specimen is a small sphere which can be produced readily using a gum additive. This has the sometimes important advantage of needing less specimen, and furthermore it produces a specimen with less preferred orientation of particles than do the cylinder methods (Hildebrand, 1953). Use of circular instead of slit collimators on some of the commonly used powder cameras results in less comparative intensity loss when spherical instead of cylindrical specimens are employed. Another procedure for dealing with extremely small samples is described by Sorem (1960).

## 3. Preferred Orientation

The degree of preferred orientation produced by most methods of specimen preparation depends upon particle shape, and is more serious for specimens with fibrous or platy crystal morphology. Flattening or rolling during specimen preparation enhances the orientation effects. Perhaps the best method of minimizing preferred orientation, which is applicable also to powder diffractometer specimens, is to prepare a thorough mixture of the crushed mineral grains with a thermo-setting plastic (see p. 283). When this has set hard it is re-ground, and it is expected that the platy or fibrous mineral grains will then occur in all orientations.

## 4. Measurement of Positions of Reflections

In order to achieve reasonable accuracy of measurement it is usually necessary to clamp the powder photograph in position on a screen illuminated by fluorescent lamps from below, and for measurement to be made by a vernier attachment that slides over a rule marked in millimetres (or half-millimetres). Instruments of this kind can be readily constructed, but are also commercially available. If a travelling microscope is used for film measurement, its magnification should be rather low (about 2×).

The interplanar spacings $d_{hkl}$ can be derived from $\theta$ or $2\theta$ values by use of published charts or tables that effectively give solutions for the Bragg equation for different X-ray wavelengths at close intervals of $\theta$ (charts – Parrish and Mack, 1963; tables – National Bureau of Standards, 1950).

### 5. Photographic Intensities

For many purposes a rough visual estimate of relative intensities is all that is needed. The crudest, but nevertheless often effective, method is to designate intensities on a scale such as very strong (vs), strong (s), moderately strong (ms), moderate (m), etc. Alternatively, a numerical scale of 0–10 (which is unlikely to be linear) can be used, corresponding to the above categories. A fair degree of accuracy can be obtained visually if comparison is made with a standard intensity scale. This can be prepared either by using a rotating step sector to obtain uniform areas of different blackening on a piece of film, or by using a suitable powder reflection recorded on film for different lengths of time. For more accurate measurements a microdensitometer should be used.

Suitable exposure times depend upon the size of specimen and camera, degree of crystallinity of the specimen, intensity of primary X-ray beam, and other factors. Commonly, exposures are of the order of one or two hours.

### 6. The Breadth of Powder Reflections

The scattered intensity in general falls off very rapidly away from the Bragg angle $\theta$ for a given $d$ value, but the spread of the reflection bears an inverse relationship with the crystallite size. Thus $\delta(2\theta) \sim \lambda/t \cos \theta$, where $\delta(2\theta)$ is as shown in Fig. 7, and $t$ is the thickness of the crystal. For $t = 1$ mm and $\theta = 60°$, $\delta(2\theta) \sim 0.001'$ but if $t = 1000$ Å, $\delta(2\theta) \sim 0.2°$. Thus the broadening of a reflection is appreciable if the crystallite size is of the order of 1000 Å or less. The above calculation gives an estimate of the broadening due to crystal size alone. For "thick" crystals other factors will be far more important, and these are features of the apparatus rather than the specimen, e.g., the beam divergence and the spectrum of the primary beam.

Another not uncommon source of line broadening is the existence of a range of $d$ values for a given set of crystal planes. This may occur through the presence of mechanical strains or through a lack of chemical homogeneity in the specimen. A powder made up of zoned crystals, for example, would give appropriately broadened reflections.

Most mineral crystals have a mosaic structure, i.e. they are composed of smaller blocks of crystal in slightly different orientations. This leads to a

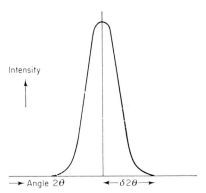

Fig. 7. Plot of intensity profile of a powder reflection. The half-breadth $\delta(2\theta)$ is influenced by crystallite size (see text).

broadening of the reflections as compared with the more rarely encountered "perfect" crystal, which lacks the mosaic structure.

## 7. Discussion of Errors for the Powder-diffraction Method in General

Differentiation of the Bragg Law equation treating $\lambda$ as a constant gives the relationship $\delta d/\delta\theta = -d \cot \theta$; thus $\delta d \to 0$ as $\theta \to 90°$. This means that a given error in measurement of the reflection angle $\theta$ will lead to only a very small error in $d$ if $\theta$ is near $90°$.

Another aspect of this relationship is shown in the reciprocal form:

$$\frac{\delta\theta}{\delta d} = -\frac{1}{d} \tan \theta$$

and:

$$\frac{\delta\theta}{\delta d} \to \infty \text{ as } \theta \to 90°$$

This means that the separation of two reflections from planes with similar $d$ values will be greater at higher Bragg angles.

An alternative differentiation of the Bragg equation, treating $d$ as a constant, gives:

$$\frac{\delta\lambda}{\delta\theta} = 2d \cos \theta$$

and from this $\delta\theta \to \infty$ when $\theta \to 90°$.

This means that the separation of reflections from a given plane for two similar wavelengths will be greatest at higher Bragg angles. The $K_{\alpha_1} - K_{\alpha_2}$ doublet, for example, is resolved readily in most cases in the back reflection region ($\theta$ approaching $90°$). Resolution of the $K_\alpha$ doublet can be a useful

274                   J. ZUSSMAN

means of distinguishing the low-angle and high-angle ends of a powder pattern. (The low-angle region also usually has greater background blackening.)

### 8. Errors in the Powder-film Method

*Absorption.* The effect of absorption on the profile of a powder reflection depends upon (a) the absorption coefficient of the specimen, (b) the diameter of the cylindrical specimen and (c) the Bragg angle of the reflection. These relationships are illustrated qualitatively in Fig. 8, from which it is seen that

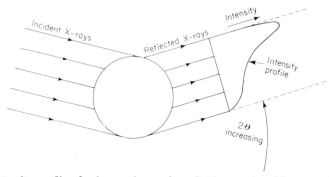

FIG. 8. Intensity profile of a low-angle powder reflection produced by a specimen with moderately high absorption.

the intensity at a given point on the film depends upon the average X-ray path lengths within the specimen for the rays reaching this point. In the case of very high absorption or a very thick specimen, the result can be that the profile goes through a minimum near its centre, and thus may be erroneously interpreted as a pair of lines. If close doublets appear on the powder photo for low-angle reflections, excessive absorption should be suspected. If the absorption error is moderate, the peak will have a shoulder (Fig. 8), and if the position of maximum intensity is observed, then the $2\theta$ value will be high and the $d$ value erroneously low.

It can be shown that the error in $d$ is proportional to $\dfrac{1}{2}\left(\dfrac{\cos^2\theta}{\sin\theta} + \dfrac{\cos^2\theta}{\theta}\right)$; thus the error is least at high Bragg angle. If estimates of the cell parameter of a cubic substance are obtained from a number of its indexed powder reflections, a more accurate value may be obtained by extrapolating against the above function to the value at $\theta = 90°$ (Fig. 9). The slope of the extrapolation line is large if the systematic absorption error is large. The scatter of points about the straight line is wide if random errors of measurement are large.

Absorption effects can be minimized by (a) reducing the specimen diameter, (b) reducing the absorption by the specimen by diluting with a low-absorbing medium, (c) using high-angle measurements, (d) using a thoroughly mixed internal standard, the $d$ values of which are accurately known. Assuming that the standard's reflections are equally affected by absorption, a correction curve can be plotted and used to apply corrections to the measured $d$ values of the specimen under investigation.

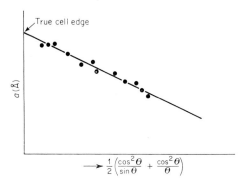

FIG. 9. Illustration of the extrapolation of the cell size of a cubic substance, as determined from various powder reflections, against a suitable function.

*Horizontal beam-divergence* (*axis of camera vertical*). Errors in $d$ produced by this are proportional to the same function as are the absorption errors, and so the single extrapolation serves for both. A thinner specimen and a narrower collimator will reduce this error but will result in loss of intensity.

*Vertical beam-divergence.* The errors produced in this case are complicated and difficult to deal with theoretically. High-angle reflections are shifted to higher, and low-angle reflections to lower angles. A practical way of minimizing these errors is to use a short length of specimen and a small collimator. A small error also arises through the use of a rod-shaped specimen and the consequent "umbrella" effect on the line profiles. This error decreases as $\theta$ approaches 90° (Lipson and Wilson, 1941).

*Rotation axis off-centre.* It is assumed that the specimen is centred accurately so that it does not move laterally on rotation. Even so, there may be errors because the rotation axis does not coincide exactly with the centre of the camera. A small displacement perpendicular to the beam produces no error in $d$ because the errors of powder arc positions on either side of the beam cancel one another. (With the Straumanis mounting, however, an error will be introduced into the estimate of the effective camera radius.) Displacement of the specimen towards the X-ray window gives erroneously low $d$ values and displacement away from the window gives high $d$ values. The error in $d$ is proportional to $\cos^2\theta$ and therefore it extrapolates to zero as $\theta$ tends towards $90°$. In order to deal with most of the above errors in $d$ values, either the extrapolation function $\dfrac{1}{2}\left(\dfrac{\cos^2\theta}{\sin\theta} + \dfrac{\cos^2\theta}{\theta}\right)$ or a simpler extrapolation against $\cos^2\theta$ can be used. The former has the advantage of a greater range of linearity, but if the back-reflection region alone is dealt with, then $\cos^2\theta$ is adequate. The more complex function is not, however, difficult to use, since it is tabulated in several text books and reference books.

The internal-standard method can be used for the correction of most of the systematic errors in powder photography, but thorough mixing and small uniform grain size should be ensured. Andrews (1951) recommends the construction of a correction curve by plotting the product of the percentage spacing error and $\sin\theta$ against $\cos^2\theta$, since this curve is easily extrapolated beyond the possibly limited range of spacings provided by the standard. Acceptable $d$ values and cell parameters for a number of simple substances suitable as standards are given in the compilation of powder patterns by Swanson et al. (1953–60).

*9. General Features of Powder Photographs*

The $d$ values and relative intensities of reflections are the main items of information but not the only ones obtainable from a powder film. More careful observation of a film is worthwhile, and the following additional features may be noted. (a) The breadth of powder reflections: these may be influenced by the condition of the specimen with respect to grain size, strains, disorder, or zoning. (b) Spottiness of powder lines: indicating crystallite size and possibly distinguishing between components in a mixture with different crystallite sizes. (c) General level of intensities of reflections and background: this may indicate whether the specimen contains atoms of high or low average atomic number. (d) Intensity of background with respect to intensity of reflections. This may give a measure of the degree of crystallinity, or may indicate that some element within the specimen is emitting fluorescent X-rays (see p. 264). An unusual mottled effect is sometimes observed on an X-ray film if it has been used in the camera in its paper wrapping. This can be produced by the fluorescent $K$ radiation from chlorine, potassium or calcium, if these elements occur in sufficient concentration in the specimen.

A rough indication of the accuracy with which $d$ values and cell parameters may be derived from a powder photograph, is given by the appearance of high-angle reflections. If the high angle $K_\alpha$ doublet is resolved, $d$ values and cell parameters can be obtained with a precision of about $0\cdot01\%$, or better. If resolution is not obtained at high angles, the precision is more likely to be of the order of $0\cdot1\%$ or worse.

## 10. Special X-ray Cameras

Various powder cameras (some available commercially), have been designed for special purposes. Among these are: focusing cameras (see p. 302), microbeam cameras (Bergmann, 1959; Carrigy and Mellon, 1964), and cameras for massive specimens (Brindley, 1955).

### B. POWDER DIFFRACTOMETRY

## 1. General

A typical X-ray diffractometer is pictured in Fig. 10 and the geometry of the diffractometer is illustrated in Fig. 11. The specimen is in the form of a flat layer of powder at the centre of a circle with radius $r$. On the circumference of the circle lies X, the source of X-rays (defined by a narrow slit perpendicular to the plane of the circle), and also a receiving slit, R, behind which is placed a Geiger or proportional counter. In the zero position, the angle $\theta$ is zero and the receiving slit is in line with the direct beam. Both specimen and counter can be driven by motor or by hand to rotate about the axis of the circle, and by means of a 2:1 reducing gear the angular velocity of the specimen is made accurately half that of the counter. Thus at any position, X-ray reflections can occur from those crystallographic planes that lie parallel to the specimen surface, as long as they have the correct $d$ value to fulfil the Bragg law. In a truly randomized powder specimen, each crystallographic plane will occur lying approximately parallel to the surface in a fair number of grains present.

An important property of diffractometer geometry is its focusing action (Fig. 11). At a given Bragg angle, if a circle is drawn passing through X, S and R, a curved specimen which conforms to this circle would give complete focusing at R of a divergent beam of X-rays. This "focusing circle" has a radius $r' = \dfrac{r}{2 \sin \theta}$. Although diffractometers in general do not use a curved specimen, the flat specimen is tangential to the focusing circle and is a close approximation to an arc of it. The focusing property means that extremely good resolution can be obtained compared with normal camera methods. With a suitable specimen the $CuK_\alpha$ doublet can be resolved at quite low angles (about 40° [$2\theta$]), whereas with the ordinary camera method, resolution is only obtained in the back-reflection region. Special specimen holders have recently come into use with diffractometers which allow the curvature of the specimen layer to a suitable radius; these are particularly useful for low-angle clay-mineral work, since they give greater accuracy in the recording of large $d$ spacings.

For diffractometry, a highly stabilized X-ray generator is required, and the X-ray source used is a line focus parallel to the axis of the goniometer. The

M

Fig. 10. Photograph of an X-ray diffractometer with vertical goniometer circle, in position against the window of a vertical X-ray tube. The beam path is normally entirely shielded, but a circular cover has been removed to show the glass plate (centre of photograph) on which the powder specimen is spread. The cable at top right of the photograph leads from the scintillation counter to the scaler and rate-meter circuits. At bottom right are shown the dial giving the Bragg angle $2\theta$, the manual control of angular setting and a cog-wheel which is part of the motorized goniometer drive.

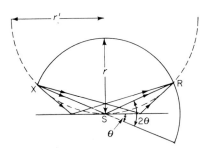

FIG. 11. Geometry of a diffractometer. X, source of X-rays; S, specimen; R, receiving slit. $r$ = radius of goniometer; $r'$ = radius of focusing circle at Bragg angle $\theta$.

take-off angle is usually about 6° so that a foreshortened view of a 1·5 × 10 mm focal line is obtained, appearing as approximately 0·15 × 10 mm. Figure 12 shows more detail of the beam-defining system of a typical diffractometer. The divergence slit for normal use passes a 1° angle, and the receiving slit is about 0·1 mm, but wider slits can be used. Intensity falls off rather rapidly with angle and can be regained at higher angles by using wider slits (say 4°). Since resolution improves with angle, it is possible to use wider slits at higher angles without losing adequate resolution, and a useful change-over position for many specimens would be at about 70° (2θ). However, in order to record very large $d$ spacings, i.e. reflections very close to the primary beam, special narrow slits are needed.

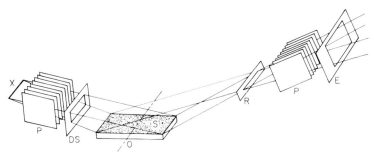

FIG. 12. The beam-defining system of a diffractometer with vertical circle. X, line focus of X-ray tube; P, vertical Soller slits to limit divergence in horizontal plane; DS, divergence slit; S, specimen; O, axis of rotation; R, receiving slit; P, vertical Soller slits; E, antiscatter slit. Distance XO = OR. (After Parrish, 1965.)

The normal procedure in diffractometry is to scan the range of Bragg angles continuously from low (near zero) to high angle, but for special purposes the counter and specimen may be made to oscillate back and forth through any desired angle. Repeated measurements can thus rapidly be made on a particularly interesting part of the powder pattern.

The pulses from the diffractometer counter are usually fed to a scaling circuit and can, if required, be totalled in a fixed time or timed for a fixed count. More usually a pattern of intensity is traced on a chart recorder via rate-meter and amplifier circuits, as the goniometer scans through the Bragg angles for the different lattice planes. The recorder used can have either linear or logarithmic response.

Although for many years a simple geiger counter has been used in most diffractometers, a considerably improved signal-to-noise ratio can be obtained by use of a proportional counter, either alone, or in conjunction with pulse-height selector circuits (see Chapter 4, p. 177). For diffractometer work, the channel width of the selector is usually set to cut off only about 5% of the total energy at each end of the pulse-height distribution curve.

If accurate intensity measurements are needed, due regard must be taken for the "dead time" of the counter and associated circuits where high intensity reflections are concerned. In such cases either the appropriate corrections must be made (Chapter 4, p. 180), or else the intensity of the incident beam reduced.

The positions of peaks on a diffractometer chart are best measured at the centres at about two-thirds of the peak height, not at the apex. The centre of an unresolved $\alpha_1$, $\alpha_2$ doublet can be used as $2\theta$ for a weighted mean wavelength $\lambda = \frac{1}{3}(2\lambda_{\alpha_1} + \lambda_{\alpha_2})$. The instrument must be set up and calibrated accurately with a standard specimen, and whatever measuring procedure is adopted for the standard's peaks should be used also for those from other specimens (see Chayes and MacKenzie, 1957). For work of the highest accuracy more sophisticated methods of determining and correcting peak positions are required (Parrish, 1960; Wilson, 1963; Parrish *et al.*, 1964).

A recorder trace will show random fluctuation in the background which may sometimes simulate a true reflection. As a rough guide, a "peak" which is three times as big as the standard deviation of the background is likely to represent a reflection. Repeated runs can of course help resolve doubts about a peak's authenticity.

## 2. Diffractometer Errors

*Flat specimen.* The use of a flat instead of a suitably curved specimen causes a negative error in peak position. The resulting error in $d$ is proportional to $\cos^2\theta$, and it therefore tends towards zero at higher angles. The error from this source is small ($\sim 0 \cdot 01°[2\theta]$), as long as the angular divergence of the X-ray beam is not large.

*Absorption.* The influence of absorption is quite the reverse from that in powder-camera work, since, with the diffractometer, the higher the absorption coefficient of the specimen, the lower the error. The shift of reflections to lower angles might be better called a transparency error. It can be shown that:

$$\delta(2\theta) = \frac{\sin 2\theta}{2\mu R} - \frac{2t\cos\theta}{R[\exp 2\mu t \cosec\theta - 1]}$$

where $\mu =$ linear absorption coefficient, $R =$ radius of diffractometer, $t =$ specimen-layer thickness.

Differentiation of the above equation shows that the error in $2\theta$ is greatest at $2\theta = 90°$ and is zero at $0°$ and $180°$. The error in $d$, however, is again proportional to $\cos^2\theta$ and so tends towards zero at high Bragg angles. The practical way of minimizing this error is to use specimens as thin as possible consistently with producing enough intensity.

*Rate-meter error.* When the rate-meter circuit is being used to give an output to the chart recorder proportional to counting rate, it is averaging the counts over a length of time depending upon the time constant of the circuit. Thus there is always a lag between the recording of a given counting rate and its occurrence at the counter. This gives a constant error in $2\theta$, the magnitude of which depends upon the time constant setting (see below).

*Zero-angle error.* Missetting of the zero of the drum recording the $2\theta$ value will cause a constant error in the apparent Bragg angle. For both rate-meter and zero-angle errors, the error in $d$ is proportional to $\cot\theta$ and so tends to zero at high angles. An error in the setting of the specimen or counter with respect to each other may be more serious, since the resulting error in $2\theta$ will vary with angle and also a serious loss of intensity will occur.

*Specimen displacement.* If the plane of the specimen does not pass through the axis of the goniometer a considerable error results. It gives an erroneously high or a low $2\theta$, depending on whether displacement is towards or away from the centre of the reflecting circle. The error in $d$ is proportional to $\cos\theta\cot\theta$ and this extrapolates to zero at high angles.

*Diffractometer errors in general.* Highly accurate results can be obtained by applying all of the various theoretical corrections to account for the errors mentioned above, and for various other errors inherent in the method. But usually, with a well designed instrument, these errors are small and can be compensated by the use of a suitable standard. In general, the more exhaustive methods of correction need be resorted to only in cases where the accuracy sought is so high that uncertainties in the cell parameter of the standard would be of importance.

## 3. Time Constant and Scanning Speed

In order to achieve an almost undistorted and immediate plot of intensity variation, the rate meter should have a very low time constant. A low time constant can be used, but this results in the recording of all the fluctuations of background due to the random way in which X-ray quanta arrive at the counter. Too large a time constant, however, means that the rate meter may average over the time during which two reflections are passed by the counter

M*

and therefore record these as one or, at best, as poorly resolved peaks. The seriousness of the latter effect clearly depends upon the speed of diffracto-meter scanning. If it is a slow scan, one can use a longer time constant without sacrificing resolution. The choice of time constant is important not only for resolving two close reflections, but also for its effect on the shape of a single peak that contains the two wavelength components $\alpha_1$ and $\alpha_2$.

A reflection is recorded during the time that the receiving slit passes through the diffracted beam. Thus the distortion of a peak will be affected by the relation of the time constant to the time of traverse. A small slit width will require a low time constant. Thus, speed, slit width, and suitable time con-stants are all inter-related. For a given slit width a suitable time constant $T$ is given by $K/S$, where $K$ is a constant and $S$ is the scan speed in degrees $2\theta$ min. If the slit width is 0·1 mm, a suitable value of $K$ is 2. A typical arrangement would thus be: slit width 0·1 mm (goniometer radius = 17 cm), $S = 1°/\text{min}$, $T = 2\,\text{sec}$.

The value of $T$ derived here is meant as a rough guide, and for particular purposes, $T$ may be adjusted to other values. Thus for example, one might wish to sacrifice some accuracy of profile and resolution to gain a smooth background by making the latter condition $S = 1$, $T = 4$.

For highest accuracy of derived $d$ values and intensity profiles, a slow scan speed should be used, e.g. $\frac{1}{2}°$ or even $\frac{1}{4}°/\text{min}$. For more routine purposes, e.g. a preliminary survey, or an A.S.T.M. identification, a faster scan is satisfactory, say 1° or even 2°/min.

The design of a diffractometer is in many respects a compromise since the two requirements of high resolution and high intensity are incompatible. The former would be increased but the latter decreased by lower take off angle from the X-ray tube, by smaller beam divergence, by smaller receiving slits, and by thinner specimen layers.

*4. Specimen Preparation*

The ideal specimen for use with an X-ray diffractometer takes the form of a thin layer of fine-grained (1–10 $\mu$m) powder. A slightly thicker layer may be acceptable, since it gives greater intensity, but "transparency" errors will be more serious. The most easily prepared, and therefore most commonly used specimen mounting is the "smear" on a glass plate. A microscope slide of suitable dimensions is used as a base for spreading a thin layer of powdered specimen. A slurry of powder can be prepared in acetone and this settles and is left behind when the acetone rapidly evaporates. If a more permanent mount is desired, a little Durofix or similar glue can be mixed with the acetone.

An alternative is the cavity mount, in which a depression is etched in a glass plate and is filled with powder, and the surface smoothed off. A variation of this is to use a metal plate containing a rectangular hole. A glass slide backing allows the hole to be filled with powder and smoothed off. All the above methods tend to produce specimens with preferred orientation, particularly if the crystallites have platy or fibrous habit. The problems with regard to preferred orientation and suitable grain size, are greater with diffractometer mounts than with powder-camera spindles. Too coarse a grain size can give very misleading relative intensities, and the existence of this trouble is not as easily noticed as it is by the spotty-ring effect in powder camera work (p. 270). For some reflections the counter may be receiving intensity from within a "spot", and for others it may be between "spots". Thus crushing to a fine grain size, though not so much as to destroy the structure, is very important. Grinding under acetone can prevent decomposition of the specimen.

The production of a smooth-topped surface by packing down from the top is bound to enhance preferred orientation, so procedures have been recommended in which specimens are packed from the rear, or side (see, for example, Rex and Chown, 1960; Niskanen, 1964). These may reduce, but do not eliminate, preferred orientation.

It is often possible by means of a special attachment, to rotate the specimen in its own plane. This can help in two respects: it can help average out inhomogeneities, since different parts of the specimen and a larger area is covered by the beam during the rotation. Also, since the beam is not parallel but slightly divergent, crystallites that are not accurately parallel to the surface may be able to reflect at some position during the rotation. More crystallites therefore can contribute, and there is a better chance that all orientations may be equally represented.

A method that seems reasonably successful in overcoming preferred orientation effects is that whereby the powdered specimen is mixed with a hard-setting plastic (e.g. Lakeside 70 dissolved in dioxan). The solid mass, after setting, is then pulverized and yields roughly equidimensional particles containing flakes or fibres of the mineral specimen that occur in all orientations (Brindley and Kurtossy, 1961). Another useful method is to prepare a specimen by sprinkling the fine powder on to a tacky adhesive surface.

## 5. Background

The use of pulse-height discrimination for improving peak-to-background ratio has already been mentioned. Another method that has been advocated to reduce background is to use as a support for the layer of specimen powder, a quartz plate instead of glass. As long as the quartz plate is not itself able to reflect (an irrational plane should be parallel to the surface), then background is considerably reduced. An alternative is to spread the specimen on a thin film that is strong enough to support it, but contains less scattering matter than does a glass plate and so contributes less to the background. A suitable film may be prepared by spreading amyl acetate on to a water surface; the film is then lifted off by using the type of metal specimen holder described above (Gude and Hathaway, 1961).

## 6. Intensity Measurement

For intensity measurement various techniques are possible. The intensity at a given Bragg angle can be measured by counting for a fixed time or by timing a fixed count with a stationary goniometer. A similar result can be obtained, but with less accuracy, by measuring the recorder-pen deflection. In both cases, background measurements have to be made and subtracted. A scanning goniometer is more usually used and the reflection profile is traced on the recorder. The correct procedure is to obtain the integrated intensity by measurement of the area under the peak, but often the peak height alone may be measured and can be assumed to be proportional to the area. If changes in peak shape are involved, however, because of variation in crystal perfection, for example, then the area should certainly be measured.

## C. COMPARISON BETWEEN CAMERA AND DIFFRACTOMETER METHODS

### 1. Advantages of Camera

(a) It is possible to obtain a good strong pattern with very little material, particularly if the technique of using a small ball of powder is used. (b) The whole pattern can generally be obtained, including the back-reflection region, whereas with the diffractometer intensity falls off drastically at high angles. (c) Broad, weak reflections seem to be more easily detected above the background by eye on a film than on a diffractometer chart. (d) Lack of random orientation due to large grain size or crystal habit is more easily recognized, and preferred orientation is less enhanced by the methods of specimen preparation. An adequately small particle size for photographic work is about 200–300 mesh, whereas for diffractometry it ought to be $<400$ mesh (approximately 5 $\mu$m).

### 2. Advantages of Diffractometer

(a) Better angular resolution is obtained than with a normal powder camera. (b) The low-angle limit is much lower than with an ordinary powder camera, and so, with a suitable slit system, reflections from planes with large $d$ values can be recorded. (c) Highly absorbing specimens do not give rise to shifted or apparently doubled reflections. (d) By oscillation back and forth over selected peaks, it is possible to follow continuously any change in pattern that occurs with time or temperature, and possibly with other variables. The most common example of this technique is that in which a specially designed diffractometer furnace is used (e.g. Skinner et al., 1962; Scott and Ruh, 1963; Levin and Mauer, 1963). (e) The diffractometer plots reflection profiles directly, whereas with the powder camera a microdensitometer is needed to measure the film blackening. (f) The diffractometer allows great versatility in use, since it may be set up either for high speed and lower accuracy or for low speed and higher accuracy. The speed advantage is particularly evident when only a small part of a diffraction pattern is needed.

## D. THE PRINCIPAL USES OF THE POWDER-DIFFRACTION METHOD

(a) Identification of unknown substances (see p. 286). (b) Quantitative estimation of compounds in mixtures (see p. 299). (c) Determination of cell parameters (see p. 288). (d) Determination of chemical composition in a solid-solution series. The structural parameters of a mineral may vary continuously over a range when the chemical composition varies in a solid-solution series (e.g. forsterite, $Mg_2SiO_4$; fayalite, $Fe_2SiO_4$). The variation may be expressed as a graph of a cell parameter against composition, and then a measured cell parameter can be used to read off the composition of an unknown. Alternatively, it often suffices to use a graph of the $d$ value of a particularly sensitive and readily measurable powder reflection, or even the separation of a pair of suitable reflections, against composition. Some examples are given in Section X, p. 320. It should be noted (i) that a binary solid-solution series does not necessarily give a strictly linear graph, and (ii) that the presence of small amounts of elements other than the end members of the series can lead to deviations from the "calibration" curve. (e) Observation of high-temperature thermal transformations. This can be done with an ordinary powder camera if the high-temperature modification is quenchable and can therefore be examined at room temperature. Alternatively a high-temperature camera (or diffractometer) may be used. The "oscillating–heating" method with a diffractometer, gives a continuous record of a transformation (see, for example, Weiss and Rowland, 1956; Graf et al., 1963). (f) Order–disorder studies. Ordered structures give sharp X-ray reflections, whereas for disordered structures some or all of the reflections may be diffuse. A continuous change in degree of order can sometimes be followed by the changing profiles of powder reflections. If ordering takes place to form a large-dimensioned superstructure, additional powder reflections may be observed. Order–disorder effects are usually better studied by single-crystal diffraction methods, but in some cases a powder study may be adequate. (g) Structure determination. This usually requires the use of single-crystal methods, but simple structures can be determined from powder data. The subject of structure determination is beyond the scope of this book, but some of the methods employed are relevant, since the relative intensities of reflections can sometimes be used as an indication of specimen composition. It is therefore necessary to know how measured intensities are affected by variations in structure, and this topic is dealt with later (p. 297).

M**

## IV. Mineral Identification

The powder pattern of a substance can be described in terms of two sets of parameters: the positions of its reflections and their relative intensities ($I$ values). Since positions depend upon instrumental factors and on $\lambda$, it is best to characterize a substance by its set of $d$ values, which are an inherent property of the crystal structure. Since powder patterns contain many reflections and their $d$ and $I$ values may cover a wide range, it is not surprising that the powder pattern is unique to a substance in much the same way as a fingerprint is to a human being. The patterns therefore are suitable as a means of identifying an unknown substance, provided that there is a reference file of patterns of all or nearly all known substances, and that there is a simple method of finding one's way in it.

A system for this purpose has been organized by the American Society for Testing and Materials (A.S.T.M.)† and it consists of two parts, the A.S.T.M. Powder Data File and the A.S.T.M. Index. The file consists of a set of cards (Fig. 13), each one containing full details of $d$ and $I$ values and other useful information about the substance to which it refers. The cards are divided into sets (an additional set is issued approximately once a year) and serially numbered within the sets. The information can be in a printed or punched (key-sort) card form, and a recent innovation is the publication of copies of

| d | 2.95 | 4.35 | 9.5 | 9.5 | $CAAL_2SI_4O_{12}\cdot6H_2O$ | $CAO\cdot AL_2O_3\cdot4SIO_2\cdot6H_2O$ |
|---|------|------|-----|-----|---|---|
| $I/I_1$ | 100 | 90 | 70 | 70 | CALCIUM ALUMINUM SILICATE HYDRATE | CHABAZITE |

Rad. CuKα  λ 1.5418        Filter Ni        Dia. 114.6MM
Cut off                 $I/I_1$ VISUAL ESTIMATE
Ref. MASON AND GREENBERG, ARKIV MINERAL. GEOL. 1 519–526
                                                        (1953)

Sys. HEXAGONAL (RHOMBOHEDRAL)  S.G. $D_{3D}^5 - \bar{R}3M$
$a_0$  13.78     $b_0$          $c_0$ 14.97     A          C 0.920
$a$              $\beta$        $\gamma$        Z 6 (HEX.) $Dx$ 2.05
Ref. WYART, COMPT. REND. ACAD. SCI.              2 (RHOMB.)
     PARIS 192 1244 (1931)

$\varepsilon\alpha$          $n\omega\beta$ 1.481     $\varepsilon\gamma$ 1.483        Sign +
2V              D 2.05 .     mp                Color COLORLESS,GREY-
Ref. MASON AND GREENBERG, IBID.                      WHITE

SAMPLE FROM SANTA CATHERINA, SOUTHERN BRAZIL.
COMMONLY SOME REPLACEMENT OF (NA, K)SI FOR CAAL.

| d Å | $I/I_1$ | hkl | d Å | $I/I_1$ | hkl |
|-----|------|-----|-----|------|-----|
| 9.5 | 70 | 101 | 1.35 | 10 | |
| 7.0 | 40 | 110 | 1.33 | 10 | |
| 6.4 | 10 | 012 | | | |
| 5.6 | 40 | 021 | | | |
| 5.0 | 40 | 003 | | | |
| 4.35 | 90 | 211 | | | |
| 3.90 | 20 | 122,300 | | | |
| 3.61 | 50 | NI | | | |
| 3.47 | 20 | 220,104 | | | |
| 2.95 | 100 | 401 | | | |
| 2.62 | 20 | 410 | | | |
| 2.51 | 30 | 125,232 | | | |
| 2.10 | 20 | 333 | | | |
| 1.82 | 30 | INDEXED | | | |
| 1.74 | 20 | BY L.G.B. | | | |
| 1.66 | 10 | | | | |
| 1.57 | 10 | | | | |
| 1.53 | 10 | | | | |
| 1.43 | 10 | | | | |
| 1.41 | 10 | | | | |

FIG. 13. A typical card in the A.S.T.M. powder-data file.

† See reference under A.S.T.M. for address.

five sets of cards in the form of a book. Sets 1–5 have been published in this way and 6–10 are to follow: these books are a very convenient compact form of the file. A cumulative-index book is published in revised form with each new edition of cards, and this is the means by which one can find the required card in the file on the basis of the $d$ values of the three strongest reflections measured from the unknown. Instructions for the use of the index and file are given in each book and need not be repeated here. Emphasis must be placed, however, firstly on the desirability of accurate $d$ values and allowance for such experimental error as remains, and secondly on the need to bear in mind some variation in relative intensities which can occur mainly through different degrees of preferred orientation in the specimen. This is particularly likely to occur with fibrous or platy minerals (see pp. 271, 283).

Some difficulty may arise in the case of a solid-solution series, where, although the members have similar crystal structures, the cell parameters and therefore the $d$ values vary continuously across the series. If the spread of $d$ values is large enough, several members of the series may be represented by separate cards in the A.S.T.M. file, but not all intermediates can be included. Relative intensities may also vary considerably with substitution in a solid-solution series. Particularly difficult minerals with regard to identification are the individual members of the feldspar group. For these the $d$ and $I$ values vary considerably, not only through extensive solid solution but also through variations in structural state (i.e. high- and low-temperature feldspars). Although a feldspar can be recognized as such by use of the A.S.T.M. index, it is doubtful whether a very precise designation can be made by this method. To accomplish this, the complete patterns and not just a selection of reflections need to be examined.

An alternative way of tracking down an A.S.T.M. card is by use of the Fink index, which is available to all users of the file. Identification by the Fink index is based upon the $d$ values of the eight strongest lines, as compared with three lines in the ordinary index. Variations of intensity as compared with that of the standard A.S.T.M. specimen because of crystallite size, strain or solid solution effects can cause difficulties when identification is based upon only the three strongest lines; the Fink index may deal more successfully with such cases.

A third method that can be used is the Matthews index, which can be purchased at extra cost. This consists of a collection of large plastic punched cards, each of which has an array of $100 \times 100$ co-ordinates representing 10,000 substances. Each card represents the presence of a strong reflection within a stated range of $d$ values, and holes are punched in it for all substances that give such reflections. Similar sets of cards represent the absence of certain $d$ values, and other sets represent the presence or absence of particular chemical elements or groups of elements. Superposition of several appropriate cards above an illuminated screen should leave illuminated only the hole corresponding to the substance which produced the powder pattern. A particularly useful feature of the Matthews index is the presence of a single card that isolates all minerals from other substances. This index promises to be of particular value in dealing with the identification of mixtures.

The limitation of the Matthews index to 10,000 substances may prove serious before very long. Computer methods of data storage and data retrieval can supplement or even replace the present methods of dealing with the problem of identification by X-ray powder patterns. (The first computer tapes are already available from the A.S.T.M.). Qualitative, and even quantitative, estimation of the components in mixtures may also be greatly facilitated by the use of computer methods.

The ordinary A.S.T.M. method may be used to identify mixtures of minerals with varying degrees of success. When only two minerals are present the task is generally not too difficult. The pattern of one mineral must be identified and its powder lines eliminated, and then the remainder can be treated as a new unknown. Several trials may of course be necessary before a correct selection of the strongest lines that all belong to one mineral is obtained. Overlapping lines from the two patterns may be troublesome, and this possibility should be allowed for. Sometimes the components of a mixture have different characteristics that may helpfully lead to a different quality in their powder patterns. For example, a harder material may grind to a coarse-grained powder and therefore give spotty powder lines.

An important instance of the use of powder patterns for identification is in high-temperature phase-equilibrium studies. The products of experimental runs are often very fine-grained, and although optical study is always made for recognition of glasses and other features if possible, X-ray powder diffraction is widely used to identify the products.

When an identification has been made by means of the A.S.T.M. powder-data file, it is often useful to compare the powder pattern directly with one given by a standard (pure) specimen of the same substance. Significant variations in intensities and in reflection profiles may be noted, which are missed when only numerical data are compared.

## V. Indexing Powder Patterns and the Determination of Cell Parameters

The mathematical relationships between the $d$ values for the different reflections and cell parameters are given below:

orthorhombic:
$$\frac{1}{d_{hkl}^2} = \frac{h^2}{a^2} + \frac{k^2}{b^2} + \frac{l^2}{c^2}$$

or:
$$Q_{hkl} = h^2A + k^2B + l^2C$$

where $\quad Q = \dfrac{1}{d_{hkl}^2}, \; A = \dfrac{1}{a^2}, \; B = \dfrac{1}{b^2}, \; C = \dfrac{1}{c^2}$

tetragonal: $\quad Q = (h^2 + k^2)A + l^2C$

cubic: $\quad Q = (h^2 + k^2 + l^2)A$

hexagonal: $\quad Q = (h^2 + hk + l^2)A + l^2C$

where $\quad A = \dfrac{4}{3a^2}, \; C = \dfrac{1}{c^2}$

monoclinic: $Q = h^2A + k^2B + l^2C + 2hl\sqrt{AC}\cos\beta$

where $\quad A = \dfrac{1}{a^2\sin^2\beta}, \; C = \dfrac{1}{c^2\sin^2\beta}, \; B = \dfrac{1}{b^2}$

It is sometimes more convenient to deal in terms of $\sin^2\theta$ values instead of $1/d^2$, and this involves only the inclusion of the constant multiplier $\lambda^2/4$ on the right-hand side of the above equations. The analogous formula for a triclinic substance would be very unwieldy, and problems concerning triclinic substances are best dealt with by using reciprocal-lattice methods (see p. 309). In terms of the reciprocal-lattice elements:

$$\frac{\lambda^2}{d_{hkl}^2} = (d_{hkl}^*)^2 = h^2a^{*2} + k^2b^{*2} + l^2c^{*2} + 2hka^*b^*\cos\gamma^*$$

$$+ 2klb^*c^*\cos\alpha^* + 2lhc^*a^*\cos\beta^*$$

(For the monoclinic system, $\alpha^* = \gamma^* = 90°$, and for orthorhombic, $\alpha^* = \beta^* = \gamma^* = 90°$, and the expression simplifies considerably).

## A. INDEXING WHEN THE CELL IS KNOWN

If the cell parameters are known accurately, the $Q$ values can be calculated for all possible reflections ($hkl$ with various values of $h$, $k$ and $l$ up to the limit where $\sin^2\theta = 1$) by using the above formulae. If the lattice type and/or space group are also known, the systematically absent reflections appropriate for the space group need not be calculated. The $Q$ values derived from the observed reflections are then compared with the complete list of $Q_{calc}$, and indices are allocated if a sufficiently good match is obtained, taking account of the errors of measurement. Computers are frequently used for this type of calculation, particularly if a large range of $2\theta$ is to be covered. The problems of ambiguities of indexing can be minimized if resolution in the powder pattern is good, and measurements are very accurate. If single-crystal data are available, the observed relative intensities of reflections may help resolve some ambiguities; the weakest reflections on single-crystal photographs are generally not discernible in powder patterns. If the crystal structure is known, the published set of calculated intensities can be useful.

B. INDEXING WHEN THE CELL IS UNKNOWN

Procedures for the solution of this problem applicable to any crystal system are described in the literature, but here it will suffice to outline the procedures that are applicable for the cubic, tetragonal and orthorhombic systems. Details of these are also available in other texts.

For cubic substances, $Q$ values are proportional to $h^2 + k^2 + l^2 (= N)$, i.e. the ratios of $Q$ values for the different reflections are those of a set of integers obtainable by summing the squares of three integers. (All of the possible values of $h$, $k$ and $l$ that give values of $N$ up to 998 are listed by Lonsdale, 1959.) When indices are derived, each observed $d$ value gives the value of the cell parameter $a$, by use of the relation $a = d\sqrt{N}$. From powder photographs, the highest angle reflections give the most accurate value for $a$, and extra-polation methods may be used (see p. 274). For diffractometer measurements with internal standard, the reflections nearest to those of the standard will give the best results.

For tetragonal minerals, those reflections that are from the basal planes (indices $00l$) will have $Q$ proportional to $l^2$, i.e. $Q$ values would be pro-portional to the squares of integers. Another sequence of reflections will be those from $hk0$ planes and for these the $Q$ values will be proportional to $h^2 + k^2$, i.e. the sum of the squares of two integers. Providing that these two sequences of lines can be recognized somehow from among the others, they can be indexed and the values of $a$ and $c$ determined. If this indexing is correct, it will then be possible to index $hkl$ reflections and to use a selection of these or all of them in order to give a more accurate redetermination of the cell parameters. For hexagonal minerals, a similar approach to the problem of indexing and cell determination can be made.

The problem of indexing cubic, tetragonal and hexagonal substances can alternatively be dealt with by graphical methods (see for example, Azaroff and Buerger, 1958; Henry et al., 1960; Myers and Davies, 1961). These methods involve specially prepared charts that show the pattern of $d$ values of a large number of reflections calculated on the basis of a continuously varying range of $a$ values for the cubic system and of axial ratios for tetragonal and hexagonal specimens.

For orthorhombic substances, a method of indexing by systematic inspection, and trial and error, is based upon the examination of the set of $Q$ values in order to find recurring differences between different pairs. For example, $Q_{321} - Q_{221} = 5A$ and more generally $Q_{3kl} - Q_{2kl} = 5A$. Similarly, $Q_{2kl} - Q_{1kl} = 3A$ and $Q_{1kl} - Q_{0kl} = A$. The differences between pairs of $Q$ values are noted systematically, and those that occur most frequently are tried as possible values for $A$, $B$, $C$, $3A$, $3B$, $3C$, etc. This method was used by Lipson (1949) and is described also by Henry et al. (1960). A similar approach has been used for the more complex case of a monoclinic pattern (Zachariasen, 1963).

No simple methods of the above kind are applicable to the triclinic system. Methods that have been developed for this are of a different kind, and aim at determining (mostly by reciprocal space considerations) a cell that describes the lattice, but is not necessarily the cell usually chosen for its description. Procedures can then be used to transform the description of the lattice from one cell to another (Ito, 1949, 1950; Lonsdale, 1952, p. 530; de Wolff, 1957). Computer methods have been devised to cope with the laborious calculations and lengthy trials of combinations involved in the various ways of dealing with this kind of problem. Although these methods have been shown to work in certain cases, they are by no means straightforward and may be unsuccessful if any impurity powder line is present, or if the positions of reflections are measured with insufficient accuracy, particularly if the unit cell is large.

## C. REFINEMENT OF CELL PARAMETERS

The problem of cell determination that the mineralogist is likely to meet most often, and which is perhaps the most important to him, is the determination of accurate cell parameters when approximate ones are already known. Thus the members of a mineral group, or a solid-solution series, will have parameters that are known to be within a particular range, and the determination of accurate values can be used to characterize the mineral more precisely.

For this problem it is assumed that the cell parameters initially known are near enough correct to give correct indices to a number of reflections. Taking as an example an orthorhombic mineral, it may be possible to select certain reflections with simple indices ($h00$, $0k0$, $00l$) to give directly the cell edges ($a = h \times d_{h00}$, etc.). The derived parameters can be checked by using reflections with more complex indices. Alternatively, if $h00$ gives $a$, then an $hk0$ reflection can be used to give $b$. Another procedure would be to use pairs of $hk0$ reflections to give simultaneous equations for determining $a$ and $b$, and then use these values together with $hkl$ reflections to solve for $c$. Also, for the general reflections, *hkl*, three equations with three unknowns could be set up and solved.

All of the above approaches have some degree of arbitrariness as to the way the parameters are determined. Separate *h*00's, pairs of *hk*0's etc., can be used, or a number of different ways can be adopted and the several results averaged. Some reflections may be measurable with greater accuracy than others; furthermore, the accuracy obtained for a given parameter by simultaneous equations will depend very much upon the indices of the particular reflections chosen. The best and most general method of accurately determining cell parameters from powder data is the method of least squares. This allows the best use of all the observed data, but it can also have built into it any weighting for or against particular observations that is thought desirable (see, for example, Lawn, 1963).

## 1. The Method of Least Squares

A simple example will illustrate the method where only two unknowns are involved, as for a tetragonal mineral. Instead of the expression for $Q$, it is more convenient to use $\sin^2\theta_{hkl} = (h^2 + k^2)A + l^2C$, which may be written $\sin^2\theta_{hkl} = \alpha A + \gamma C$, where $\alpha = h^2 + k^2$, $\gamma = l^2$, $A = \dfrac{\lambda^2}{4a^2}$, $C = \dfrac{\lambda^2}{4c^2}$.

An equation may be written for each observed reflection:

$$\alpha_1 A + \gamma_1 C = \sin^2\theta_1 \tag{1}$$

$$\alpha_2 A + \gamma_2 C = \sin^2\theta_2 \tag{2}$$

$$\cdot \qquad \cdot \qquad \cdot$$
$$\cdot \qquad \cdot \qquad \cdot$$

$$\alpha_i A + \gamma_i C = \sin^2\theta_i \tag{3}$$

If $A$, $C$ and $\sin\theta$ are absolutely correct, then for each reflection $\alpha A + \gamma C - \sin^2\theta = 0$. However, because of experimental errors in $\theta$ we will have, for the $i$th reflection:

$$\alpha_i A + \gamma_i C - \sin^2\theta_i = \Delta_i \tag{4}$$

The method of least squares (un-weighted) gives the values of $A$ and $C$ that minimize $\sum_i \Delta_i^2$. The procedure is as follows.

Multiply all "equations" by their $\alpha$'s and add them:

$$A \sum \alpha_i^2 + C \sum_i \alpha_i \gamma_i = \sum_i \alpha_i \sin^2\theta_i \tag{5}$$

Multiply all equations by their $\gamma$'s and add them:

$$A \sum \alpha_i \gamma_i + C \sum \gamma_i^2 = \sum_i \gamma_i \sin^2\theta_i \tag{6}$$

Equations (5) and (6) are called "the normal equations" and can be solved for $A$ and $C$.

If there are three unknowns (orthorhombic) there will be three normal equations, and for the monoclinic system there will be four, proceeding from:

$$\sin^2\theta = h^2A + k^2B + l^2C - 2hl\,AC\cos\beta$$

The solution of the four equations will give $A$, $B$, $C$, and $AC\cos\beta$, and so eventually, $a$, $b$, $c$ and $\beta$.

## 2. Drift Constant

When an internal standard is used, there should be no appreciable systematic error in the measured $\theta$ values. If no standard is used, or if a systematic error is known to occur and it is a known function of $\theta$, then a better solution can be obtained by including a drift constant, $D$. Thus for diffractometry generally it can be shown that a systematic error in $\sin^2\theta$ occurs which is proportional to $\sin^2 2\theta$. The general equation can then be written (e.g. for tetragonal minerals) $\alpha A + \gamma C + \sin^2 2\theta D = \sin^2\theta$. This introduces one more unknown, the drift constant, but if many reflections (equations) are available, this does not matter, and the results for $A$ and $C$ should be improved.

When a least-squares solution has been obtained by using a number of reflections that have been indexed with fair confidence, further indices may be obtained, and more data can be put into the next cycle of least squares refinement. Some of the reflections assumed to have been correctly indexed may, after the least-squares solution, show a large $\Delta$ between observed and calculated values of $\sin^2\theta$ ($\Delta$ can also be listed in terms of $\theta_{obs.}$ and $\theta_{calc.}$ or $1/d^2_{obs.}$ and $1/d^2_{calc.}$). This may indicate an error of measurement or calculation, or an incorrect index assignment. New indices should be assigned if possible, and a further cycle of refinement undertaken, since the incorrectly indexed reflection will have influenced the previous solution adversely. Most least-squares calculations are done with the aid of computers.

For more detailed treatment of the least-squares method and for discussion of errors, the reader is referred to Smith (1956) and Kelsey (1964).

### D. THE RELATION BETWEEN CELL PARAMETERS, DENSITY AND CHEMICAL FORMULA

Measurement of cell parameters leads to knowledge of the volume of the unit cell, and this can be usefully correlated with the density of the mineral as follows:

$$D = \frac{\text{Mass of unit cell content}}{\text{Volume of unit cell}}$$

$$= \frac{\sum(\text{atomic weights}) \text{ of formula unit} \times Z \times 1\cdot660_4}{[V \text{ (in Å)}]^3}$$

where $Z$ = number of formula units per unit cell.

The above expression contains three variables, the density, the "molecular weight" of the formula unit, and the cell volume, and a fourth factor $Z$, which must be an integer. If $Z$ is known, then measurement of two of the three variables enables one to solve the equation for the third. For example, if the formula and $Z$ are known and $V$ is determined, then $D$ may be calculated. This $D_{\text{calc.}}$ can perhaps be compared with a measured density as a check on the correctness of the chemical formula.

As another example, if the "molecular weight" is known approximately and then $V$† and $D$ are measured accurately, this leads to an approximate value for $Z$, which will be nearly an integer. If the nearest integer is re-substituted in the equation, a more correct "molecular weight" can be determined.

Lastly, if a chemical analysis of a mineral is given in terms of oxide percentages, then the normal method of deriving the chemical formula yields only the relative number of the different atoms present and not the absolute numbers in a unit cell. If, however, the cell volume and density have been measured accurately then the percentages of the various oxides can be converted into absolute numbers of atoms in the cell (Hey, 1939, 1954; Nicholls and Zussman, 1955).

## VI. X-ray Diffraction Intensities

### A. VARIOUS GEOMETRIC AND PHYSICAL FACTORS

If use is to be made of relative intensities of diffraction, either for quantitative estimates of mixtures or as an indication of chemical composition, it is necessary to appreciate the geometric and physical factors that influence these intensities. For powder diffraction $E = K.I_0.L.P.p.A.F^2$, where $E$ = energy of diffracted beam, $K$ = constant, $I_0$ = intensity of incident beam, $L$ = Lorentz factor, $P$ = polarization factor, $p$ = multiplicity, $A$ = absorption factor, $F$ = structure factor.

The various factors are discussed in the following pages.

† The general formula for the volume is:

$$V = abc(1 + 2\cos\alpha\cos\beta\cos\gamma - \cos^2\alpha - \cos^2\beta - \cos^2\gamma)^{\frac{1}{2}}$$

## 1. Lorentz Factor

This makes allowance for the fact that for different reflections the intensity of the powder arc measured represents a different fraction of the total intensity spread over the complete cone of reflection. Also, the probability that a particular set of planes will occur in a reflecting position is a function of $\theta$.

For the powder method, $L = \dfrac{1}{\sin \theta \sin 2\theta}$

## 2. Polarization Factor

The X-ray beam incident on a face of a crystal in the specimen is unpolarized (except when a monochromator is used) but the total amplitude $G$ can be regarded as resolved into two components at right angles; $G_x$ perpendicular to the plane containing the X-ray beam and the normal to the crystal face, and $G_y$ parallel to this plane (Fig. 14). The component $G_x$ is unchanged

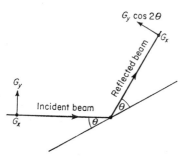

FIG. 14. Resolution of the incident and reflected X-ray beams in and normal to the plane of incidence for derivation of the polarization factor.

on reflection but the component at right angles to $G_x$ in the reflected beam is $G_y \cos 2\theta$. The resulting intensity is thus proportional to $G_x^2 + G_y^2 \cos^2 2\theta$. Since $G_x = G_y = G/\sqrt{2}$ the reflected intensity is proportional to $\frac{1}{2}(1 + \cos^2 2\theta)$.

For the powder method the combined Lorentz and polarization factors is thus $(1 + \cos^2 2\theta)/(\sin^2 \theta \cos\theta)$ and this function can be found tabulated at small intervals of $\theta$ in various texts. It is useful to appreciate that qualitatively the $LP$ factor enhances the observed intensities at both low and high Bragg angles with respect to those at medium angles.

The expression for the combined Lorentz and polarization factors for a single-crystal experiment in which the crystal is rotated about an axis normal to the incident beam is $(1 + \cos^2 2\theta)/\sin 2\theta$. The single-crystal formula may sometimes be appropriate in powder diffractometry, since specimens that have a marked preferred orientation may effectively behave as a mosaic single crystal.

## 3. Multiplicity

Reflections from crystallographically equivalent planes occur at the same Bragg angle, and their intensities are superimposed in powder patterns. As an example in the orthorhombic system, $00l$ planes are doubly recorded ($00l$ and $00\bar{l}$), whereas $hk0$, $0kl$, and $h0l$ intensities are quadrupled (e.g. $hk0$, $\bar{h}k0$, $h\bar{k}0$, $\bar{h}\bar{k}0$), and $hkl$'s are recorded eight-fold.

It should be noted that nonequivalent planes may also in certain cases have identical $d$ values and therefore give exactly superimposed reflections on powder patterns (e.g. $hk0$ and $kh0$ in a tetragonal crystal with space group P4). [See Lonsdale, 1952, p. 31.]

## 4. Absorption Factor

For the case of X-ray reflection from a specimen in the form of a flat layer, as for example in normal diffractometry, it can be shown that the absorption factor is simply proportional to $1/\mu$ where $\mu$ is the linear absorption coefficient of the specimen.

For a powder photograph with a cylindrical or spherical specimen, the correction is more complicated, but it has been determined and tabulated for specimens with different radii (Peiser *et al.*, 1955).

The linear absorption coefficient is represented by $\mu$ in the expression $I = I_0 e^{-\mu t}$, where $I_0 =$ intensity of incident beam and $I =$ intensity of beam transmitted through thickness of specimen $t$.

The linear absorption coefficient depends upon the physical state of the specimen; a coefficient independent of state is the mass-absorption coefficient, $\mu_m$, which is equal to $\mu/\rho$ (where $\rho$ is the density). $\mu$ can be calculated for a compound or mixture in terms of the values of $\mu_m$ for the elements present. Thus $\mu = \rho \sum_i x_i \mu_{m_i}$, where $x$ and $\mu_m$ are the fractions by weight and the mass-absorption coefficients of the several elements present, respectively. (An example of this calculation for diopside is given in Chapter 4, p. 169.)

An alternative calculation for $\mu$ can be made in terms of the gram-atomic absorption coefficient, $\mu_g$, of the elements present: $\mu_g = \mu_m w$, where $w$ is the atomic weight: $\mu = \rho . \sum_i n_i \mu_{gi} / \sum_i n_i w_i$, where $n$ is the number per formula unit of each particular kind of atom.

## 5. Diffractometer Intensities at Low Bragg Angles

With a normal size of smear or cavity mount, the X-ray beam divergence is such that the beam lies entirely within the specimen for all but the lowest angles of incidence. If, however, the specimen area is unusually small, or a reflection at an extremely low Bragg angle is being recorded, the above condition may not hold. The limiting angle is $\theta_0$, and $\sin\theta_0 \sim \dfrac{R\delta}{W}$ where $R =$ radius of goniometer circle, $\delta =$ divergence angle of incident beam (in radians), $W =$ width of specimen.

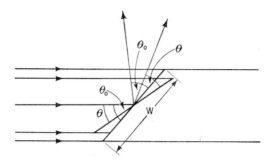

FIG. 15. Effect of specimen width on intensity at low Bragg angles. The width W of the specimen is such that it does not intersect the whole of the X-ray beam at Bragg angles lower than $\theta_0$, and there is consequently a fall-off in intensity. The diagram uses a large $\theta_0$ for ease of illustration.

If $\theta < \theta_0$, the area of the beam intercepted by the specimen is proportional to $\sin\theta$ (see Fig. 15). In order to put all reflections on a comparable scale, intensities of reflections up to $\theta_0$ should be multiplied by $\sin\theta_0/\sin\theta$. Below the limiting angle, $\theta_0$, however, the X-ray beam may hit the end of the specimen holder and give rise to increased background, so that it is advisable to use a narrower divergence slit for recording reflections at very low Bragg angles. Scaling of intensities can in this case be accomplished by recording certain reflections in an overlapping range with both apertures.

### B. THE STRUCTURE FACTOR

The structure factor (more strictly the "structure amplitude") depends upon (1) the number and kinds of atoms present; (2) their positions in the unit cell. A single atom scatters X-rays in the forward direction in direct proportion to the number of electrons it contains. The amplitude scattered falls off with increasing angle and in the way indicated by the atomic-scattering factor ($f$) curve.†

† The $f$ curves are usually given for atoms or ions at rest, but the atoms in a crystal are in thermal vibration, the effect of which is to make $f$ fall away more rapidly with increasing Bragg angle. A temperature factor that takes account of this can be applied to the $f$ curve for stationary atoms and this has the form: $\exp - B\sin^2\theta/\lambda^2$, where $B$ is a constant. $B$ may vary from one atom to another depending upon its nature and its structural environment, but a typical average value for a specimen at room temperature is unity. The thermal-vibration effect can usually be allowed for empirically when sets of calculated and observed intensities are compared.

Tables of atomic-scattering factors for different Bragg angles, or the constants of an analytical expression that gives the scattering curves, are available (Lonsdale, 1962).

The intensity of a reflection $hkl$ is proportional to the square of the structure amplitude $F$, where $F^2_{hkl} = A^2_{hkl} + B^2_{hkl}$, and:

$$A_{hkl} = \sum_j f_j \cos 2\pi(hx_j + ky_j + lz_j)$$

$$B_{hkl} = \sum_j f_j \sin 2\pi(hx_j + ky_j + lz_j)$$

$x_j$, $y_j$, $z_j$ are the atomic co-ordinates of the $j$th atom expressed as fractions of the $a$, $b$, and $c$ cell dimensions, respectively. The summation is over all the atoms in the unit cell, but the presence of symmetry elements allows the summation to be done over a fraction of the cell with appropriate modification of the trigonometric expression. If the structure concerned is centrosymmetric, then $B = 0$. When some atoms lie on special positions, in order to avoid errors it is as well to think of the calculation in terms of the whole unit cell. Computer programmes are generally available for structure-amplitude calculations.

### C. SYSTEMATIC ABSENCES

Application of the formula for the structure amplitude to the content of unit cells that are centred or contain certain symmetry elements (glide planes and screw axes) will demonstrate that reflections from particular kinds of lattice plane will be systematically of zero intensity. The following systematic absences indicate the lattice type:

$hkl$ reflections absent when $(k + l)$ is odd —— $A$-centred lattice

$hkl$ reflections absent when $(h + l)$ is odd —— $B$-centred lattice

$hkl$ reflections absent when $(h + k)$ is odd —— $C$-centred lattice

$hkl$ reflections absent when $(h + k + l)$ is odd —— $I$-centred lattice

$hkl$ reflections absent when indices are mixed
$\qquad\qquad\qquad\qquad$ odd and even —— $F$-centred lattice

If a rhombohedral lattice is treated by using Miller-Bravais axes and indices $(hk * l)$, then only reflections with $(-h + k + l) = 3n$ or $(h - k + l) = 3n$ will be present.

The presence of glide planes and screw axes of symmetry can also sometimes be detected by systematic absence rules that apply to particular kinds of reflection rather than to general $hkl$ reflections.

Thus:

$hk0$ absent when $h$ is odd —— glide of $\dfrac{a}{2} \perp c$.

$hk0$ absent when $k$ is odd —— glide of $\dfrac{b}{2} \perp c$.

$hk0$ absent when $(h + k)$ is odd —— glide of $\dfrac{a + b}{2} \perp c$.

and similar rules applying to $0kl$ and $h0l$ reflections.

Also:    $h00$ absent when $h$ odd —— screw axis parallel to $a$.

       $0k0$ absent when $k$ odd —— screw axis parallel to $b$.

       $00l$ absent when $l$ odd —— screw axis parallel to $c$.

The assemblage of symmetry elements deduced from the systematic absences of a complete set of diffraction data can indicate that the space group of a mineral must be one of relatively few alternatives. Usually, however, such information can be obtained from single-crystal data, and only to a limited extent from powder data.

### D. QUANTITATIVE ESTIMATION OF MIXTURES BY X-RAY DIFFRACTION

Considerable difficulties arise in using X-ray diffraction for accurate quantitative estimates of the amounts of different minerals present in mixtures, even if one restricts the problem to a two-component mixture.

In a two-component mixture we can regard one component as the mineral to be determined, and the other as a matrix. The intensity of an X-ray reflection will be affected by absorption in the specimen as follows:

$$I \sim \frac{Kx}{\rho[x\mu + (1 - x)\mu']}$$

where $x$ = weight fraction of component to be determined, $I$ = intensity of reflection, $\rho$ = density of mixture, $\mu$ = mass-absorption coefficient of component to be determined, $\mu'$ = mass-absorption coefficient of matrix.

If $\mu = \mu'$ (e.g. in the case of polymorphs such as calcite and aragonite), then $I$ is directly proportional to $x$, and a straight-line relationship can be plotted. If $\mu \neq \mu'$, the relationship between $I$ and $x$ will be expressed by a curve as shown in Fig. 16. If the matrix is known then standard mixtures can be made up and the calibration curve plotted and used for unknown samples.

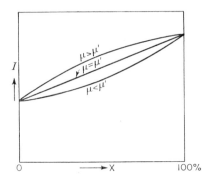

FIG. 16. Graph of diffracted intensity $I$ against concentration $x$ for a compound with mass-absorption coefficient $\mu$ in a matrix with coefficient $\mu'$.

If the matrix is not known, an internal standard may be used. It is assumed that the intensities from the standard are affected in the same way by absorption as are those from the unknown. If this is true, $x = kI/I_s$ where $I$ is the intensity of a given reflection of the unknown and $I_s$ is the intensity of a given reflection of the standard, assuming a constant weight of standard in all samples. This approximation to a linear relationship will be more closely attained if absorption is not greatly affected by changes in composition of the specimen; the latter condition can be more nearly achieved by diluting the specimen with an amorphous substance (a gum or a glass). For all methods of quantitative estimation of minerals, good results cannot be achieved unless the particle size in the specimen is very small (of the order of $10^{-4}$cm).

Mitchell (1960) has described an internal-standard method and has given a "universal scale" of intensity values for certain strong reflections for a number of pure substances. The expression he uses is:

$$x = 100 \cdot \frac{s}{w} \cdot \frac{I_s}{I_{s_m}} \cdot \frac{I_N}{I_{N_m}}$$

where $x =$ weight% of mineral $N$ being determined, $w =$ weight of sample used, $s =$ weight of internal standard added, $I_N =$ universal scale intensity of line of $N$, $I_{N_m} =$ measured intensity of line of $N$, $I_s =$ universal scale intensity of line of standard, $I_{s_m} =$ measured intensity of line of standard.

The method should give reasonable results (probable error about $\pm$ 5%) for both powder-camera and diffractometer methods of recording intensities, but important factors to be controlled are crystallite size, preferred orientation, homogeneous mixing of sample and standard, uniform specimen mounting and instrumental stability.

An important limitation to the usefulness of the method is imposed when there is the possibility of solid-solution replacements in the chemical formulae of the minerals concerned. Substitutions will affect relative intensities, so the universal-scale mineral used should ideally have exactly the same composition as the mineral being determined. An iron-rich olivine, for example, cannot be determined in a mixture with reference to the universal-scale intensity for forsterite, $Mg_2SiO_4$.

An alternative method of correcting for absorption errors in quantitative determinations by diffractometry, is to prepare a tablet sample of the mixture concerned, with known thickness, and actually measure the absorption of an X-ray beam passing through it (Norrish and Taylor, 1962). This method is described more fully in Chapter 4, p. 194, where it is applied to X-ray fluorescence analysis. A method of using X-ray diffraction patterns for the modal analysis of some felsic rocks is described by Tatlock (1966).

The sensitivity of the X-ray powder method to the detection of a small amount of a particular mineral in a mixture depends a great deal on the minerals concerned. A well crystallized mineral for which a particular reflection happens to be very strong (e.g. quartz or calcite) can often be detected at below the 1% level, whereas quite large concentrations of a poorly crystallized mineral may be undetected.

Various methods for determining mixtures of clay minerals and non-clay minerals in clays have been described in the journals dealing with clays and ceramics (see below).

## VII. Clay Minerals

X-ray powder-diffraction work with clay minerals involves a number of difficulties in addition to those encountered with other minerals. Clay minerals are less perfectly crystallized and they are extremely fine grained, so that reflections are broader, intensities are lower and there are fewer measurable reflections. The structures of clay minerals are often disordered, so that reflections have very variable profiles and some are so diffuse as to be unobservable; the positions of asymmetrically broadened reflections are not easily designated.

Some clay minerals have a very long repeat distance perpendicular to their structural layers, so that important basal reflections may occur at very low Bragg angles: in order to be able to record reflections with $d$ values up to about 20 Å a Bragg angle of about 2° ($2\theta$ Cu radiation) must be attainable. Most ordinary cameras are not suitable for this purpose unless specially designed narrow collimators are available (see Aruja, 1961; Cole, 1961), and the use of a focusing camera (p. 302) will generally give far better results.

Diffractometers with special narrow slit systems are capable of dealing wi
spacings up to about 30 Å (Cu radiation).

Difficulties may arise also because many clay minerals have platy mo
phology, and so there is an enhancement of preferred orientation of th
specimen particles. This tendency can sometimes be turned to good accou
(see, for example, Bloxam, 1963), since it is often the basal reflections tha
are those most usefully studied. Furthermore, problems of indexing may t
simplified if it is known that a particular series of stronger reflections a
likely to be from 00l planes.

A further complication in dealing with clay minerals is that many specime
have a variable content of inter-layer water molecules and of exchangeab
cations, which leads to variation in the powder pattern $d$ values. (The inte
layer spacings are sometimes affected by atmospheric humidity.) Also mixe
layer clay minerals occur with varying proportions of two components, e.;
illite–montmorillonite or chlorite–kaolinite; these result in broadened refle
tions, some of which occur at non-Bragg positions.

In view of the above mentioned special characteristics of X-ray diffractio
from clay minerals, special techniques of specimen preparation, film an
diffractometer recording, and interpretation are needed. Auxiliary techniqu
such as size fractionation, dissolution by acids, treatment with organic liquid
thermal treatment or cation exchange, are often employed with great advar
tage. Detailed information on the investigation of clay minerals by X-rays
given in the volume edited by Brown (1961), and many papers involving th
use of X-rays have been published in the annual volumes of "Clays and Cla
Minerals", in "Clay Minerals Bulletin" and in journals dealing with ceramic
Diffractometer patterns of reference clay minerals were presented by Mollo
and Kerr (1961). Quantitative estimation of mixtures of clay minerals ha
also been the subject of many investigations and presents many difficultie
(see, for example, Brindley and Kurtossy, 1961, 1962; van der Marel, 1962
Lapham and Jaron, 1964).

## VIII. Focusing Cameras

The geometry of a typical focusing camera is illustrated in Fig. 17. Th
horizontal line focus of an X-ray tube at $A$ yields a divergent beam which i
reflected by the curved crystal of a monochromator $B$ with focus on th
film $D$ at the circumference of the vertical circle of the camera. The curvatur
of the monochromator crystal is adjustable for various X-ray wavelengths

The specimen is in the form of a paste smeared on to a sample holder, $C$
which also lies on the circumference of the camera. This arrangement allow
the recording of fully focused powder reflections up to $2\theta = 90°$, and in som

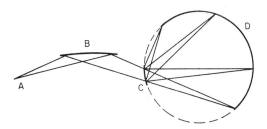

Fig. 17. The geometry of a typical focusing camera. A, horizontal focus of an X-ray tube; B, curved crystal monochromator; C, powder specimen; D, photographic film.

cameras four specimens can be dealt with simultaneously. Monochromator and camera can be enclosed in a gas-tight casing, so that vacuum can be employed and background scattering reduced to a very low level.

Focusing cameras are capable of producing photographs of very high quality with respect to precision, resolution of closely spaced reflections, sensitivity for weak reflections, and ability to record reflections at very low Bragg angles. Fuller descriptions of these instruments can be referred to elsewhere (Guinier, 1964; Brindley, 1961), and examples of applications to clay mineral studies are given by Porrenga (1958).

## IX. Single-Crystal X-ray diffraction

Although many mineralogical problems may be tackled by the method of X-ray powder diffraction, which is rapid and which can yield very accurate results for $d$ spacings and cell parameters, there are circumstances in which single-crystal methods are preferable or even essential. The main difficulty with powder patterns is that several reflections may be superimposed, and the weaker reflections are often not observable. Single-crystal X-ray photographs contain more complete and more readily analysed information about the crystal structure of the specimen, since the reflections from all crystal planes can, by suitable techniques, be recorded separated from one another. All reflections can be indexed, their positions can be used to determine symmetry, cell parameters and space group, and their intensities can be measured to give other structural information.

Single-crystal methods may of course be essential if only a few small grains of specimen are available.

### A. OSCILLATION AND ROTATION PHOTOGRAPHS

If a single crystal is set with an edge of its unit cell (say $c$) vertical and a horizontal beam of X-rays impinges upon it, the diffracted X-rays will lie on cones of semi-vertical angle $(90° - \phi)$ where $n\lambda = c \sin \phi$ (Fig. 18). A cylindrical film with vertical axis passing through the crystal will record the diffracted beams on horizontal lines ("layer lines"), which are the intersections of these cones with the film. If the "crystal" had only one-dimensional regularity (parallel to $c$), the layer lines would be lines of continuous blackening but regularity in other directions restricts diffraction intensities to two other sets of cones in space, and only where all three sets of cones intersect at a point on the film can there be a "reflection". Thus reflections may occur as

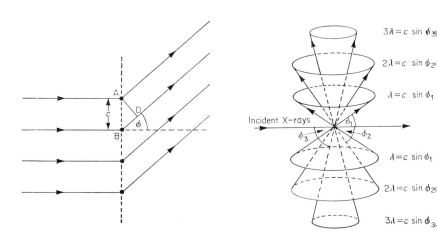

Path difference $= BD = c \sin \phi = n\lambda$

FIG. 18. Formation of layer lines in an oscillation or rotation photograph. The conditions are similar to those for diffraction from a vertical row of atoms with repeat distance $c$. First-, second- and third-order cones are illustrated on the right, and the dependence of cone angle on $c$ is indicated on the left.

spots on the layer lines, but with a strictly monochromatic and parallel beam and a stationary crystal, only very few reflections will occur. More reflections are recorded if the crystal is oscillated or rotated about the vertical axis, since the condition for the layer lines remains unchanged, but the angles of the other cones change and more intersections of cones can occur giving a number of reflections on the layer lines (Fig. 19).

The geometry and theory described above is that used in a common instrument for recording single-crystal data, the oscillation or rotation camera (Fig. 20). A cell edge that is parallel to the camera axis can readily

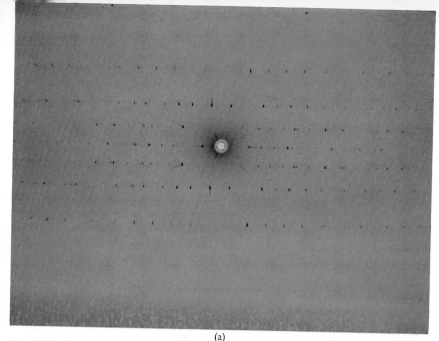

(a)

FIG. 19(a). Oscillation photograph (15°) from single crystal of natrolite taken with nickel-filtered CuK radiation. The crystal has been mounted with the z axis vertical, and so layer lines of reflections are produced corresponding to the repeat distance $c$ (= 6·60 Å). The symmetry of this photograph above and below the equator indicates that the z axis is one of even rotation symmetry, or that there is a mirror plane perpendicular to z. Reflections on the zero-, first-, second- and third-layer lines have indices $hk0$, $hk1$, $hk2$ and $hk3$, respectively. The streaks are produced by white radiation.

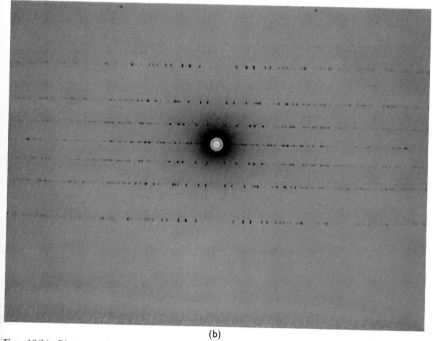

(b)

FIG. 19(b). Photograph produced as above, but with crystal rotating about z instead of oscillating. Many more reflections are recorded and there is now additional symmetry as between right and left.

FIG. 20. Photograph of a single-crystal X-ray goniometer. At the centre is the goniomete
arc system for orientating the crystal; on top of the arcs is a piece of Plasticine in whicl
the glass fibre holding the crystal can be mounted.The entrance collimator, alignin
telescope and cam system for oscillating the crystal can be seen, and at the bottom rigl
is the cylindrical pot containing X-ray film, which, in operation, sits in the well at th
centre of the instrument. The "X-rays on" sign is a safety device which flashes when th
X-ray tube is energized.

be determined using the equations:

$$n\lambda = c \sin \phi \text{ and } h_n = R \tan \phi$$

where $n$ = order of layer line ($n = 0$ for equatorial line, $n = 1$ for 1st layer above or below equator, etc.), $h$ = height of layer line above or below the equator, $R$ = radius of camera.

Thus if $R$ is known and $h_n$ is measured, $c$ can be deduced. For an approximate cell parameter, a Bernal chart[†] (Fig. 30) may be used superimposed on the film. Layer-line positions can be read from this directly to give a value for $\sin \phi$ known as $\zeta$. Then:

$$c = \frac{n\lambda}{\zeta}$$

It is important to note that the distance given by the layer-line measurements is always that repeat distance of the Bravais lattice which is parallel to the camera axis. In the simplest case this corresponds to a cell edge, but it may be a face diagonal, body diagonal, or indeed any direction through lattice points. Only when the crystal has a prominent lattice row vertical will layer lines appear prominently and relatively widely spaced. If a face diagonal passes through a face-centring lattice point, or a body diagonal through a body-centring point, then the repeat distance derived will be half of the corresponding face or body diagonal. Some typical situations are illustrated in Fig. 21. For the non-special orientation (v), layer lines will correspond to the large distance $p'$ and so they will be very closely spaced with relatively few spots on each, and they will therefore be more difficult to discern on the photographs.

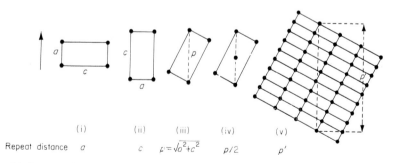

FIG. 21. Some typical situations in which lattice-repeat distances can give rise to layer lines on oscillation or rotation photographs. The repeat distance determined by the layer-line spacing is given below each of the figures (i) to (v).

[†] Bernal charts and other charts used in X-ray diffraction work can be purchased from the Institute of Physics and Physical Society, 47 Belgrave Square, London, W.1.

The setting of the crystal with each cell edge vertical in turn can yield a complete set of cell parameters with accuracy of about 1%. The angle $\beta$ of a monoclinic cell can be determined, albeit inaccurately, by measurement of the [101] or [10$\bar{1}$] diagonal and use of the relationship:

$$p^2 = a^2 + c^2 \pm 2ac \cos \beta.$$

A method of using rotation or oscillation photographs for determining cell parameters more accurately is described later (p. 319).

*1. Crystal Mounting*

The crystal (size usually of the order of 1 mm or less) may be attached to the end of a thin glass fibre by means of a small amount of adhesive (Seccotine, Durofix, shellac dissolved in alcohol) or Vaseline. The crystal should first be examined under a polarizing microscope in order to see whether it is simple or composite, twinned or untwinned, and also to determine if possible which morphological direction is parallel to the axis desired as rotation axis. This examination can sometimes be done conveniently in benzene as immersion medium, its rate of evaporation giving long enough to make observations and avoiding the difficulties of removing a less volatile liquid. The crystal is positioned conveniently on the microscope slide and the glass fibre held in tweezers (a little Plasticine in the tweezer tips prevents crushing the fibre). The very tip of the fibre is dipped in glue and then rapidly touched on the appropriate part of the crystal (Fig. 22). The fibre diameter should be small, and a suitable length is about 1 cm.

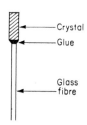

FIG. 22. Mounting of crystal on glass fibre for single-crystal X-ray diffraction.

*2. Crystal Orientation*

The crystal should be set on the arcs of the camera with its axis as nearly vertical as possible: the crystal morphology and the use of a polarized-light attachment may help in this preliminary setting. A photograph is then taken with the specimen oscillating through a 15° angle. This is unlikely to have exactly straight layer lines, and the manner and extent of their deviation from the horizontal can be used to apply corrections to the crystal orientation. A simple method of correction (Davies, 1950, 1961) is as follows. For the first photograph the arcs of the goniometer are set to make an angle of 45° with the X-ray beam. The displacements of the zero-layer line from the equator are measured at $2\theta = 90°$ on each side of the film, and these give the corrections to be applied to each of the arcs in the manner shown by the example in Fig. 23. The extent of each correction in degrees is given by dividing the displacement in mm by 0·71 (assuming camera diameter 5·73 cm).†

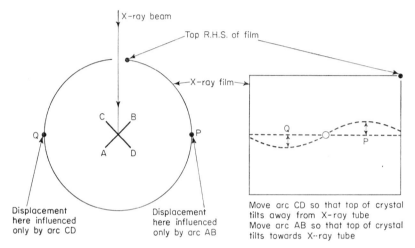

FIG. 23. Method of aligning crystal to give horizontal layer lines by measurements of zero-layer displacements.

### B. THE RECIPROCAL LATTICE

The methods described so far for relating X-ray diffraction patterns to the geometry and content of the unit cell of a mineral are adequate for dealing with certain types of problem, but they are unduly cumbersome for many others. The diffraction patterns have a reciprocal relationship to the Bravais lattice and therefore they can be more readily interpreted in terms of a lattice that is reciprocal to the Bravais lattice. The reciprocal relationship is seen immediately in the Bragg law $\lambda = 2d \sin \theta$, where smaller lattice spacings give reflections at larger Bragg angles. Similarly, in a rotation photograph widely spaced layer lines correspond to a small repeat distance in the unit cell and vice versa. The concept of the reciprocal lattice is described below.

† The methods for setting a crystal in a desired orientation on a precession camera are described by Buerger (1964).

A set of planes of spacing $d$ can be represented both in orientation and $a$ value by a row of points at regular intervals along a line normal to the planes. In Fig. 24, for example, the set of planes (100) is represented by the point A along the normal to the planes such that $OA = \dfrac{\lambda}{d_{100}}$. (The constant $\lambda$ is introduced here because of its subsequent usefulness.) The next point along the line, B, represents a set of planes with the same orientation but with half the spacing since $OB = \dfrac{2\lambda}{d} = \dfrac{\lambda}{d/2}$. Thus the points A and B can be given the indices 100 and 200 of the lattice planes which they represent.

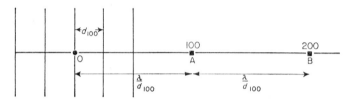

FIG. 24. Representation of a set of planes with spacing $d$ by a series of points O-A-B- etc. in reciprocal space. $\lambda$ is a constant, usually the X-ray wavelength.

### 1. Reciprocal Lattice for Orthorhombic Crystals

Suppose we consider only $hk0$ planes of an orthorhombic lattice. In Fig. 25 the rectangular cell OPQN has sides $a$ and $b$. Along the direction $x^*$, perpendicular to (100), are marked distances $\dfrac{\lambda}{a}$, $\dfrac{\lambda}{2a}$, etc. corresponding to the spacings 100, 200, etc., and similarly along $y^*$ perpendicular to (010) are marked a series of points corresponding to the spacings of $d_{010}$, $d_{020}$, etc. A network of lines is constructed parallel to $x^*$ and $y^*$ through the axial points already marked, giving intersections such as that labelled R. It can be shown that OR is perpendicular to PN, the lattice plane (110), and that $OR = \dfrac{\lambda}{d_{110}}$ so that the lattice plane (110) is represented by R in the same way that $0k0$ and $h00$ planes are represented by axial points. A general proof can show that a similar relation holds for all $hk0$ planes, and so the network of points can be labelled with the indices of the planes which they represent.

The array of points such as R form a lattice, which is called the reciprocal lattice of the two-dimensional Bravais lattice. The above construction can be extended to three dimensions. From a three-dimensional set of planes of a Bravais lattice one can derive a set of points forming a three-dimensional reciprocal lattice.

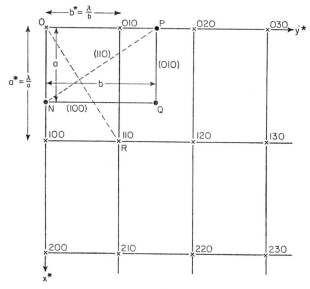

Fig. 25. Construction of a two-dimensional reciprocal lattice ($a^* \times b^*$) corresponding to the rectangular projection of an orthorhombic unit cell ($a \times b$). Each point $hk0$ of the reciprocal lattice represents the corresponding set of planes $hk0$ in the Bravais lattice.

The reciprocal lattice bears a very close relationship to an X-ray diffraction photograph. Thus, for example, a set of planes in a crystal gives a reflection at a point on the X-ray film and the planes are represented by a point in the reciprocal lattice; all reflections from planes $hk0$ lie on the equator ("zero layer") of a $c$ axis rotation photograph and all points $hk0$ lie on the "zero layer" of the reciprocal lattice; the points of the reciprocal lattice of a cell with $c$ vertical are on levels separated by $\lambda/c$, and the layer-line separation on a $c$ axis rotation photograph bears an inverse relationship to the $c$ parameter.

## 2. Reciprocal Lattice for Monoclinic Crystals

y *axis vertical.* In Fig. 26(a) is shown the construction of the zero layer of the reciprocal lattice, i.e. the representation of $h0l$ planes. The reciprocal-lattice axis $x^*$ is drawn perpendicular to (100) planes and the repeat distance, $a^*$, is marked off along it such that $a^* = \dfrac{\lambda}{d_{100}}$, i.e. $a^* = \dfrac{\lambda}{a \sin \beta}$. Similarly, $z^*$ is drawn perpendicular to (001) planes and $c^*$ is marked off, where $c^* = \dfrac{\lambda}{c \sin \beta}$. The network is completed by lines through axial points parallel to $x^*$ and $z^*$. The interaxial angle $x^*: z^*$ is called $\beta^*$, and $\beta^* = 180° - \beta$. In this orientation the three-dimensional reciprocal lattice will consist of planes of

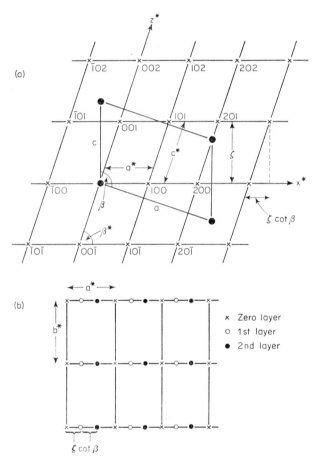

FIG. 26. (a) Construction of the zero layer of the reciprocal lattice for a monoclinic crystal
mounted with $b$ vertical ($\zeta$ is $\lambda/b$, and is not shown in the diagram for this orientation). This diagram also serves to illustrate a vertical section (in the plane $xz$) of the
reciprocal lattice when $c$ is vertical ($\zeta$ is $\lambda/c$, as shown in the diagram). (b) Horizontal
layers of the reciprocal lattice for a monoclinic crystal with $c$ vertical. Each layer is a
rectangular array, and successive layers are displaced by $\zeta \cot \beta$.

points superimposed vertically above the zero-layer points and the separa-

tion of these planes will be $b^* = \dfrac{\lambda}{b}$, since $y$ is perpendicular to the $xz$ plane.

The separation of horizontal layers of reciprocal lattice points is defined as $\zeta$

and in the present example $\zeta = \dfrac{\lambda}{b}$.

$z$ *axis vertical*. Figure 26(a) can also be used to show a vertical section of the reciprocal lattice which lies in the plane $xz$ (also $x^*z^*$). In this orientation:

$$\zeta = c^* \sin \beta = \frac{\lambda}{c}$$

(If $a$ is vertical $\zeta = \lambda/a$.)

Each horizontal layer of the reciprocal lattice consists of a rectangular network $a^* \times b^*$ (Fig. 26b), but it is clear from Fig. 26(a) that successive layers will be displaced in the $x^*$ direction by a distance $c^* \cos \beta$, i.e. $\zeta \cot \beta$.

If the angle $\beta = 90°$, then successive layers are superimposed without displacement.

### 3. Triclinic Crystals

When $\alpha \neq \beta \neq \gamma \neq 90°$, for any axis of rotation, all higher layers are displaced in both of the horizontal axial directions by appropriate amounts. Layers again, however, have a vertical separation of $\zeta = \frac{\lambda}{p}$ where $p$ is the repeat distance along the rotation axis in real space.

### 4. The Ewald (or Reflecting) Sphere

Graphical construction of a scale drawing of the reciprocal lattice is an aid to determining the geometrical relationships between the X-ray reflections from all lattice planes and the incident X-ray beam. It is necessary, however, to use in addition the concept of the Ewald (or reflecting) sphere. This is a sphere of unit radius (on the same scale as that used for the reciprocal-lattice construction), which passes through the origin of the reciprocal lattice. A two-dimensional illustration is given in Fig. 27, in which a set of vertical planes with spacing $d$ in a crystal is represented by the point P in the reciprocal lattice and the circle is the equator of the reflecting sphere. If P falls on the circle as shown, then:

$$OP = 2 \cdot OC \sin \theta = 2 \sin \theta, \text{ and } OP = \frac{\lambda}{d}$$

$$\therefore \frac{\lambda}{d} = 2 \sin \theta$$

Thus the set of planes makes an angle $\theta$ with CO for which the Bragg Law is obeyed, i.e. the incident beam is represented by CO and the Bragg reflection from the planes represented by P is given by the direction CP.

A similar proof holds for the three-dimensional case, and the general statement may be made that only if a reciprocal-lattice point falls on the reflecting sphere are the reflection conditions for its set of planes satisfied; the converse of this is also true.

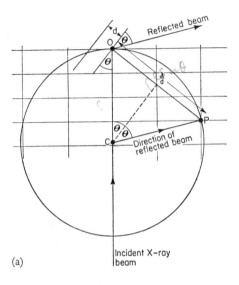

(a)

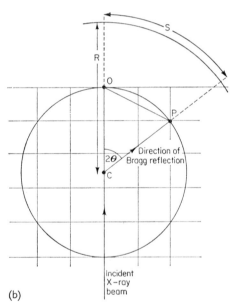

(b)

FIG. 27. (a) Reflecting circle (radius unity), superimposed on a reciprocal lattice. Reflection can occur from a set of planes if their reciprocal lattice point lies on the reflecting circle. (b) The geometry of the reflecting condition for a reflection on the zero layer of a rotation photograph. Radius of camera = R; distance of reflection spot from meridian = S.

It should be noted that if λ is changed, the point P moves along OP and thus moves off the reflecting sphere. Similarly if θ changes, i.e. the crystal is moved with respect to the X-ray beam (the reciprocal lattice goes with it), then P moves off the reflecting sphere. (This latter contingency can be expressed equally well by moving the reflecting sphere and keeping the reciprocal lattice stationary.)

The above concepts enable one to visualize, and to analyse graphically in some cases, the directions of reflections from specific crystal planes when the crystal takes up any orientation with respect to the X-ray beam. As the crystal is moved in the X-ray beam, various reciprocal-lattice points in succession cut the reflecting sphere, the corresponding reflections emanate from the crystal, and the pattern in which they are recorded depends upon the position and configuration of the X-ray film at the time each reflection occurs.

It is possible, by a suitably coupled motion of crystal and flat film, to record a diffraction pattern that is an undistorted picture of a section of the reciprocal lattice. This is done by a precession camera, the principle of which is illustrated in Fig. 28. An advantage of this method is that the reflections can be indexed by inspection, since they occur in regular sequences along straight lines as do the rows of the reciprocal lattice points (see Buerger, 1964).

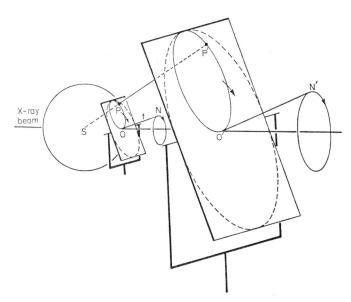

Fig. 28. Illustration of the principle of the precession camera. The crystal is at O and the centre of the film at O'. P is a point in a central section of the reciprocal lattice the normal to which (ON) is made to precess about the direction of the X-ray beam. A similar coupled precession motion of the film ensures an undisturbed picture of the parts of the reciprocal lattice intersected by the reflecting sphere (centre S). (After Buerger, 1964.)

A more simple mechanism is employed in the Weissenberg camera (crystal rotated, cylindrical film moved parallel to cylinder axis), and in the resulting diffraction pattern, sequences along reciprocal-lattice rows are recorded along recognizable curves and are therefore not difficult to identify and index. (See Henry *et al.*, 1960.) In both the precession and Weissenberg methods, reflections corresponding to a single selected layer of the reciprocal lattice are isolated by means of a suitable screen, and recorded on the film.

A still more simple mechanism is that of a rotation (or oscillation) camera (rotating crystal, stationary cylindrical film), and with this the relation between the diffraction pattern on the film and the reciprocal lattice is more complex. This latter method is, however, an extremely useful one for obtaining preliminary information about the nature and state of a crystal so that further details will be given here about its application.

*5. Formation of a Rotation Photograph*

As the crystal is rotated about a vertical cell edge the reciprocal-lattice points pass through the reflecting sphere and intersect it in a series of horizontal circles. The reflections therefore lie on the surfaces of a set of cones (see Fig. 29), and so they intersect a cylindrical film at points along horizontal layer lines (see p. 304).

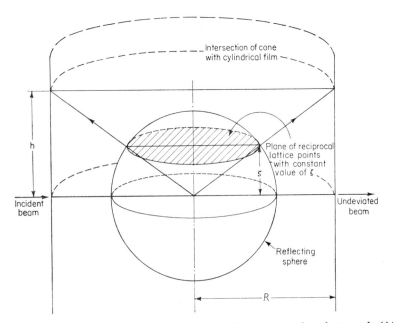

Fɪɢ. 29. Diagram illustrating the formation of layer lines in a rotation photograph. (After Henry *et al.*, 1960.)

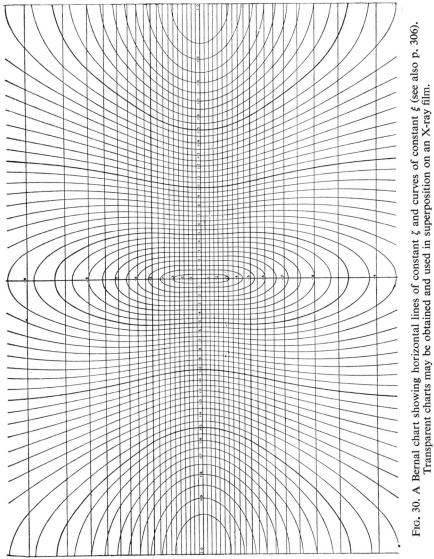

FIG. 30. A Bernal chart showing horizontal lines of constant $\zeta$ and curves of constant $\xi$ (see also p. 306). Transparent charts may be obtained and used in superposition on an X-ray film.

*Indexing reflections on the zero layer.* In Fig. 27(a), OP $= 2 \sin \theta = \xi$. The symbol $\xi$ is used to denote the perpendicular distance of a reciprocal lattice point to the vertical axis through O.

The geometry of the reflecting condition is shown in Fig. 27(b), where it is seen that $S = 2\theta R$, where $S$ is the distance of a reflection from the meridian of the film and $R$ is the camera radius.

N*

Thus, from a measured value of *S*, *θ* and hence *ξ* can be determined for the reflection concerned.

*Indexing reflections on higher layer lines.* The geometry is now such that a reflection with the same *ξ* value as one on the zero layer may lie at a different distance from the meridian of the film. The Bernal chart (Fig. 30) mentioned on p. 307 is constructed with curves of constant *ξ* (known as "row lines") as well as lines of constant *ζ*, so that by superposition of the chart on the film, values of *ξ* and *ζ* can be read off for reflections on any layer line. Thus, the process of indexing in principle is: from a rotation film to measure *ζ* and *ξ* for each reflection; *ζ* indicates on which layer of the reciprocal lattice the reflection occurs, and *ξ* indicates the distance of the reciprocal lattice point from the rotation axis.

However, for both zero and higher layers it is very likely that two or more reciprocal-lattice points will have the same or nearly the same *ξ* value, and this will lead to ambiguities of indexing. This difficulty can be overcome by use of an oscillation instead of a rotation photograph.

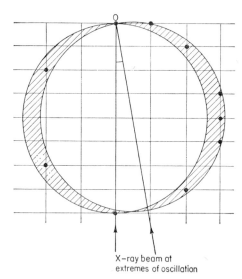

X–ray beam at
extremes of oscillation

FIG. 31. Illustration of the principle of the oscillation photograph. Heavy points of the reciprocal lattice represent reflections that can occur during oscillation of the crystal through a small angle.

*6. Indexing an Oscillation Photograph*

A graphical method is often employed for indexing reflections on an oscillation photograph. For equatorial reflections, the zero layer of the reciprocal lattice is drawn to scale, and the reflecting circle (diameter equal to that of the reflecting sphere) is drawn on a transparent sheet and superimposed with its circumference passing over the origin, where a pin is inserted through both sheets. Oscillation of the crystal through a small angle may be simulated by oscillating the overlay through the same angle. A second reflecting circle can be drawn 15° away from the first circle, and the lune-shaped areas between the two circles then represents the area of reciprocal space swept out during an oscillation (Fig. 31). In practice an oscillation angle of 15° is usually sufficiently small to allow unambiguous indexing. If the lunes are orientated correctly with respect to the reciprocal lattice, only those reciprocal-lattice points within the lunes can represent planes that give reflections on the oscillation photograph.

For other than zero-layer lines, the lunes will be formed by circles of smaller radius $r$ where $r = \sqrt{1 - \zeta^2}$ (see Fig. 32). For orthogonal cases, the upper layers are dealt with by pinning at O as for the zero layer (the reflecting circles will not then pass through O). For other cases (e.g. monoclinic, $c$ or $a$ axis vertical), O should be displaced by the appropriate distance (see p. 312).

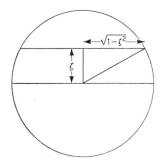

FIG. 32. Vertical section through the reflecting sphere, showing the radius of the reflecting circle for upper-layer lines.

## 7. Accurate Determination of Cell Parameters With Oscillation Photographs

A single example will serve to illustrate the principle of the method. An orthorhombic crystal is set with $c$ vertical and an oscillation photograph is taken with the beam parallel to $a$ in the middle of the oscillation. The reciprocal-lattice construction shows that pairs of $hk0$ and $h\bar{k}0$ high-angle reflections will occur on the film. The distance between these can be measured along the equator and can be used as in powder photography to give accurate values of $a$ and $b$ by using extrapolation or least-squares procedures. For a less symmetrical situation, reflections can be produced at the same Bragg angle on both sides of the film by a suitably arranged double exposure with the crystal in two different orientations. For a monoclinic crystal with $c$ vertical, values can be found for $b^*$ and $a^*$, and with $b$ vertical for $a^*$, $c^*$ and $\beta^*$. The above method was described by Weisz et al. (1948).

Other single-crystal methods for the precise determination of cell parameters are given by Christ (1956), Bond (1960), Herbstein (1963) and Main
and Woolfson (1963).

### C. USES OF OSCILLATION AND ROTATION PHOTOGRAPHS

*1. Oscillation Photographs*

(a) For indexing reflections. (b) For studying symmetry. If the equator of
the photograph is a mirror line, either the oscillation axis is 2-, 4-, or 6-fold,
or there is a horizontal mirror plane.† (In a rotation photograph the equator
and meridian are *always* mirror lines.) A study of symmetry can also be made
by arranging that the X-ray beam goes accurately along a particular axis, or
by taking a series of photographs with the crystal set at regular intervals
(60°, 90°) about a vertical symmetry axis. (c) For a general survey of a crystal
to show whether composite or single, and whether twinned or not. (d) To
determine the orientation of the crystal lattice with respect to morphology.
(e) To obtain accurate cell parameters from high-angle reflections. (f) For
the determination of possible space groups (moving-film methods are better
for this purpose).

*2. Rotation Photographs*

(a) For indexing, if the cell is very simple. (b) For comparison of two crystals suspected to be of the same compound. Oscillation photographs of two
crystals of the same substance will differ if their orientations about the camera
axis differ, but rotation photographs about the same axis will be identical.
(c) For determination of the vertical repeat distance by using layer-line separations. The layer lines can sometimes be better defined through the presence
of more reflections, but the crystal must be aligned very accurately.

## X. Examples of Applications of X-ray Diffraction
## Methods to the Study of Minerals

Below are listed in alphabetical order, the principal minerals or mineral
groups to which X-ray diffraction determinative methods have been applied.

### ALUMINOSILICATES

Powder photographs have been used for the quantitative estimation of the
aluminosilicates mullite, kyanite, andalusite and sillimanite (Johnson and
Andrews, 1962).

† The presence of these symmetry elements will result in a symmetrical relationship
between reflections with respect to both position on the film and intensity. Geometrical
symmetry across the equator without symmetry of intensity can result in the case of some
trigonal crystals with triad vertical.

## AMPHIBOLES

Determinations of the cell parameters of amphiboles from powder and single-crystal X-ray data have been correlated with composition (amphiboles in general: Whittaker, 1960; ferrohastingsites: Borley and Frost, 1963; arfvedsonites and ferrohastingsites: Frost, 1963; cummingtonite–grunerite series: Klein, 1964).

## ANALCITE

The unit cell edge and $2\theta_{639}$ have been related to the compositions of analcites in the nepheline–albite compositional range (Wilkinson, 1963).

## ARAGONITE

Aragonite is often partially inverted to calcite in fossil shells. The proportion of aragonite in a mixture of aragonite and calcite can be determined by X-ray diffraction methods, and is a comparatively simple problem since the absorption coefficient of the sample does not depend on the proportions of the two components (see p. 300). (Lowenstam, 1954; Davies and Hooper, 1963).

## ARSENOPYRITE

The determination of the arsenic: sulphur ratio from the cell constants was described by Morimoto and Clark (1961).

## CALCITE

The substitution of magnesium for calcium in calcite reduces the cell parameters, and the variation in cell parameters has been studied by Goldsmith *et al.* (1955) and by Waite (1963). The former authors also studied the variation of lattice parameters with manganese content.

The percentage of calcite in a calcite–dolomite rock has been determined by comparing the relative intensities of the strongest X-ray powder reflections from the two minerals (Tennant and Berger, 1957; Gulbrandsen, 1960).

## CHLORITE

In the chlorites the basal spacing $d_{001}$ appears to be influenced principally by substitution of aluminium for silicon, and the $d_{060}$ value is a useful guide to iron content. (See, for example, Hey, 1954; Brindley and Gillery, 1956.)

For chlorites also, the intensity relationships among basal reflections have been used determinatively (Brown, 1955; Brindley and Gillery, 1956; Lapham, 1958; Drits, 1962; Schoen, 1962; Petruk, 1964).

## DOLOMITE (see calcite)

### FELDSPARS

### 1. Alkali Feldspars

The potassium content of an alkali feldspar can be determined by measurement of the $\bar{2}01$ spacing (Bowen and Tuttle, 1950; Orville, 1958). If the feldspar is unmixed into sodium and potassium-rich regions, the otherwise single $\bar{2}01$ peak is generally split into two, although absence of a split peak is not conclusive evidence of homogeneity. The $d$ values of the separate $\bar{2}01$ peaks cannot always reliably be used to estimate the composition of the separate phases, but a perthite can be homogenized by heating to 1050°C, and the $\bar{2}01$ spacing will then give the bulk composition. (See also Kuellmer, 1959.)

It appears that a moderate calcium content does not affect the potassium determination by the above method (Carmichael and MacKenzie, 1964), but higher calcium content does (Kempe, 1966). Carmichael and MacKenzie used $d_{\bar{2}01}$ together with $\alpha^*$ to determine ternary composition.

Another informative method of examining perthitic feldspars is by means of their single-crystal X-ray patterns (see, for example, Smith and MacKenzie, 1955).

An X-ray powder method for estimating potassium and barium feldspars was given by Gay (1965).

### 2. Plagioclase Feldspars

X-ray determination of the plagioclases is difficult for two reasons. First that the sodium and calcium ions are not very different in size (it would seem also that the accompanying Si,Al substitution has rather little effect on the cell parameters). Any determinative curves will therefore have a rather small slope, and so the accuracy obtainable by their use will not be high. The second difficulty is that the determinative curves differ for low- and high-temperature plagioclases (plagioclases occur also in intermediate temperature states). For low-temperature plagioclases in the range $An_0$–$An_{30}$ the parameter $2\theta$ $(\bar{1}31) - 2\theta\,(131)$ gives good correlation with composition. (Smith and Yoder, 1956; Smith and Gay, 1958; Bambauer et al., 1965.) Synthetic plagioclases in this range give quite different values. The distinction between low- and high-temperature forms is, however, less marked for compositions more calcic than $An_{30}$.

## FeO – TiO$_2$ – Fe$_2$O$_3$ SYSTEM

The variations in cell parameters in the various solid-solution series within this system have been investigated by Akimoto (1957), Vincent *et al.* (1957), and Wright (1959).

### GARNETS

Formulae have been given relating cell sizes to known substitutions in garnets (Gnevushev *et al.*, 1956; Mikheev, 1957; Bertaut and Forrat, 1957), but because of the number of variables involved, these cannot be used in reverse to determine composition from cell edge. For determining a garnet composition, use can be made of the measured cell edge together with other properties such as density and refractive index. (See, for example, Sriramadas, 1957; Henriques, 1958; Winchell, 1958.)

For the determination of the cell edge of a garnet use can be made of several resolved $\alpha_1$ and $\alpha_2$ high-angle strong reflections on powder photographs, or else diffractometry may be used to record lower-angle indexed peaks, making use of an internal standard. Some useful high-angle reflections are those with $N = 116$ (indices 10,4,0 or 8,6,4); $N = 120$ (10,4,2); $N = 128$ (880); $N = 152$ (12,2,2 and 10,6,4); $N = 216$ (14,4,2). Copper radiation is suitable except for the most iron-rich garnets, for which Mo, Co or Fe radiation can be used.

HEMATITE (see FeO – TiO$_2$ – Fe$_2$O$_3$ system)

ILMENITE (see FeO – TiO$_2$ – Fe$_2$O$_3$ system)

### MELILITE GROUP

The variation of lattice parameters for the series åkermanite (tetragonal *a* 7·84, *c* 5·01 Å) – gehlenite (tetragonal *a* 7·69, *c* 5·08 Å), has been plotted by Neuvonen (1952). The influence of ferric iron and sodium on cell parameters is also discussed.

### MICAS

The powder patterns of di- and tri-octahedral micas can generally be distinguished by means of the 060 reflection, which for di-octahedral micas is close to 1·50 Å and for tri-octahedral micas is between 1·53 and 1·55 Å. In addition, the basal reflection at $d \sim 5$ Å is strong for most di-octahedral, and weak for most tri-octahedral micas (exceptions are lepidolite and glauconite).

Sodium and calcium micas can be distinguished from potassium micas by having smaller basal spacings.

In the study of micas and of other layered silicates, an important problem is the distinction between different polymorphs. The powder patterns of $1M$, $2M$ and $3T$ polymorphs of di-octahedral micas are in general distinguishable from each other (Yoder and Eugster, 1955), but for the tri-octahedral micas the powder patterns of $1M$ and $3T$ polymorphs are identical. Weissenberg photographs show differences in this case, but even these are rather subtle. The part of the powder pattern which is most useful for distinguishing poly-morphs is that for spacings between 4·4 and 2·6 Å. $1M$ and $1Md$ (disordered micas) are distinguished by the absence of $hkl$ reflections from the latter.

Various relationships between the cell parameters and chemical compo-sition have been published, but they usually depend upon so many variables as to be of limited use for determinative purposes. Furthermore, recent work on the structures of micas has cast doubt on the validity of some of these relationships. (Radoslovich, 1962.)

For the phlogopite–biotite series of micas, a way of determining the iron : magnesium ratio is by measurement of the relative intensities of certain basal reflections (Berkhin, 1954; Gower, 1957; Engel and Engel, 1960; Schiaffino, 1962).

## NEPHELINE

In the nephelines the principal solid-solution substitution is one of Na $\rightleftharpoons$ K. This replacement affects the cell parameters and can be followed by variations in the $2\theta$ values of appropriate reflections (Smith and Sahama, 1954; Hamilton and MacKenzie, 1960).

The substitution of calcium for sodium accompanied by aluminium for silicon is associated with a linear increase in the cell volume (Miyashiro and Miyashiro, 1954; Donnay et al., 1959). At the kalsilite end of the Na–K range, the hexagonal cell dimensions are $a$ 5·1597, $c$ 8·7032 Å for $Ne_0$ and $a$ 5·1485, $c$ 8·6428 Å for $Ne_{10}$ (Smith and Tuttle, 1957): the separation of reflections $10\bar{1}2$ and $10\bar{1}1$ on a powder pattern can be used to estimate composition (Sahama et al., 1956).

## OLIVINES

An estimate of the magnesium : iron ratio of an olivine can be made by measurement of the $d$ value of the 130 reflection (Yoder and Sahama, 1957; but see also Smith and Stenstrom, 1965). An alternative method uses the reflection 062 and is recommended particularly for olivines in the range $Fo_{80}$–$Fo_{90}$ (Jackson, 1960). Although the accuracy claimed for this method is greater than for that in which 130 is used, there are the disadvantages that the 062 is a weaker reflection and also that there is a greater chance of interference from peaks from associated minerals. For both of these reasons, therefore, the former method is more likely to be useful in dealing with unseparated olivines. Other studies concerned with the determination of olivine by X-ray diffraction are: Eliseev (1957); Sahama and Hytönen (1958); Jambor and Smith (1964).

## PYROXENES

### 1. Clinopyroxenes

The most common clinopyroxenes are those that lie in the field diopside–hedenbergite–ferrosilite–clinoenstatite. The variation of cell parameters over this field was studied by Brown (1960). See also Viswanathan (1966). The effect of the substitution of aluminium for magnesium and silicon on cell parameters was discussed by Sakata (1957).

The cell parameters of clinopyroxenes can be obtained from indexed powder patterns; $a \sin \beta$ can be deduced more directly than $a$ and $\sin \beta$ separately, and it can be a useful parameter for determinative purposes. Linear regression graphs of cell parameters against composition have been calculated by Winchell and Tilling (1960) for clinopyroxenes, which may be of use in estimating approximate compositions that could produce a given set of lattice parameters.

Determinative curves for synthetic clinopyroxenes from $Di_{95}En_5$ to $Di_{40}En_{60}$ with $2\theta$ $(31\bar{1})$ and for $Di_{40}En_{60}$ to $En_{100}$ with another identifiable reflection have been given by Schairer and Boyd (1957).

X-ray data for the acmite–diopside system were given by Nolan and Edgar (1963).

### 2. Orthopyroxenes

The effect of magnesium–iron replacement on cell parameters and also the influence of calcium and aluminium content have been studied by Hess (1952), Kuno (1954) and Howie (1962).

The cell parameters of orthopyroxenes can be derived from single-crystal photographs but unless special techniques are used in these, more accurate values can be obtained from powder patterns by using a silicon or quartz internal standard. Some prominent reflections that can be used for this purpose are 12,0,0; 060; 14,5,0; 650; 250; 610; 420; 202; 502; 521; 531; 11,3,1. Single reflections (e.g. 12,0,0, 060), and selected $hkl$ reflections, may be used to determine $a$, $b$ and $c$, and the values obtained can be checked or re-determined by means of observations on other reflections or groups of reflections. Alternatively, all reflections that can be reliably indexed and accurately measured can be used to give a solution by the method of least squares (see p. 292).

Powder and single-crystal X-ray techniques have been used to study ordered and disordered enstatite (meteoritic and terrestrial) by Pollack and Ruble (1964).

## PYRRHOTINE

Pyrrhotines often have the non-stoicheiometric composition $Fe_{1-x}S$ with $x$ varying from 0 to 0·125. The value of $x$ can be correlated with the value of the lattice spacing $d_{102}$ (Arnold and Reichen, 1962). See however, Buseck (1964), Groves and Ford (1963, 1964) and Arnold (1966).

## SCAPOLITE

Scapolites are tetragonal, and range in composition between marialite $Na_4Al_3Si_9O_{24}Cl$ and meionite $Ca_4Al_6Si_6O_{24}CO_3$, the cell dimensions of which are $a$ 12·075, $c$ 7·516 Å and $a$ 12·13, $c$ 7·69 Å, respectively, for the synthetic end members (Eugster and Prostka, 1960). A high content of sulphate or potassium ions can, however, give values outside the above range. Burley *et al.* (1961) have shown the variation of the separation $2\theta_{400}-2\theta_{112}$ with composition.

## SERPENTINES

The serpentine minerals are usually very fine grained and the three principal polymorphs are often difficult to identify by optical or other methods. By means of X-ray powder-diffraction patterns, though these are often poor because of disordered structures, a specimen can usually be referred to either the chrysotile, lizardite or antigorite serpentine sub-groups. (Whittaker and Zussman, 1956.)

## SPHALERITE

Substitution of iron for zinc in the structure of sphalerite results in a linear increase in the length of the cubic unit cell edge. (Skinner *et al.*, 1959; Skinner, 1961; Skinner and Bethke, 1961.)

Natural specimens of sphalerite usually contain mixtures of cubic and hexagonal (wurtzite type) stacking sequences. Short and Steward (1959) have described an X-ray method for determining the proportions of each.

## SPINELS

There are too many possible variations in the compositions of spinels for the size of the cubic unit cell to be used determinatively. Combined knowledge, however, of cell size, density and refractive index may be useful (see

Deer *et al.*, 1962, volume 5, p. 61). An X-ray method for determining spinels in the system $MgAl_2O_4 - MgCr_2O_4 - MgFe_2O_4$ has been described by Allan (1966).

## ZIRCON

Zircon is tetragonal with $a \sim 6.60$, $c \sim 5.98$ Å. The cell parameters are changed by metamictization (Holland and Gottfried, 1955; Ueda, 1956). See also Ueda (1957) and Lima de Faria (1964) for data on other metamict minerals.

### BIBLIOGRAPHY

(Full details are given in the list of references)

*Theory and Methods*
  *General.* Azaroff and Buerger (1958). Buerger (1942). Guinier (1964). Henry, Lipson and Wooster (1960). Klug and Alexander (1954). Lonsdale (1952, 1959, 1962). Mueller *et al.* (1960, etc.). Peiser, Rooksby and Wilson (1955).
  *Special.* Brown (1961) – Clay minerals. Buerger (1964) – Precession camera. Parrish (1965) – Diffractometry.
*Mineral Data*
  A.S.T.M. Powder Data File and Index. Berry and Thompson (1962) – Ore minerals. Bragg and Claringbull (1965) – Crystal structures. Brown (1961) – Clay minerals. Chukhrov and Bonstedt-Kupletskaya (1960–65) – General. Deer *et al.* (1962, 1963) – Rock-forming minerals. Donnay and Donnay (1963) – Crystal data. Frondel (1962) – General. Heller and Taylor (1956) – Calcium silicates. Lima de Faria (1964) – Metamict minerals. Mikheev (1957) – X-ray determinative tables. Acad. Sci. U.S.S.R. publication. Palache, Berman and Frondel (1944, 1951) – General. Strunz (1966) – Chemical and crystallographic data. Swanson *et al.* (1953–60) – Standard powder patterns.
*Handbooks, Tables, Etc.*
  Croft (1965), $\sin^2\theta - 2\theta$ table. Lonsdale (1952, 1959, 1962). Mirkin (1964). National Bureau of Standards (1950), $d - \theta$ tables. Parrish and Mack (1963), $d - \theta$ charts.

### REFERENCES

Akimoto, S. (1957). Magnetic properties of ferromagnetic oxide minerals as a basis of rock magnetism. *Adv. Phys.* **6**, 288.
Allan, W. C. (1966). An X-ray method for defining composition of a magnesium spinel. *Am. Miner.* **51**, 239.
Andrews, K. W. (1951). The determination of interplanar spacings and cell dimensions from powder photographs using an internal standard. *Acta cryst.* **4**, 562.

Arnold, R. G. (1966). Mixtures of hexagonal and monoclinic pyrrhotite and the measurement of the metal content of pyrrhotite by X-ray diffraction. *Am. Miner.* **51**, 1221.

Arnold, R. G. and Reichen, L. E. (1962). Measurement of the metal content of naturally occurring, metal-deficient, hexagonal pyrrhotite by an X-ray spacing method. *Am. Miner.* **47**, 105.

Aruja, E. (1961). A new collimator and beam trap for the Philips 114.83 mm. diameter X-ray diffraction powder camera. *Clay Miner. Bull.* **4**, 307.

A.S.T.M. Index to the X-ray powder data file. A.S.T.M. Diffraction Data Sales Dept., 1916 Race Street, Philadelphia, Penna. 19103, U.S.A.

Azaroff, L. V. and Buerger, M. J. (1958). "The Powder Method in X-ray Crystallography." McGraw-Hill, New York.

Bambauer, H. U., Corlett, M., Eberhard, E., Gubser, R., Laves, F., Nissen, H. U. and Viswanathan, K. (1965). Variations in X-ray powder patterns of low structural state plagioclases. *Schweiz. miner. petrogr. Mitt.* **45**, 327.

Bergmann, M. E. (1959). X-ray micro-diffraction cameras. *Norelco Reptr* **6**, 96.

Berkhin, S. I. (1954). Rontgenograms of iron-magnesia micas. *Dokl. Akad. Nauk S.S.S.R.* **95**, 145.

Berry, L. G. and Thompson, R. M. (1962). "X-ray Powder Data for Ore Minerals: the Peacock Atlas." Geological Society of America.

Bertaut, F. and Forrat, F. (1957). Étude des paramètres des grenats. *C. r. hebd Séanc. Acad. Sci., Paris* **244**, 96.

Bloxam, T. W. (1963). X-ray diffraction using rock chips. *Mineralog. Mag.* **33**, 619.

Bond, W. L. (1960). Precision lattice constant determination. *Acta cryst.* **13**, 814.

Borley, G. and Frost, M. T. (1963). Some observations on igneous ferrohastingsites. *Mineralog. Mag.* **33**, 646.

Bowen, N. L. and Tuttle, O. F. (1950). The system $NaAlSi_3O_8 - KAlSi_3O_8 - H_2O$. *J. Geol.* **58**, 489.

Bragg, W. L. and Claringbull, G. F. (1965). "Crystal Structures of Minerals." Bell, London.

Brindley, G. W. (1955). *In* "Glancing-angle and flat-layer techniques in X-ray Diffraction by Polycrystalline Materials." Ed. H. S. Peiser, H. P. Rooksby and A. J. C. Wilson. Institute of Physics, London.

Brindley, G. W. (1961). *In* "The X-ray Identification and Crystal Structures of Clay Minerals." Ed. G. Brown. Mineralogical Society (Clay Minerals Group), London.

Brindley, G. W. and Gillery, F. H. (1956). X-ray identification of chlorite species. *Am. Miner.* **41**, 169.

Brindley, G. W. and Kurtossy, S. S. (1961). Quantitative determination of kaolinite by X-ray diffraction. *Am. Miner.* **46**, 1205.

Brindley, G. W. and Kurtossy, S. S. (1962). Quantitative determination of kaolinite by X-ray diffraction. A reply to H. W. van Der Marel. *Am. Miner.* **47**, 1213.

Brown, G. (1955). The effect of isomorphous substitutions on the intensities of 00*l* reflections of mica- and chlorite-type substitutions. *Mineralog. Mag.* **30**, 657.

Brown, G. (Ed.) (1961). "The X-ray Identification and Crystal Structures of Clay Minerals." Mineralogical Society (Clay Minerals Group), London.

Brown, G., Dibley, G. C. and Farrow, R. (1956). An extrusion method for bonded powder specimens. *Clay Miner. Bull.* **3**, 19.

Brown, G. M. (1960). The effect of ion substitution on the unit cell dimensions of the common clinopyroxenes. *Am. Miner.* **45**, 15.

Buerger, M. J. (1942). "X-ray Crystallography." Wiley, New York.

Buerger, M. J. (1964). "The Precession Method in X-ray Crystallography." Wiley, New York.

Burley, B. J., Freeman, E. B. and Shaw, D. M. (1961). Studies on scapolite. *Can. Mineralogist* **6**, 670.

Buseck, P. R. (1964). Discussion of "Pyrrhotite measurement" by Groves and Ford. *Am. Miner.* **48**, 911; **49**, 1491.

Carmichael, I. S. E. and MacKenzie, W. S. (1964). The lattice parameters of high-temperature triclinic sodic feldspars. *Mineralog. Mag.* **33**, 949.

Carrigy, M. A. and Mellon, G. B. (1964). Authigenic clay mineral cements in Cretaceous and Tertiary sandstones of Alberta. *J. sedim. Petrol.* **34**, 461.

Chayes, F. and MacKenzie, W. S. (1957). Experimental error in determining certain peak locations and distances between peaks in X-ray (powder) diffractometer patterns. *Am. Miner.* **42**, 534.

Christ, C. L. (1956). Precision determination of lattice constants of single crystals using the Weissenberg camera. *Am. Miner.* **41**, 569.

Chukhrov, F. V. and Bonstedt-Kupletskaya, E. M. (1960, 1963, 1965). "Minerals", Vols. 1, 2.1. 2.2. I.G.E.M. Acad-Nauk U.S.S.R.

Cole, W. F. (1961). Modifications to standard Philips powder cameras for clay-mineral work. *Clay Miner. Bull.* **4**, 312.

Croft, W. J. (1965). "Tables of $\sin^2 \theta$ vs $2\theta$ in Intervals of $0.01°2\theta$." Sperry Rand Corp. Sudbury, Mass.

Davies, P. T. (1950). The setting of single crystals from zero layer curve photographs. *J. scient. Instrum.* **27**, 338.

Davies, P. T. (1961). The setting of single crystals. *Acta cryst.* **14**, 1295.

Davies, T. T. and Hooper, P. R. (1963). The determination of the calcite:aragonite ratio in mollusc shells by X-ray diffraction. *Mineralog. Mag.* **33**, 608.

Deer, W. A., Howie, R. A. and Zussman, J. (1962, 1963). "Rock Forming Minerals." Vols. 1 to 5. Longmans Green, London.

Donnay, G., Schairer, J. F. and Donnay, J. D. H. (1959). Nepheline solid solution. *Mineralog. Mag.* **32**, 93.

Donnay, J. D. H. and Donnay, G. (1963). Crystal data: part II. determinative tables. ACA Monogr. **5**.

Drits, V. A. (1962). The quantitative X-ray phase analysis of clay minerals. *Soviet Phys. Crystallogr.* **6**, 423. (*Kristallografiya*, 1961, **6**, 530.)

Eliseev, E. N. (1957). X-ray study of the minerals of the isomorphous series forsterite–fayalite. *Mém. Soc. Russe Min.* **86**, 657.

Engel, A. E. J. and Engel, C. G. (1960). Progressive metamorphism and granitization of the Major paragneiss, northwest Adirondack Mountains, New York. Part II. Mineralogy. *Bull. geol. Soc. Am.* **71**, 1.

Eugster, H. P. and Prostka, H. J. (1960). Synthetic scapolites. *Bull. geol. Soc. Am.* **71**, 1859 (abst).

Frondel, C. (1962). "Dana's System of Mineralogy." 7th ed., Vol. 3. Wiley, New York.

Frost, M. T. (1963). Amphiboles from Younger Granites of Nigeria. Part II. X-ray Data. *Mineralog. Mag.* **33**, 377.

Gay, P. (1965). An X-ray powder method for the estimation of (K,Ba) feldspars. *Mineralog. Mag.* **34**, 204.

Gnevushev, M. A., Kalinin, A. I., Mikheev, V. I. and Smirnov, G. I. (1956). Changes in the cell dimensions of the garnets as a function of their chemical composition. *Mém. Soc. Russe Min.* **85**, 473.

Goldsmith, J. R., Graf, D. L. and Joensuu, O. I. (1955). The occurrence of magnesian calcites in nature. *Geochim. cosmochim. Acta* **7**, 212.

Gower, J. A. (1957). X-ray measurement of the iron–magnesium ratio in biotites. *Am. J. Sci.* **255**, 142.

Graf, R. B., Wahl, F. M. and Grim, R. E. (1963). Phase transformations in silica-alumina–magnesia mixtures as examined by continuous X-ray diffraction. II. Spinel-silica compositions. *Am. Miner.* **48**, 150.

Groves, D. I. and Ford, R. J. (1963, 1964). Note on the measurement of pyrrhotite composition in the presence of both hexagonal and monoclinic phases. *Am. Miner.* **48**, 911: **49**, 1496.

Gude, A. J. and Hathaway, J. C. (1961). A diffractometer mount for small samples. *Am. Miner.* **46**, 993.

Guinier, A. (1964). Théorie et Technique de la Radiocristallographie. Dunod, Paris.

Gulbrandsen, R. A. (1960). Petrology of the Meade Peak phosphatic shale member of the Phosphoria formation at Coal Canyon, Wyoming; and a method of X-ray analysis for determining the ratio of calcite to dolomite in mineral mixtures. *Bull. U.S. geol. Surv.* 1111–C: D, 71.

Hamilton, D. L. and MacKenzie, W. S. (1960). Nepheline solid solution in the system $NaAlSiO_4$–$KAlSiO_4$–$SiO_2$. *J. Petrology.* **1**, 56.

Heller, L. and Taylor, H. F. W. (1956). "Crystallographic Data for the Calcium Silicates." H.M.S.O., London.

Henriques, Å. (1958). On the determination of garnet without chemical analyses. *Ark. Miner. Geol.* **2**, 349.

Henry, N. F. M., Lipson, H. and Wooster, W. A. (1960). "The Interpretation of X-ray Diffraction Photographs," 2nd Edn. Macmillan, New York.

Herbstein, F. H. (1963). Accurate determination of cell dimensions from single-crystal X-ray photographs. *Acta cryst.* **16**, 255.

Hess, H. H. (1952). Orthopyroxenes of the Bushveld type, ion substitution and changes in unit cell dimensions. *Am. J. Sci.*, Bowen vol. 173.

Hey, M. H. (1939). On the presentation of chemical analyses of minerals. *Mineralog. Mag.* **25**, 402.

Hey, M. H. (1954). A new review of the chlorites. *Mineralog. Mag.* **30**, 277.

Hildebrand, F. A. (1953). Minimizing effects of preferred orientation in X-ray diffraction powder patterns. *Am. Miner.* **38**, 1050.

Holland, H. D. and Gottfried, D. (1955). The effect of nuclear radiation on the structure of zircon. *Acta cryst.* **8**, 291.

Howie, R. A. (1962). Some orthopyroxenes from Scottish metamorphic rocks. *Mineralog. Mag.* **33**, 903.

Ito, T. (1949). A general powder X-ray photography. *Nature, Lond.* **164**, 755.

Ito, T. (1950). "X-ray Studies on Polymorphism." Maruzen, Tokyo.

Jackson, E. D. (1960). X-ray determinative curve for natural olivines of composition $Fo_{80-90}$. *Prof. Pap. U.S. geol. Surv.* 400B, 432.

Jambor, J. L. and Smith, C. H. (1964). Olivine composition determination with small-diameter X-ray powder cameras. *Mineralog. Mag.* **33**, 730.

Johnson, W. and Andrews, K. W. (1962). Quantitative X-ray examination of aluminosilicates. *Trans. Br. Ceram. Soc.* **61**, 724.

Kelsey, C. H. (1964). The calculation of errors in a least squares estimate of unit-cell dimensions. *Mineralog. Mag.* **33**, 809.

Kempe, D. R. C. (1966). A note on the $20\bar{1}$ spacing of some lime-rich alkali feldspars from Kangerdlugssuaq, East Greenland. *Mineralog. Mag.* **35**, 704.

Klein, C. (1964). Cummingtonite–grunerite series: a chemical, optical and X-ray study. *Am. Miner.* **49**, 963.

Klug, H. P. and Alexander, L. E. (1954). "X-ray Diffraction Procedures." Wiley, New York.

Kuellmer, F. J. (1959). X-ray intensity measurements on perthitic materials. I. Theoretical considerations. (1960). II. Data from natural alkali feldspars. *J. Geol.* **67**, 648: **68**, 307.

Kuno, H. (1954). Study of orthopyroxenes from volcanic rocks. *Am. Miner.* **39**, 30.

Lapham, D. M. (1958). Structural and chemical variation in chromium chlorite. *Am. Miner.* **43**, 921.

Lapham, D. M. and Jaron, M. G. (1964). Rapid, quantitative illite determination in polycomponent mixtures. *Am. Miner.* **49**, 272.

Lawn, B. R. (1963). On the weighting of reflexions in least-squares calculation of non-cubic unit-cell dimensions. *Acta cryst.* **16**, 1256.

Levin, E. M. and Mauer, F. A. (1963). Improved sample holder for X-ray diffracto-meter furnace. *J. Am. Ceram. Soc.* **46**, 59.

Lima de Faria, J. (1964). "Identification of Metamict Minerals by X-ray Powder Photographs." Junta de Investigações de Ultramar, Lisbon.

Lipson, H. (1949). Indexing powder photographs of orthorhombic crystals. *Acta cryst.* **2**, 43.

Lipson, H. and Wilson, A. J. C. (1941). The derivation of lattice spacings from Debye-Scherrer photographs. *J. scient. Instrum.* **18**, 144.

Lonsdale, K. (Gen. Ed.) "International Tables for X-ray Crystallography." Vols. 1–3. I. "Symmetry Groups" (1952); II. "Mathematical Tables" (1959); III. "Physical and Chemical Tables" (1962). International Union of Crystallography. Kynoch Press, Birmingham, England.

Lowenstam, H. A. (1954). Factors affecting the aragonite–calcite ratios in carbonate-secreting marine organisms. *J. Geol.* **62**, 284.

Main, P. and Woolfson, M. M. (1963). Accurate lattice parameters from Weissen-berg photographs. *Acta cryst.* **16**, 731.

van der Marel, H. W. (1962). Quantitative determination of kaolinite by X-ray diffraction. A reply to G. W. Brindley and S. S. Kurtossy. *Am. Miner.* **47**, 1209.

Mikheev, V. I. (1957). X-ray determination tables for minerals. *Acta cryst.* **10**, 768 (abst.).

Mirkin, L. I. (1964). "Handbook of X-ray Analysis of Polycrystalline Materials." (Transl. by J. E. S. Bradley). Consultants Bureau, New York.

Mitchell, W. A. (1960). A method for quantitative mineralogical analysis by X-ray powder diffraction. *Mineralog. Mag.* **32**, 492.

Miyashiro, A. and Miyashiro, T. (1954). Unit cell dimensions of synthetic nepheline. *J. Fac. Sci. Tokyo Univ.* (Sect. 2) **9**, 267.

Molloy, M. W. and Kerr, P. F. (1961). Diffractometer patterns of A.P.I. reference clay minerals. *Am. Miner.* **46**, 583.

Morimoto, N. and Clark, L. A. (1961). Arsenopyrite crystal-chemical relations. *Am. Miner.* **46**, 1448.

Mueller, W. M., Mallett, G. and Fay, M. (Eds.). (1960–    ). "Advances in X-ray Analysis." (Reports of annual conferences held at Denver, Colorado.) Plenum Press, New York.

Myers, E. J. and Davies, F. C. (1961). A direct graphical method for the precise determination of lattice parameters of tetragonal or hexagonal crystals. *Acta cryst.* **14**, 194.

National Bureau of Standards (1950). "Tables for Conversion of X-ray Diffraction Angles to Interplanar Spacings." U.S. Government Printing Office, Washington, D.C.

Neuvonen, K. J. (1952). Thermochemical investigation of the åkermanite-gehlenite series. *Bull. Commn geol. Finl.* **26**, 1.

Nicholls, G. D. and Zussman, J. (1955). The structural formula of a hydrous amphibole. *Mineralog. Mag.* **30**, 717.

Niskanen, E. (1964). Reduction of orientation effects in the qualitative X-ray diffraction analysis of kaolin minerals. *Am. Miner.* **49**, 705.

Nolan, J. and Edgar, A. D. (1963). An X-ray investigation of synthetic pyroxenes in the system acmite–diopside–water at 1,000 kg/cm$^2$ water-vapour pressure. *Mineralog. Mag.* **33**, 625.

Norrish, K. and Taylor, R. M. (1962). Quantitative analysis by X-ray diffraction. *Clay Miner. Bull.* **5**, 98.

Orville, P. M. (1958). Feldspar Investigations. Carnegie Inst. Washington, *Ann. Rept. Dir. Geophys. Lab.* p. 206.

Palache, C., Berman, H. and Frondel, C. (1944, 1951). "Dana's System of Mineralogy," 7th Edn, Vols. 1, 2. Wiley, New York.

Parrish, W. (1960). Results of the I.U.Cr. Precision Lattice-parameter Project. *Acta cryst.* **13**, 838.

Parrish, W. (1965). X-ray Analysis Papers. Centrex, Eindhoven, Holland.

Parrish, W. and Mack, M. (1963). "Data for X-ray Analysis," 2nd Edn., Vols. 1, 2, 3. Charts for the Solution of Bragg's Equation. Philips Technical Library.

Parrish, W., Taylor, J. and Mack, M. (1964). Dependence of lattice parameters on various angular measures of diffractometer line profiles. *Adv. X-ray Analysis* **7**, 66. Plenum Press, New York.

Peiser, H. S., Rooksby, H. P. and Wilson, A. J. C. (Ed.). (1955). "X-ray Diffraction by Polycrystalline Materials." Institute of Physics, London.

Petruk, W. (1964). Determination of the heavy atom content in chlorite by means of the X-ray diffractometer. *Am. Miner.* **49**, 61.

Pollack, S. S. and Ruble, W. D. (1964). X-ray identification of ordered and disordered ortho-enstatite. *Am. Miner.* **49**, 983.

Porrenga, D. H. (1958). The application of a multiple Guinier camera (after P. M. de Wolf) in clay mineral studies. *Am. Miner.* **43**, 770.

Radoslovich, E. W. (1962). The cell dimensions and symmetry of layer-lattice silicates. II. Regression relations. *Am. Miner.* **47**, 617.

Rex, R. W. and Chown, R. G. (1960). Planchet press and accessories for mounting X-ray powder diffraction samples. *Am. Miner.* **45**, 1280.

Sahama, Th. G. and Hytönen, K. (1958). Calcium bearing magnesium-iron olivines. *Am. Miner.* **43**, 862.

Sahama, Th. G., Neuvonen, K. J. and Hytönen, K. (1956). Determination of the composition of kalsilites by an X-ray method. *Mineralog. Mag.*, **31**, 200.

Sakata, Y. (1957). Unit cell dimensions of synthetic aluminian diopsides. *Jap. J. Geol. Geogr.* **28**, 161.

Schairer, J. F. and Boyd, F. R. (1957). Pyroxenes: the join MgSiO$_3$–CaMgSi$_2$O$_6$. Carnegie Inst. Washington, Ann. Rept. Dir. Geophys. Lab., p. 223.

Schiaffino, L. (1962). L'effetto dell'orientazione preferenziale sulle intensita, osservate al diffratometro, delle reflessioni basali dei minerali micacei. *Periodico Miner.* **31**, 7.

Schoen, R. (1962). Semi-quantitative analysis of chlorites by X-ray diffraction. *Am. Miner.* **47**, 1384.

Scott, R. K. and Ruh, E. (1963). Design and performance of a high-temperature X-ray diffractometer furnace. *J. Am. Ceram. Soc.* **46**, 513.

Short, M. A. and Steward, E. G. (1959). Measurement of disorder in zinc and cadmium sulphides. *Am. Miner.* **44**, 189.

Skinner, B. J. (1961). Unit-cell edges of natural and synthetic sphalerites. *Am. Miner.* **46**, 1399.

Skinner, B. J. and Bethke, P. M. (1961). The relationship between unit-cell edges and composition of synthetic wurtzites. *Am. Miner.* **46**, 1382.

Skinner, B. J., Barton, P. B. and Kullerud, G. (1959). Effect of FeS on the unit cell edge of sphalerite. A revision. *Econ. Geol.* **54**, 1040.

Skinner, B. J., Stewart, D. B. and Morgenstern, J. C. (1962). A new heating stage for the X-ray diffractometer. *Am. Miner.* **47**, 962.

Smith, J. R. and Yoder, H. S. (1956). Variations in X-ray powder diffraction patterns of plagioclase feldspars. *Am. Miner.* **41**, 632.

Smith, J. V. (1956). The powder patterns and lattice parameters of plagioclase feldspars. I. The soda-rich plagioclases. *Mineralog. Mag.* **31**, 47.

Smith, J. V. and Gay, P. (1958). The powder patterns and lattice parameters of plagioclase feldspars. *Mineralog. Mag.* **31**, 744.

Smith, J. V. and MacKenzie, W. S. (1955). The alkali feldspars: II. A simple X-ray technique for the study of alkali feldspars. *Am. Miner.* **40**, 733.

Smith, J. V. and Sahama, Th. G. (1954). Determination of the composition of natural nephelines by an X-ray method. *Mineralog. Mag.* **30**, 439.

Smith, J. V. and Stenstrom, R. C. (1965). Chemical analysis of olivines by the electron microprobe. *Mineralog. Mag.* **34**, 436.

Smith, J. V. and Tuttle, O. F. (1957). The nepheline–kalsilite system. I. X-ray data for the crystalline phases. *Am. J. Sci.* **255**, 282.

Sorem, R. K. (1960). X-ray diffraction technique for small samples. *Am. Miner.* **45**, 1104.

Sriramadas, A. (1957). Diagrams for the correlation of unit cell edges and refractive indices with the chemical composition of garnets. *Am. Miner.* **42**, 294.

Strunz, H. (1966). "Mineralogische Tabellen." Akademische Verlagsgesellschaft Geest & Portig K.—G. Leipzig.

Swanson, H. E. and others (1953–1960). "Standard X-ray Diffraction Powder Patterns." N.B.S. Circular 539. U.S. Govt. Printing Office, Washington D.C.

Tatlock, D. B. (1966). Rapid Modal Analysis of Some Felsic Rocks from Calibrated X-ray Diffraction Patterns. *Bull. U.S. geol. Surv.*, 1209.

Tennant, C. B. and Berger, R. W. (1957). X-ray determination of the dolomite–calcite ratio of a carbonate rock. *Am. Miner.* **42**, 23.

Ueda, T. (1956). On the biaxialization of zircon. *Mem. Coll. Sci., Kyoto Univ.* Ser. B **23**, 297.

Ueda, T. (1957). Studies on the metamictization of radioactive minerals. *Mem. Coll. Sci., Kyoto Univ.* Ser. B, **24**, 81.

Vincent, E. A., Wright, J. B., Chevallier, R. and Mathieu, S. (1957). Heating experiments on some natural titaniferous magnetites. *Mineralog. Mag.* **31**, 624.

Viswanathan, K. (1966). Unit cell dimensions and ionic substitutions in common clinopyroxenes. *Am. Miner.* **51**, 429.

Waite, J. M. (1963). Measurement of small changes in lattice spacing applied to calcites of a Pennsylvanian age limestone. *Am. Miner.* **48**, 1033.

Weiss, E. J. and Rowland, R. A. (1956). Oscillating-heating X-ray diffractometer studies of clay mineral dehydroxylation. *Am. Miner.* **41**, 117.

Weisz, O., Cochran, W. and Cole, W. F. (1948). The accurate determination of cell dimensions from single-crystal X-ray photographs. *Acta cryst.* **1**, 83.

Whittaker, E. J. W. (1960). The crystal chemistry of amphiboles. *Acta cryst.* **13**, 291.

Whittaker, E. J. W. and Zussman, J. (1956). The characterization of serpentine minerals by X-ray diffraction. *Mineralog. Mag.* **31**, 107.

Wilkinson, J. F. G. (1963). Some natural analcime solid solutions. *Mineralog. Mag.* **33**, 498.

Williams, J. M. (1963). Small diameter thin-walled gelatin capillaries for powder X-ray diffraction use. *Rev. scient. Instrum.* **34**, 1430.

Wilson, A. J. C. (1949). Straumanis' method of film shrinkage correction modified for use without high angle lines. *Rev. scient. Instrum.* **20**, 831.

Wilson, A. J. C. (1963). "Mathematical Theory of X-ray Powder Diffractometry." Philips Technical Library.

Winchell, H. (1958). The composition and physical properties of garnet. *Am. Miner.* **43**, 595.

Winchell, H. and Tilling, R. (1960). Regressions of physical properties on the compositions of clinopyroxenes. *Am. J. Sci.* **258**, 529.

de Wolff, P. M. (1957). On the determination of unit-cell dimensions from powder diffraction patterns. *Acta cryst.* **10**, 590.

Wright, J. B. (1959). Some further heating experiments on natural titaniferous magnetites. *Mineralog. Mag.* **32**, 32.

Yoder, H. S. and Eugster, H. P. (1955). Synthetic and natural muscovites. *Geochim. cosmochim. Acta* **8**, 225.

Yoder, H. S. and Sahama, Th. G. (1957). Olivine X-ray determinative curve. *Am. Miner.* **42**, 475.

Zachariasen, W. H. (1963). Interpretation of monoclinic powder X-ray diffraction patterns. *Acta cryst.* **16**, 784.

# CHAPTER 7

# Electron Microscopy and Electron Diffraction

## J. D. C. McCONNELL

*Department of Mineralogy and Petrology*
*University of Cambridge, England*

## I. Introduction

This Chapter has been written with the object of demonstrating the general theory and application of electron microscopy in the field of mineralogical studies. In this sphere the electron microscope is at present used principally in the study of morphology and diffraction phenomena in finely crystalline material. The potential of the electron microscope in the study of microstructure and defects in mineralogical materials is such that this aspect of electron microscopy is likely to become more important in the future. Since this is the case, the present Chapter includes aspects of the theory of electron diffraction and contrast phenomena, which are relevant to such study.

In the investigation of diffraction effects in minerals, the electron microscope provides a technique that is complementary to conventional X-ray diffraction studies. It is seldom convenient to undertake a complete determinative study of cell dimensions and crystal structure by using electron-diffraction effects alone, though this may be necessary in the case of very fine-grained materials (Gard and Taylor, 1957a). One factor that makes structure determination extremely difficult is the existence of multiple elastic scattering. Owing to the strong interaction between the electron beam and crystalline material, even thin crystalline flakes produce diffracted spectra with anomalous intensities. Since this is the case, it is seldom possible to deal with diffracted intensities in terms of the simple kinematic theory that is adequate for the treatment of X-ray diffraction by small single-crystals. In the kinematic treatment one assumes a single elastic scattering incident as the rule.

Single-crystal diffraction and resolution experiments on mineralogical materials are almost invariably made on random and unorientated crystalline fragments obtained by crushing. Limited facilities for the manipulation of such grains by tilting are available on the modern electron microscope, but suitably orientated single crystals can be selected by trial and error, since it is possible to observe the diffraction pattern of each grain directly on the fluorescent screen of the electron microscope by using the technique of selected-area electron diffraction. Recently a technique for the preparation of ultra thin sections of ceramic materials and rocks for the electron microscope has been described (see p. 345).

The use of electron-diffraction techniques is particularly convenient where the cell dimensions of the phase studied are already known. In this case the orientation of the single crystal can be readily defined, and it is possible to study the distribution and character of weak or anomalous diffraction effects. The situation is particularly favourable, since the radiation is highly monochromatic and high-resolution diffraction patterns can be obtained.

The possibility of combining diffraction studies and direct observation in the electron microscope make it an extremely important tool in the direct study of solid-state reactions. In this case it is possible to utilize the heating effect of the electron beam, or to use a heated electron-microscope stage.

In the direct study of the microstructure of minerals on a scale on 10 Å or greater, there is at present no alternative to the use of the electron microscope. The resolution of microstructure in the electron microscope involves an appreciation of the origin of contrast effects in the electron-optical image. Contrast effects arise in a number of different ways depending on the nature of the material being examined.

In the case of well crystallized materials contrast arises, in the main, owing to strong diffraction effects in the specimen. Contrast that arises in this way is described as electron-diffraction contrast. It is used very extensively in the study of defects in metals and permits the observation of twinning and exsolution phenomena in mineralogical material. The interpretation of electron-diffraction contrast effects requires an analysis of the associated diffraction effects in the imperfect, or composite, single crystal.

In the case of amorphous or semi-amorphous materials, contrast is usually extremely weak and is due essentially to variation in scattering power in the specimen.

Where the specimen used in the electron microscope is particularly inhomogeneous, as for example when porous materials are examined, strong phase contrast results in the focused image. Such materials show intense contrast in out-of-focus images, and the situation is analogous to that in which the light microscopist uses phase-contrast techniques. Similar techniques have been tried with limited success in electron microscopy (Heidenreich, 1964, p. 140).

Strong contrast also results in electron microscopy, where it is possible to utilize a pair of diffracted rays due to a periodicity in well crystalline material. This technique yields an image of the corresponding lattice repeat and can be used for the direct study of dislocations and other mistakes in the periodic structure of the phase concerned. In principle, an image of the atomic structure would be formed if all the diffracted radiation could be recombined in this way with correct phase and amplitude. Unfortunately, the spherical and other aberrations associated with even the best modern electron microscope lenses render this technique impracticable. The maximum resolution currently achieved in this way is of the order of 1 Å under the most favourable experimental conditions.

## II. General Principles of Electron Microscopy

### A. THE CONDITIONS NECESSARY FOR HIGH RESOLUTION

In order to appreciate the use and current limitations of the electron microscope in resolution experiments, it is necessary to deal briefly with the theory of image formation, and the nature of the imperfections in current electron-microscope lenses.

In any microscope, the ultimate limits of resolution, with a perfect imaging system, depend on the wavelength of the radiation used and the numerical aperture of the objective lens. This follows from the fact that radiation diffracted by a periodicity in the object must be made to recombine by interference to produce an image. Where a single point source of radiation is imaged by a perfect lens of aperture of half-width $\alpha$, the image is not a perfectly defined point but a disc, the Airy disc, whose dimensions depend on the wavelength of the radiation used. This means that two sources of radiation separated by a distance $d$ cannot be resolved separately unless:

$$d > 0\cdot61\lambda/n \sin \alpha \qquad (1)$$

In this expression $\lambda$ is the wavelength of the radiation, $n$ is the refractive index of the immersion medium† and $\alpha$ is half the angular aperture of the objective lens.

In a light microscope it is possible to make an objective lens with a numerical aperture of $1\cdot4$ by using an immersion medium of refractive index $1\cdot5$. Hence, with light of wavelength 4000 Å the ultimate limit of resolution is approximately $0\cdot2$ $\mu$m (or 2000 Å) under the most favourable conditions. Higher resolution may be achieved in light microscopy by using light of shorter wavelength. X-rays have a wavelength comparable with atomic dimensions and, in principle at least, might be used in microscopy, but in this case the refractive index for solids is very close to unity and X-rays are heavily absorbed. Hence a practical imaging system is not possible.

Accelerated electrons have an associated wavelength which may be defined in terms of the accelerating voltage $V$ by the relationship:

$$\lambda(\text{Å}) = \frac{12\cdot27}{V^{\frac{1}{2}}}\left[1 + 0\cdot978\times10^{-6}\,V\right]^{-\frac{1}{2}} \tag{2}$$

Since electrons are charged particles they can be deflected and focused by using either an electrostatic or an electromagnetic field. At an accelerating voltage of 100 kV the associated wavelength of electrons is $0\cdot03702$ Å. This wavelength is much smaller than atomic dimensions, implying that a perfect imaging system in which electrons are used should permit resolution of subatomic detail in the object examined. In modern electron microscopes, electromagnetic lenses are used and have the advantage that the strength of the lenses can be varied conveniently by changing the energizing current. They suffer, however, from spherical-aberration effects, since it is impossible to prepare lenses of this type with the correct distribution of field intensity at a distance from the optical axis of the lens. Spherical-aberration effects thus limit the useful apertures of such lenses to divergences of the order of $10^{-2}$ radian. This defines a practical limit to possible resolution in the range 5–10 Å under normal conditions.

Chromatic-aberration effects in the electron microscope will also be important if the accelerating voltage is not highly stabilized. Similarly the lens currents must be highly stabilized for high-resolution microscopy. On modern electron microscopes, stabilization of the high voltage and the lens current is adequate and need not be the determining factor in high resolution.

Interaction of the electron beam and the magnetic field of the imaging lens in the electron microscope involves rotation of the beam about the optic axis of the imaging system, in addition to deflection towards the optic axis of the system. This means that the image in the electron microscope is rotated with

† $n \sin \alpha$ is described as the "numerical aperture" of the lens.

respect to the object and the rotation is dependent on the overall magnification. This is a point that must be borne in mind when the image of a single crystal and its diffraction pattern are compared, since the two images are examined at different settings of the lens currents. It is usual to take standard photographs of a fibrous crystalline material, and its diffraction pattern, in order to determine the effective rotation at several convenient magnifications of the electron-optical image.

## B. IMAGE FORMATION AND ELECTRON DIFFRACTION

The electron microscope may be used for bright- and dark-field imaging and for electron diffraction. In order to illustrate the use of the instrument for these techniques, it is necessary at this point to consider schematically the several possible functions of an oversimplified single-lens system. It will be assumed that the lens is free from aberration and that the object is a one dimensional grating as illustrated in Fig. 1. The incident radiation will be assumed to be plane parallel and monochromatic.

Under these experimental conditions the object grating will produce a series of diffracted spectra due to the periodicity of the grating. These spectra will be separately resolved in the back focal plane of the objective lens, i.e. in the plane labelled B in Fig. 1. If a photographic plate is inserted in this plane, the diffraction pattern may be recorded, and will comprise zero, first, second, etc. spectra as indicated in Fig. 1. It is convenient at this point to note that the individual spectra provide information on the gross features of the object grating. If, for example, the number of elements in the grating is small, or the grating contains irregularities, the individual spectra will be broadened or otherwise modified.

Under the experimental conditions defined above, the single lens will produce a magnified image of the object grating, as in plane C of Fig. 1. If this is to be a faithful image of the original grating all the diffracted radiation must be returned to the image plane and allowed to interfere with correct phase and amplitude.

One may now consider the effects of limiting the number of spectra that are utilized in the formation of the image. If a small aperture is inserted in the back focal plane of the objective lens and this obstructs all but one of the diffracted spectra, the image observed will no longer show the periodicity of the grating and only the gross features of the grating will be observed. Owing to the strong spherical aberration in the electron microscope, it is essential to insert a limiting aperture of this kind in the instrument in order to increase the resolution, but in so doing it follows that detail associated with the diffracted spectra must be lost.

A bright-field image of the object is obtained in the electron microscope when the limiting, or objective, aperture in the back focal plane of the objective lens is used to isolate the zero order of the diffraction pattern. A dark-field image results when the objective aperture is placed so that it permits the passage of a non-zero diffraction maximum. In such a dark-field image, only regions of the object that possess the corresponding diffracting periodicity will be observed as regions of illumination against a dark background. It is possible

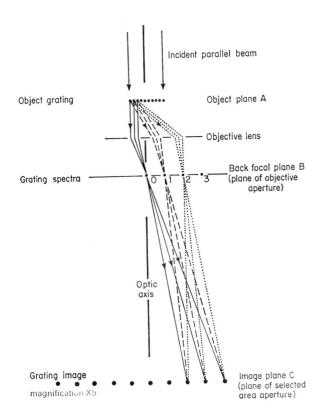

FIG. 1. Diagram illustrating, in a very general way, the main functions of the objective lens system of an electron microscope. The plane of the specimen, in the diagram a grating, is indicated at A. Diffraction spectra from the grating, labelled 0, 1, 2 and 3 in the diagram, are formed in the back focal plane of the objective lens as indicated at B. The magnified image of the grating which is formed by recombining the diffracted radiation is indicated at C. In the modern electron microscope it is possible to insert and manipulate an objective aperture in the plane at B, hence selecting specific diffraction maxima for image formation. Small areas of the specimen are isolated for diffraction study by inserting a selected-area aperture in the image plane at C, where the dimensions of the object are considerably magnified.

to increase the resolution in such a dark-field image by tilting the incident illumination so that the diffracted ray, which has been isolated in this way, passes along the optical axis of the imaging lenses.

The possible results of using a single pair of diffracted spectra for image formation remain to be considered. In order to obtain an image in this way the spectra must be close to the optical axis of the instrument and have small angular separation. It is also preferable that the incident illumination be tilted so that both spectra are equally inclined about the optical axis of the microscope. The final image obtained with a pair of diffracted spectra will show, in addition to the gross features of the object, a suite of $cos^2$ intensity fringes corresponding to the periodicity that produced the spectra. Since gross irregularities in the periodicity will be resolved in the fringe pattern, this technique may be used in the direct observation of dislocations and other lattice imperfections (Heidenreich, 1964, p. 304; Hirsch et al., 1965, p. 353).

The basic principles of selected-area diffraction may be explained by reference to Fig. 1. The single crystals examined in the electron microscope may be smaller than 1 $\mu$m and may occur in close proximity to other grains. It is physically impossible to prepare a mask or aperture that is sufficiently small to isolate such grains in the original object plane, A, of Fig. 1. Effective isolation of the single crystal is essential, however, if one wishes to study the diffraction pattern from a single grain or part of a crystal. This problem has been solved in modern electron microscopes by inserting a limiting aperture in the image plane of the objective lens where the dimensions of the area of interest are considerably magnified. In this way, by using a 25 $\mu$m aperture it is possible to isolate a region less than 1 $\mu$m in diameter in the object plane. Care is necessary in the study of diffraction effects from areas selected in this way, since both spherical-aberration effects and mis-focusing of the objective lens can lead to diffraction contributions from outside the selected area. This problem has been examined in some detail by Hirsch et al. (1965, p. 18).

In order to observe the diffraction pattern from the selected area on the fluorescent screen of the microscope, it is necessary to weaken one of the additional lenses between the objective and the final image. It is usual to do this by defocusing the intermediate lens, which allows the diffraction pattern in the back focal plane of the objective lens to fall in the object plane of the defocused intermediate lens.

In modern instruments it is possible to change over quickly and conveniently from direct observation of a selected area to the corresponding diffraction pattern by operating a single switch. In order to observe the whole of the diffraction pattern it is also necessary to remove the objective aperture from the back focal plane of the objective lens.

Considerable attention has been devoted to the functions of the various components of the objective lens system in the above paragraphs, since most

o

problems in electron microscopy, and in particular those involving electron-diffraction contrast, demand examination of both the direct image and the corresponding diffraction pattern and in most cases information from both is necessary in interpretation.

### C. GENERAL CHARACTERISTICS OF THE ELECTRON MICROSCOPE

The positions of the main components of a modern electron microscope are illustrated in Fig. 2, from which it will be obvious that the layout is similar to that used in a light microscope. The column of the electron microscope is maintained under high-vacuum conditions by means of a diffusion pump, and the final image is observed through a lead glass window, on a fluorescent screen. In modern electron microscopes facilities are available for photography, and both the photographic plates and the specimen can be

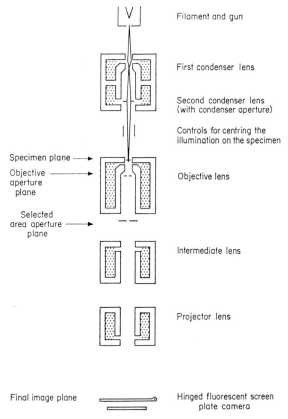

FIG. 2. Diagram showing schematically the positions of the principle components of a modern electron microscope.

changed without letting the main column of the microscope up to atmospheric pressure.

Modern electron microscopes employ two lenses in the condensing system. The initial lens produces a demagnified image of the electron source, which is a heated filament, and the second lens is used to illuminate the specimen with a beam of relatively low divergence. The condensing-lens system also contains an aperture, which limits the general irradiation of the sample, and a spot size a few microns in diameter can be obtained. For most applications, except possibly very high resolution, a condenser aperture of 250 $\mu$m diameter is convenient. The ability to produce a small and well defined region of illumination in the electron microscope is extremely important, since the sample, when irradiated by the direct beam, rapidly builds up a contamination layer with associated rapid loss of contrast detail. The heating effects of the direct beam are also important when materials that react at low temperatures are examined. Under such conditions, very low beam currents must be used during the examination, and it is also desirable to work with the condenser lens considerably defocused.

Facilities also exist on the modern electron microscope for correcting the astigmatism of the condenser-lens system. The use of these facilities is important, not only in direct-resolution experiments, but also in the study of diffraction effects, since the individual diffraction maxima will be distorted if appreciable astigmatism exists.

In recording diffraction photographs with the selected-area technique, extremely high resolution in the diffraction pattern may be obtained if the condenser lens is extensively defocused (weakened). Under these conditions, the diffraction pattern need not be visible on the fluorescent screen and exposure times of the order of three minutes are not unusual. The diffraction resolution achieved in this way is extremely high and considerably superior to conventional X-ray single-crystal techniques. This situation is illustrated by the study of additional intensity maxima in nepheline (McConnell, 1962).

The main functions of the objective lens in an electron microscope have already been discussed in dealing earlier with the theory of image formation. The ultimate performance of the complete instrument depends critically on the objective lens. Residual astigmatism in this component can be corrected for as in the condenser system, and methods of testing both the limits of resolution and the residual astigmatism have been dealt with fully in Kay (1965). The objective lens carries aperture slides in both the back focal plane and the image plane, and in each case a number of apertures can be selected and centred conveniently. The objective lens is followed usually by two lenses, described as the intermediate and projector lenses, as illustrated in Fig. 2. The use of the intermediate lens in obtaining selected-area diffraction has already been noted.

High-resolution photographs are normally taken at a magnification greater than 50,000. Exposure times are of the order of ten seconds.

A facility that is sometimes required in electron microscopy, and which has already been considered above, is the possibility of tilting the illumination, so that a selected diffracted ray from a single crystal passes through the objective lens parallel to its axis. Not all modern instruments lend themselves conveniently to this manipulation but the technique has been used with considerable effect for high resolution in dark-field operation (Glossop and Pashley, 1959; Hirsch *et al.*, 1965, p. 300).

One facility that must have increased importance in the future in mineralogical studies is the use of a cooling stage. In its simplest form this comprises a cold finger of copper which dips into liquid air and terminates at, or near to, the specimen in the microscope (Kay, 1965). This facility reduces the rapid build up of contamination on the specimen and also makes it possible to study mineral specimens, such as hydrates, that would normally dehydrate readily in the electron beam.

Facilities for tilting a single-crystal specimen in the electron microscope are available for most modern instruments. These have been described and illustrated by Kay (1965). The use of a tilting stage is useful in the study of diffraction phenomena.

## III. Specimen Preparation

A very complete account of the methods used in specimen preparation for the electron microscope has been given by Kay (1965). Techniques may be divided into those in which a replica of specimen grains, or of the surface of a bulk specimen is used, and methods that utilize the actual specimen itself within the electron microscope.

### A. REPLICATION TECHNIQUES

In the replication method it is possible only to obtain information on the topography of the specimen surface insofar as this is preserved in the replication process. In this case a very thin film is deposited on the surface of the specimen or on a mount of material particles. This film may be deposited from solution or by vacuum evaporation. The film is subsequently stripped off and examined in the electron microscope.

Since thin films of suitable material, such as carbon, have low contrast in the electron microscope, it is usual to accentuate their topography by a shadowing technique in which a metal is vacuum-deposited on the surface of the replica from an oblique angle. The regions covered by the metal show up with strong contrast in the electron microscope and appear as shadows. In certain replication processes, the specimen itself is shadowed and the supporting layer is added subsequently.

Details of the use of replication technique in the study of clays has been given by Comer and Turley (1955). A further example of the use of replicas is contained in a study of the breakdown products of kaolinite at high temperature by Brindley and Nakahira (1959). In this case a pre-shadowed carbon replica of the surface of a heated crystal of kaolinite is used to show the distinctive morphology of the surface, which is clearly due to the development of mullite crystals in strongly preferred orientation. In this case the specimen was shadowed with platinum, and the replica was stripped by removing the silicate in hydrofluoric acid.

Replication technique is also illustrated in the study of the morphology of montmorillonoids by Nixon and Weir (1957).

### B. DIRECT METHODS OF OBSERVATION

Traditionally, it has been usual to examine mineralogical specimens by making use of finely ground material, which may be deposited from suspension on a suitable supporting film on a standard electron microscope grid. In this technique the specimen is finely ground in absolute alcohol, or freshly distilled water, to form a visibly cloudy suspension. This is then placed on a carbon-coated grid with a micropipette and the liquid is allowed to evaporate. Some experience is required in order to produce a grid that is correctly loaded, and experience is also required in the selection of carbon films of suitable thickness. If these are too thin they tend to rupture. If they are too thick contrast detail in the specimen itself is impaired. Suitable carbon films are transparent and almost colourless. The preparation of carbon films by vacuum deposition is described by Kay (1965). Such films may be used to support material on a heating stage in the electron microscope.

In the direct methods of examination described above, it is not possible to study the microfabric of a ceramic material or a fine-grained rock. The ability to prepare thin sections of such materials has recently been demonstrated by Doherty and Leombruno (1964). In this technique, which is essentially an extension of normal thin-section-making procedure, sections approximately 1000 Å thick were prepared. Such sections are sufficiently thin for transmission studies in the electron microscope.

## IV. The Scattering of Electrons by Crystalline Solids

### A. THE CHARACTERISTICS OF ELASTIC AND INELASTIC ELECTRON SCATTERING

Electrons are very strongly scattered by solids, and in consequence crystals only a few hundred Ångströms in thickness yield scattering patterns that are directly visible on the fluorescent screen of the electron microscope. Two types of scattering are involved in the case of crystalline materials and may be distinguished as follows. Inelastic scattering results when the electron loses energy in the scattering incident with consequent change in wavelength. The energy may appear as heat in the specimen or may be responsible, for example, for the production of X-ray photons. The latter effect is exploited in the electron-probe microanalyser. Elastic scattering of electrons by the crystalline specimen is produced by periodicities in the scattering object and occurs without change in the wavelength. The presence of inelastic-scattering effects in the electron microscope is inconvenient, since they lead to loss of contrast in the image and occur, in addition to normal elastic diffraction effects, in the diffraction pattern.

Owing to the strong interaction between the electron beam and even thin crystalline specimens multiple elastic scattering is usually the rule. This means that a detailed treatment of elastic scattering involves the use of the dynamical theory of electron diffraction, which considers the interaction, or coupling, of all the diffracted radiation. In this respect, electron diffraction differs from X-ray diffraction, in which similar dynamical effects are important only in dealing with relatively large and perfect single crystals.

Although in the present Chapter it would be unwise to attempt to deal in any detail with the dynamical theory of electron diffraction, a brief mention of the salient features of the theory is necessary, since the direct observation of effects due to the dynamical theory is unavoidable in practice. Perhaps the most obvious of these effects is the existence of thickness interference contours.

In the dynamical theory it is necessary to consider the coupling of the direct beam and the diffracted spectra. Where a single diffracted spectrum is operative, this coupling implies that the relative intensity of the diffracted beam and the undeviated beam is a function of the thickness of the specimen. For a critical thickness of the specimen, which is described as the extinction thickness, the diffracted beam has zero intensity. At half this thickness, the incident beam has zero intensity and the whole of the intensity appears in the diffracted beam. The phenomena is illustrated in Fig. 3. With increase in thickness the condition for maximum intensity in the diffracted beam is simply:

$$t = \tfrac{1}{2}(2n + 1)\, t_o \tag{3}$$

where $t$ is the thickness, $t_o$ the extinction thickness and $n$ is an integer.

The numerical value of the extinction thickness for any given crystalline material depends on the structure factor for the reflection concerned and may be calculated directly from the atomic-scattering amplitudes for electrons where the structure of the compound is known. Atomic-scattering amplitudes for electrons are given in the International Tables for Crystallography and

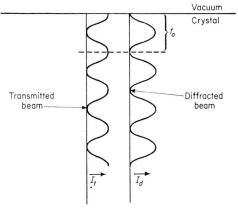

FIG. 3. Periodic variation of intensity in the transmitted and diffracted beams with depth of penetration in the crystal. The extinction thickness is indicated by the distance $t_o$. Note that, at half the extinction distance, the intensity of the diffracted beam is at its maximum.

are also listed in Hirsch *et al.* (1965, p. 489). The atomic-scattering amplitudes require to be corrected for the appropriate energy of the accelerated electrons (Hirsch *et al.*, 1965, p. 495), and relativistic correction for the associated wavelength should be used in determining the extinction thickness in the following relationship:

$$t_o = \frac{\pi V_c \cos \theta}{\lambda F_g} \qquad (4)$$

In this expression $t_o$ is the extinction thickness, $\lambda$ the wavelength, $V_c$ the volume of the unit cell, $\theta$ the Bragg angle and $F_g$ the structure factor for the reflection concerned.

In metals, where the atomic-scattering amplitudes are large and the final structure factors also large, calculated extinction distances are small. For example, in the case of gold, the extinction distance for the 111 reflection is only 159 Å. This low value for extinction thickness implies that dynamical effects are important in considering the intensities of reflections from films that are considerably less than 100 Å thick.

In the case of silicate structures, where the atomic-scattering amplitudes are in general small and the final structure factors are also relatively small, extinction distances are larger. Calculations by the present author indicate that for the strong $24\bar{1}$ and $20\bar{1}$ reflections from adularia the corresponding extinction thicknesses are of the order of 1200 Å. This means that extremely strong diffraction effects will be observed in suitably orientated single-crystal flakes of approximately half this thickness. The implications of this condition will be mentioned later in the present Chapter in dealing with electron-diffraction contrast effects in adularia.

The presence of thickness-extinction contours in the electron-optical image is, as already noted, an obvious demonstration of dynamical diffraction effects. Thickness-extinction contours are illustrated in Fig. 4. The presence or absence of these contours forms a very useful guide to the actual thickness

FIG. 4. Thickness-extinction contours at the edges of a thick crystal of plagioclase feldspar.

of the single-crystal flake. In the study of simple diffraction effects on mineralogical material, it is desirable to choose only thin crystalline flakes. Where these are considerably thinner than the extinction thickness, it will be satisfactory to interpret the diffraction effects in terms of the simple kinematic approach, which assumes a single elastic-scattering incident as the rule. An account of the dynamical theory of electron diffraction is presented by Hirsch et al. (1965, p. 195).

It is reasonable to suppose that, in the examination of certain mineralogical materials, particularly where heavy elements are involved, the implications of the dynamical theory may be much more important than in the case of the simpler silicate structures.

## B. INELASTIC SCATTERING

Inelastically scattered electrons, as already noted, suffer a change in wave-length in the scattering incident. Inelastically scattered electrons must there-fore be affected by chromatic-aberration effects in the imaging system and hence give rise to a loss in contrast detail. In the diffraction pattern, inelastically scattered electrons are concentrated about the incident beam (000), and about the directions of the strong elastic maxima. The most strik-ing effect produced by these inelastically scattered electrons is observed in thick and relatively perfect single crystals, where they give rise to Kikuchi lines in the diffraction pattern. These lines result where the inelastically scattered radiation travels in the correct direction to be elastically scattered by a periodicity in the crystal. They are diffracted outwards from this direction producing a dark, or deficiency line, and appear as a light, or excess line, further from the origin of the diffraction pattern. This effect is illustrated in Fig. 5, and the diffraction pattern presented as Fig. 6. The exact positions of

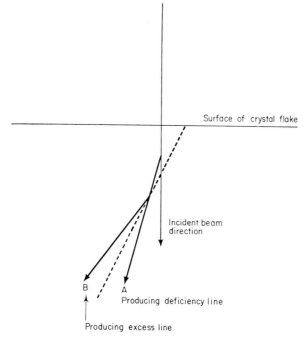

FIG. 5. Geometry of the formation of Kikuchi lines. Direction A represents the inelastic-scattering direction appropriate to the Bragg plane shown dashed. A deficiency line occurs at this position. The diffracted radiation produces an excess line corresponding to the direction B on the diagram.

O*

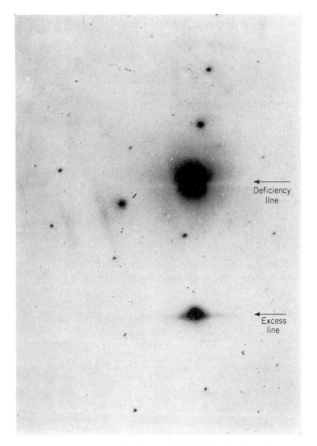

FIG. 6. Electron-diffraction photograph from a thick flake showing the presence of Kikuchi lines. The deficiency line and the excess line in one case have been indicated.

the Kikuchi lines may be used to determine the precise orientation of the single-crystal specimen (Hirsch *et al.*, 1965, p. 123). The presence of Kikuchi lines in a diffraction pattern usually implies that the effects of both inelastic and multiple elastic scattering are important, and that the specimen is too thick for the study of simple kinematic diffraction effects.

## C. ELASTIC SCATTERING

### 1. Single-crystal Diffraction Effects

Since the main uses of the electron microscope in mineralogical studies involve diffraction and high-resolution microscopy, the role of elastic scattering is fundamental. Not only is the diffraction pattern of use in its own right, but, for well crystalline materials, useful contrast effects in the direct image are due primarily to diffraction effects in the specimen. Contrast produced in this way is described as electron-diffraction contrast and requires a detailed analysis of the diffraction effects in the single crystal concerned.

The simple kinematic approach to the interpretation of electron diffraction overlaps to a considerable extent the treatment for single-crystal X-ray diffraction (Chapter 6). Accounts of the theory and practice of electron diffraction have been provided in the mineralogical literature (Ross and Christ, 1958; Gard, 1964). A very brief review will be given here. The treatment of contrast phenomena in both crystalline and semi-amorphous materials is less adequate elsewhere and will be presented here in more detail.

Crystalline material is characterized by the existence of three-dimensional regularity, which can be described formally in terms of a lattice-point array. Diffraction effects occur under these conditions when Bragg's law is satisfied, a condition that implies that all the lattice points scatter in phase. In the study of elastic scattering from crystals it is usual to make use of the Ewald construction, which defines the conditions for diffraction for any given suite of lattice planes.

Use of the Ewald construction in the interpretation of electron-diffraction effects from single crystals is particularly straightforward for the following reason. The wavelength associated with electrons accelerated at 100 kV is approximately 0·037 Å, as already noted earlier. In this case the reciprocal-lattice construction for single crystals with basic lattice translations of the order of 10 Å involves elementary reciprocal repeats that are very much smaller than the radius of the Ewald sphere. This means in turn that many diffraction maxima occur within a small angle about the direct beam. Hence it is not important to distinguish between the value of the angle of diffraction as measured in radians, and its sine or tangent. This important simplification means that the actual diffraction pattern as recorded on a flat plate normal to the electron beam can be regarded as a direct section of the reciprocal-lattice construction, and the dimensions and angles characteristic of the reciprocal-lattice section can be measured directly on the plate with a ruler and protractor to a very good approximation.

It is necessary to standardize the scale of the diffraction pattern. This may be done conveniently by using the recorded diffraction pattern of a substance with known cell dimensions. It is also possible to shadow the specimen with a thin film of gold and to use the powder pattern of this material for calibration. The accurate determination of $d$ spacings and cell dimensions cannot be made on the electron microscope without the use of such an internal standard, since although the accelerating voltage is highly stable over short periods, appreciable fluctuations exist from day to day. It is convenient to calibrate the diffraction plate in terms of the length, in centimetres, that is equivalent to one reciprocal Ångström unit. It is also possible to define the scale of the diffraction pattern in terms of the effective distance between the specimen and the image plane, as in an electron-diffraction camera, but since imaging lenses are used in the electron microscope for the observation of selected-area diffraction, this distance does not correspond to the actual distance as measured in the microscope. Calibration of the diffraction plate in terms of the apparent camera length involves the relationship:

$$\lambda L = dr \qquad (5)$$

where $\lambda$ is the wavelength, $L$ the effective camera length, $d$ the spacing of Bragg planes and $r$ the distance from the origin of the diffraction pattern of the corresponding diffraction maximum.

The use of a reciprocal-lattice model is essential in the interpretation of single-crystal electron-diffraction patterns. One of the most obvious features of such patterns is the existence of zones of reflections that correspond to the intersection of planes of reciprocal-lattice points with the Ewald sphere in the reciprocal-lattice construction. These distinctive zones are described as Laue zones. Their formation is illustrated in Fig. 7.

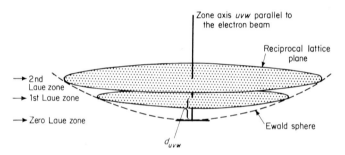

FIG. 7. Schematic diagram illustrating the formation of Laue zones in single-crystal electron-diffraction photographs. Individual planes of reciprocal lattice points normal to a zonal direction that is parallel to the axis of the microscope have symmetrical circular intersections with the Ewald sphere which is shown dashed.

Where the single crystal examined has a prominent zone axis $[uvw]$ parallel to the optical axis of the microscope, the Laue zones are arranged concentrically about the origin of the diffraction pattern. The individual maxima within the Laue zones may be indexed very easily if the corresponding zone axis has simple indices and the cell dimensions of the crystalline phase are known. The process of indexing an electron-diffraction pattern in this way is illustrated in Fig 9. These diagrams also indicate the effects on the diffraction pattern of limited thickness in the single-crystal flake, a topic that will be dealt with in more detail later in this Chapter.

Maxima in the zero Laue zone must have indices that satisfy the zone law, $hu + kv + lw = 0$, and additional Laue zones satisfy the more general relationship $hu + kv + lw = N$, where $N$ is an integer. The unit repeat along the zone axis may be determined with reasonable accuracy from the relationship:

$$R = \lambda L \, (2N \, d/\lambda)^{\frac{1}{2}} \qquad (6)$$

where $R$ is the radius of the non-zero Laue zone, $L$ the effective camera length, $\lambda$ the wavelength, $N$ the value of the integer in the zone law and $d$ the spacing along the zone axis.

If a prominent zone axis in the single crystal examined is not quite parallel to the optical axis of the microscope, the corresponding Laue zones are no longer symmetrically arranged about the origin of the diffraction pattern. This effect is illustrated in the diffraction photograph provided as Fig. 8(a).

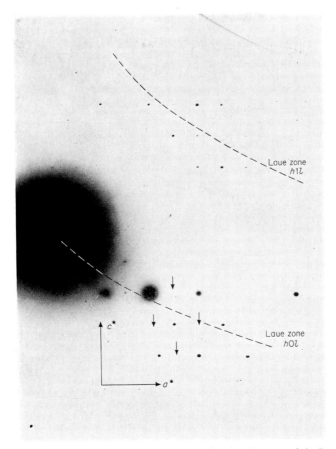

FIG. 8(a). Electron-diffraction photograph of sillimanite showing part of the Laue zone $h1l$ at the top of the photograph and part of the Laue zone $h0l$ with its distinctive absences at the bottom of the photograph. The positions of several systematically absent maxima have been indicated by arrows.

The geometrical analysis of single-crystal diffraction patterns has been dealt with by Hirsch *et al.* (1965, p. 108), and by Gard (1964, p. 256).

## 2. *The Observation and Study of Single-crystal Diffraction Patterns*

Since single-crystal electron-diffraction photographs can be interpreted simply and directly in terms of the reciprocal-lattice construction, as noted earlier, the distribution of intensity maxima throughout reciprocal space may be used in the recognition of specific reciprocal-lattice sections in unorientated material. The recognition of the principal reciprocal-lattice sections is particularly easy where the symmetry of the distribution is high. Orthogonal reciprocal-lattice sections are particularly easy to recognize. Another aspect of reciprocal space distribution which provides considerable aid in the recognition of specific reciprocal-lattice sections is the existence of systematic absences among the recorded intensity maxima. Systematic absences occur throughout reciprocal space in the case of non-primitive direct lattices, and additional systematic absences occur owing to the presence of certain space-group symmetry operators (see Chapter 6, p. 299). Thus glide planes produce systematic absences in certain zero Laue zones, and screw axes produce systematic absences along isolated rows of reciprocal-lattice points. The data on space groups and their associated systematic absences are tabulated in the International Tables for Crystallography.

Systematically weak intensity maxima also occur in the single-crystal diffraction pattern where it is possible to describe the structure to a good approximation in terms of a small unit cell with translation vectors that are submultiples of the true unit cell. In this case, which occurs quite frequently in mineralogical material, the small cell is described as a sub-cell and is responsible for the strong intensity maxima. The weak additional maxima exist owing to differences in the sub-cells, and may be described in terms of the larger unit cell. When the single-crystal diffraction pattern is indexed in terms of the large unit cell, the weak additional maxima satisfy simple selection rules that operate throughout reciprocal space and are analogous to systematic absences due to lattice type.

The recognition of systematically weak, or absent, diffraction maxima in a given single-crystal electron-diffraction pattern provides an invaluable aid in the recognition of specific single-crystal orientations. In certain cases, such effects may be used to identify general orientations, and this may frequently be done from observations on the screen of the electron microscope without recording and measuring the diffraction pattern. These techniques have a direct bearing on the problem of distinguishing between several crystalline phases in a single preparation. An example of the use of this technique is provided by the study of coexisting mullite and sillimanite in a metamorphic rock (Smith and McConnell, 1966).

The principles outlined above have been illustrated in the diffraction photography of sillimanite provided in Fig. 8(a), which also illustrates the formation of Laue zones.

Sillimanite is orthorhombic with space group symmetry *Pbnm*, and has the following systematic absences: $h + l$ odd for reflections $h0l$; $k$ odd for reflections $0kl$; $h$ odd for reflections $h00$; and finally $k$ odd for reflections $0k0$. In addition, sillimanite shows systematically weak reflections for all maxima with $l$ odd. The corresponding distribution of possible intensity maxima is shown in the reciprocal-lattice model illustrated in Fig. 8(b), which provides the key

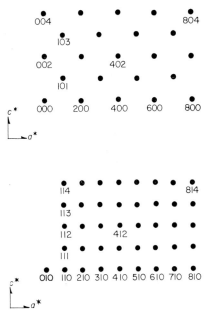

FIG. 8(b). Reciprocal-lattice plots for the Laue zones $h0l$ and $h1l$ for sillimanite. Note the orthogonal character of the plots and the systematic absences in the Laue zone $h0l$.

to the analysis of the diffraction pattern of sillimanite provided as Fig. 8(a). In this case it is possible to be certain about the nature of the single crystal examined and also to know its precise orientation from a single diffraction photograph.

It is often convenient to prepare in advance a reciprocal-lattice drawing, preferably to scale, as an aid to the identification of reciprocal-lattice sections that may be observed in any given problem involving single crystals in random orientation.

The use of single-crystal electron-diffraction technique in mineralogical studies can be illustrated from a large number of references. Particularly good

examples are provided by the study of some vanadium minerals by Ross
(1959) and by the study of the minerals rhodesite and mountainite by Gard
and Taylor (1957a). Examples of the study of the distribution of weak additional maxima in the diffraction pattern are provided by Gard and Taylor
(1957b), and Gard (1964).

## 3. Diffraction Effects Associated with Lattice Imperfections

In dealing earlier in the present Chapter with the theory of image formation
and electron diffraction, it was noted that the character of individual diffraction maxima depends on the dimensions and perfection of the object grating.
In the same way the dimensions of the single crystal examined and the presence
of lattice defects define the character of the diffraction maxima produced by
a single crystal.

The most obvious lattice defect that must be considered in this connection
is the limited thickness of the crystalline flake that can be used in single-crystal
electron-diffraction studies owing to the presence of inelastic and multiple-
elastic scattering. This defect in the three-dimensional lattice grating means
that the distribution of permitted intensity in the diffraction pattern is not
confined to the reciprocal-lattice point but occurs as a spike which is orientated
normally to the single crystal flake. Fuller treatment of the diffraction
problem in this case shows that the intensity along this spike is modulated by
a function whose period is determined by the reciprocal of the thickness of
the flake, and that the permitted intensity is zero at a series of points along
the spike. This problem is dealt with by Hirsch et al. (1965, pp. 97–100) and
also by Heidenreich (1964, p. 218).

This relaxation of the diffraction conditions in the case of a thin flake has a
very important bearing on the character of the single-crystal diffraction
pattern obtained from thin crystals by using electron diffraction. The Ewald
construction in this case is no longer concerned with the coincidence of a
sharply defined reciprocal-lattice point and the Ewald sphere, and diffraction
will occur over a finite range of settings of the single crystal (corresponding to
the intersection of the spike of permitted intensity and the Ewald sphere).
This effect results in simultaneous diffraction from a large number of planes
in the crystal for a single stationary setting. This effect is illustrated schematically in Fig. 9(a,b) and is responsible for the large number of diffraction
maxima observed in Fig. 8(a).

With continuously decreasing thickness, the three-dimensional diffraction
pattern grades into a two-dimensional pattern owing to increase in the length
of the reciprocal-lattice spikes. Where there is reason to suspect that the single
crystal studied is very thin, caution must be exercised in determining repeat
distances from the diffraction pattern.

Other types of lattice defect also yield distinctive diffraction effects. The

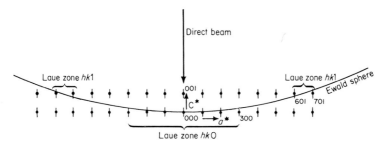

FIG. 9(a). Schematic section through the origin of the reciprocal-lattice construction in a direction parallel to the primary electron beam. Note that the spikes due to the thin flake, which pass through each reciprocal-lattice point, intersect the Ewald sphere over a considerable area of the reciprocal-lattice section.

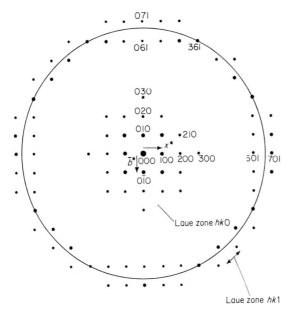

FIG. 9(b). Schematic section of the same reciprocal-lattice construction passing through the origin of the reciprocal lattice and indicating the points of intersection of the spikes with the Ewald sphere. Note the presence of Laue zones $hk0$ and $hk1$.

existence of a simple sinusoidal lattice perturbation, with wavelength much greater than the basic lattice repeats, leads to additional intensity at a distance from the main reciprocal-lattice points. The intensity of this additional scattering depends on the vector amplitude of the distortion wave, and the orientation of the Bragg planes. It is possible to study the nature of a complex lattice perturbation by examining the distribution and intensity of such additional scattering throughout reciprocal space (McConnell, 1965). Since the recorded single-crystal diffraction pattern may be regarded as a nearly planar section through reciprocal space, analysis of the distribution of additional scattering is particularly straightforward.

Single-crystal electron-diffraction techniques may also be used to study the mutual orientation of the components of composite crystals. In a twinned crystal, for example, duplicate reciprocal distributions occur, and these must be related by the twinning operator. Similarly, orientated exsolution effects may be studied in terms of the composite electron-diffraction pattern. Since it is possible to use both bright- and dark-field direct microscopy in addition to selected-area diffraction in the electron microscope, the shape and disposition of inclusions can be studied in addition to their mutual orientations. In this field the electron microscope provides a technique that is superior to single-crystal X-ray studies. Good examples of the study of orientated exsolution effects by electron microscopy are provided by Fleet and Ribbe (1963, 1965).

It is important to note at this point that the contrast observed in the electron-optical images of twinned and composite crystals is due almost entirely to effects associated with diffraction rather than differences in the mass thickness, or absorption, in the specimen. Contrast effects that arise in this way are described as electron-diffraction contrast and will be dealt with in some detail in the following Section.

## V. Electron Optical Contrast Phenomena

### A. PHASE- AND AMPLITUDE-CONTRAST EFFECTS IN THE ELECTRON MICROSCOPE

Both phase and amplitude contrast occur in focused electron-optical images and arise in exactly the same way as in the normal light-optical microscope. Amplitude contrast occurs where the specimen is inhomogeneous and contains regions of relatively opaque material. Platinum-shadowed replicas produce strong contrast in this way, but in the study of normal mineralogical specimens in the electron microscope, amplitude contrast is not particularly important.

Phase-contrast effects in the electron microscope are of considerable importance and result, as in the ordinary light microscope, from appreciable optical path differences in different parts of the specimen. Strong phase-contrast effects occur in porous materials and originate in this way. As in the normal light microscope, phase contrast in the final focused electron-optical image cannot be observed. Under these conditions the light-optical microscopist uses phase-contrast techniques, and analogous techniques have been used with limited success in electron microscopy (Heidenreich, 1964, p. 142). In the observation of phase-contrast effects in the electron microscope it is usual to record out of focus images in which the phase contrast appears as amplitude contrast. The extent of defocus required depends on the scale of the corresponding repeat distance in the specimen. If the repeat distance is $a$, the amount of defocus required for maximum amplitude contrast is given by:

$$\Delta f = a^2/2\lambda \tag{7}$$

where $\Delta f$ is the defocus term and $\lambda$ is the wavelength. Where the specimen contains a number of different repeat distances, those that most nearly satisfy equation (7) will be emphasized most. Contrast that arises in this way is described as phase contrast by defocus and has been discussed in some detail by Heidenreich (1964, p. 140, p. 357).

### B. ELECTRON-DIFFRACTION CONTRAST

The phenomenon of electron-diffraction contrast is peculiar to the electron-optical images of well crystalline material and has no direct analogue in light microscopy. It is due essentially to the existence of very strong elastic scattering from lattice periodicities in the single-crystal specimen. Earlier in the

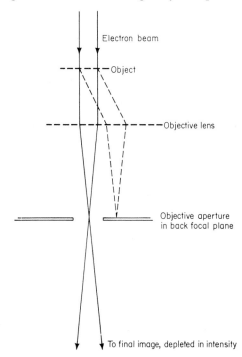

FIG. 10. Schematic diagram to illustrate the origin of electron-diffraction contrast effects. Strong elastic scattering by the object, or a certain region of the object, diverts intensity from the direct beam into a diffracted ray, which is not permitted to reach the final image plane owing to the presence of a limiting aperture, the objective aperture, in the back focal plane of the objective lens. In the diagram the diffracted radiation is shown dashed.

present Chapter, in dealing with the dynamical theory of electron diffraction, it was noted that, under suitable conditions, the whole of the intensity in transmission can appear in the diffracted beam. If radiation that is diffracted in this way falls outside the objective aperture it will not be returned to the plane of the final image and the crystal, or that portion of it that is strongly diffracting, will appear dark in the bright-field image. When the corresponding strong diffraction spectrum is selected under conditions of dark-field operation, the contrast will be reversed, and the diffracting single crystal will appear light on a dark ground. The mechanism of formation of electron-diffraction contrast in the bright field image is indicated in Fig. 10.

The use of electron-diffraction contrast effects provides an extremely powerful technique in the study of defects in crystalline solids. Extensive study of defects such as dislocations in thin metal foils has been made in this way (Hirsch *et al.*, 1965, p. 247).

The direct observation of twinning effects in the electron microscope is due

FIG. 11(a). Photograph of a flake of moonstone showing alternate lamellae of albite and orthoclase due to diffraction contrast. This result is due to the slight difference in crystallographic orientation of the two components. (After Fleet and Ribbe, 1963.)

entirely to electron-diffraction contrast and may be explained simply as follows. If the twinned single crystal is in an arbitrary orientation, this will normally mean that different diffraction conditions will hold for the two components of the twin. Where strong diffraction occurs in only one of the two lattice orientations in the twin, the corresponding set of twin lamellae will appear darker in the bright-field image. If the twinning is cyclic alternate lamellae will show similar contrast effects.

Exactly similar conditions apply in the observation and study of orientated exsolution effects in the electron microscope since the host phase and the exsolved lamellae will normally have related, but independent, lattice orientations. Electron-diffraction contrast effects in exsolved lamellae of sodium feldspar in a moonstone are illustrated in Fig. 11(a), and the corresponding composite electron-diffraction pattern from the same crystal is shown as Fig. 11(b) (Fleet and Ribbe, 1963). By using both bright- and dark-field observation in association with the analysis of the composite diffraction pattern, it is possible, in this case, to identify the individual lamellae of sodium and potassium feldspar.

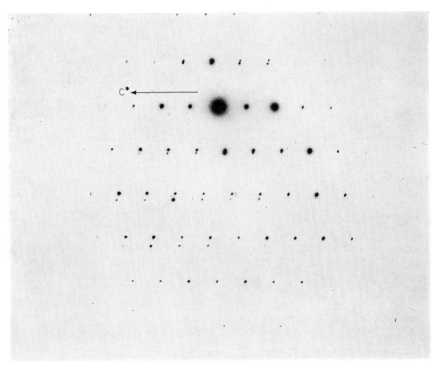

FIG. 11(b). Composite single-crystal diffraction photograph showing $h0l$ reflections for both the orthoclase and the albite components. (After Fleet and Ribbe, 1963.)

## C. ELECTRON-DIFFRACTION CONTRAST EFFECTS DUE TO LATTICE DEFECTS

Electron-diffraction contrast effects also occur, as already noted, where the single crystal is deformed. A particularly simple example of this phenomena occurs where the single crystal is bent. In this case, the conditions for strong diffraction by a single suite of Bragg planes will be satisfied only locally in the single-crystal flake. In the corresponding reflecting-sphere construction the reciprocal lattice may be regarded as rotating about the origin of reciprocal space in moving from one part of the bent crystal to another. Where the conditions for strong diffraction are satisfied in the flake, a dark band will be observed in the bright-field image of the flake. This dark band is described as an extinction-bend contour, and its width depends on the thickness of the crystal flake, the indices of the corresponding Bragg reflection and the radius of curvature of the bent single crystal. The origin of the extinction-bend contour may readily be demonstrated by selecting the diffraction maximum responsible, using dark-field technique. In the dark-field image-reversed contrast will be observed with the contour appearing light against a dark background. By using the dark-field technique in this way it is possible to

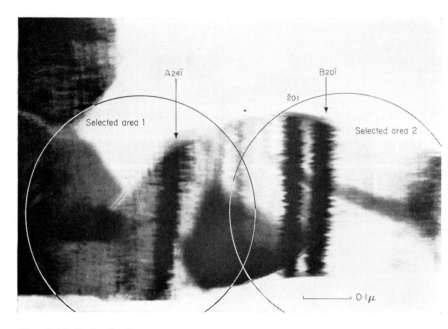

Fig. 12(a). Extinction-bend contours in a thin crystalline flake of adularia. Diffraction photographs from the two selected areas (1) and (2) are shown in Figs. 12(d) and (e) respectively. The extinction contours indicated at A and B correspond to the very strong diffraction maxima that have been similarly labelled and ringed on the selected-area diffraction photographs.

index each of the extinction-bend contours present in the deformed single crystal, and hence define the deformation precisely.

Extinction-bend contours in a single-crystal flake of adularia are illustrated in Fig. 12(a), and the individual intensity maxima responsible for each of the contours are indicated in the corresponding selected-area diffraction patterns provided in Figs. 12(d) and (e). Dark-field photographs obtained by selecting certain of these maxima are also given as Figs. 12(b) and (c). The extremely

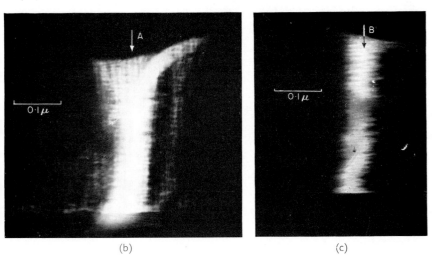

(b)                                     (c)

FIG. 12(b). Dark-field image produced by selecting diffraction maximum A of Fig. 12(d) with a 25 μm objective aperture. Note the reversed contrast and the distinctive cross-hatching on this extinction contour.
(c). Dark-field image obtained by using the ringed diffraction maximum B of Fig. 12(e) for image formation.

strong electron-diffraction contrast that is observed in the principal extinction-bend contours illustrated in Fig. 12 is due to the fact that this particular flake is close to half the extinction thickness for the strong maxima 20Ī and 24Ī that are concerned. For this flake it is possible to obtain an independent check on thickness, since extinction-bend contours 20Ī and 2̄01 are both present. By using this pair of extinction-bend contours, it is possible to calculate the radius of curvature of the flake and hence to determine the thickness of the flake by measuring the width of the individual contours. This technique has been described by Hirsch et al. (1965, p. 525).

Electron-diffraction contrast also occurs where the single crystal is deformed in other ways. In the case of a simple edge dislocation, the dislocation introduces a relative rotation of the Bragg planes. Where such a dislocation crosses an extinction-bend contour, the dislocation line is marked by a

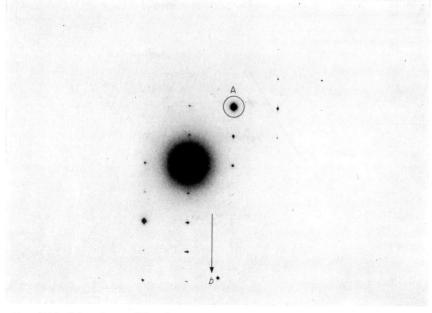

FIG. 12(d). Selected-area diffraction photograph from the area (1) defined in Fig. 12(a).

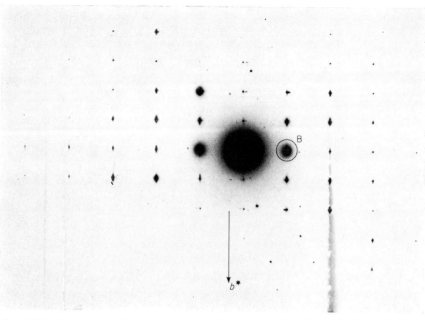

FIG. 12(e). Selected-area diffraction photograph of the area (2) indicated in Fig. 12(a). Note the relative change in orientation of the flake along its length as indicated by the change in the diffraction pattern.

relatively sharp change in contrast. Similarly other lattice perturbations lead to a modification of the contrast within an extinction bend contour. Contrast will be absent where the amplitude vector for the perturbation is normal to the reciprocal lattice vector for the extinction contour concerned. This condition may be used to define the Burgers vector of dislocations.

Use of electron-diffraction contrast theory is directly applicable to the study of lattice defects in mineralogical material. A more detailed examination of the diffraction photographs of adularia given in Fig. 12 shows that the individual Bragg maxima have associated streaks of intensity. This additional intensity distribution is due to the presence of a system of very-fine-scale lattice perturbations associated with the onset of inversion in adularia to the triclinic state. Further inspection of the extinction-bend contours of Fig. 12 shows that, in both the bright- and dark-field images, the corresponding lattice perturbation is resolved as a modulation of the intensity distribution within the contours themselves. The existence of two perturbations at right angles to one another is evident in the $24\bar{1}$ contour, labelled A in Fig. 12. The absence of the second modulation in contour B indicates that in this case the amplitude vector for the perturbation lies in the corresponding lattice planes $20\bar{1}$. A complete analysis of the lattice perturbation in adularia can be made in this way (McConnell, 1965).

The above example of the use of electron-diffraction contrast technique not only indicates the extreme facility with which lattice defects may be studied but also emphasizes the close relationship that exists, in general, in electron microscopy, between study of the diffraction pattern and direct imaging technique in both dark- and bright-field operation.

## VI. High Resolution Electron Microscopy

Earlier in the present Chapter it was noted that where a single diffraction maximum is used in resolution experiments on crystalline material, only the gross features of the crystal are resolved. The resolution of defects in both dark- and light-field microscopy due to electron-diffraction contrast effects must be included under this heading. In such experiments the fine structure of crystalline solids is not resolved. As already noted earlier, a further possibility exists in the case of crystalline solids with long lattice spacings. This involves the direct imaging of such spacings by including two or more diffraction spectra within the objective aperture of the microscope. Under these conditions the corresponding lattice spacings are resolved as a fringe pattern, and the images may be described as periodic. Such images are capable of revealing lattice defects directly.

This technique was first successfully applied by Menter (1957) in the resolution of the $20\bar{1}$ planar spacing in platinum phthalocyanine. Examples of the subsequent use of this technique are provided by the resolution of long lattice spacings in antigorite (Chapman and Zussman, 1959), by the resolution of antiphase domain structures in alloys (Glossop and Pashley, 1959), and the resolution of similar antiphase domain structures in the plagioclase feldspars (McConnell and Fleet, 1963). A detailed account of periodic images is given by Hirsch *et al.* (1965, p. 353). Fringe patterns in antigorite are illustrated in Fig. 13.

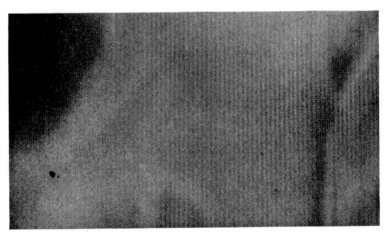

FIG. 13. Direct resolution of the lattice spacing (approx. 40Å) in antigorite. (After Chapman and Zussman, 1959.)

In dealing with periodic images produced in this way, it is important to realize that the fringe pattern, as observed, does not correspond directly to a mass-thickness relationship in the single-crystal specimen. As noted earlier, maximum contrast in the fringe pattern can be obtained by a defocusing technique and the relative positions of the dark and light fringes can be reversed, as, for example, where the fringe pattern crosses a thickness extinction contour. Considerable care must also be exercised in interpretating irregularities in the spacing of the fringes, since this is dependent on uniformity in the thickness of the specimen over the region examined. This topic is dealt with in some detail by Heidenreich (1964, p. 236).

## VII. The Direct Study of Phase Transformations in the Electron Microscope

The importance of adequate control over the illumination conditions in the electron microscope, particularly when hydrate minerals are being studied, has been stressed earlier in this Chapter. The heating effects of the direct beam can be usefully employed in certain cases to study solid-state reactions, in

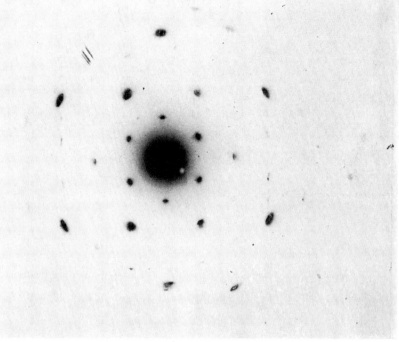

FIG. 14(a). Electron-diffraction photograph of norstrandite [Al(OH)₃].

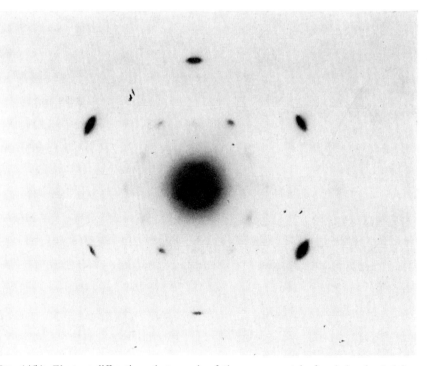

FIG. 14(b). Electron-diffraction photograph of the same crystal after being heated for several minutes in the electron beam. Dehydration is associated in this case with the development of a defect spinel phase.

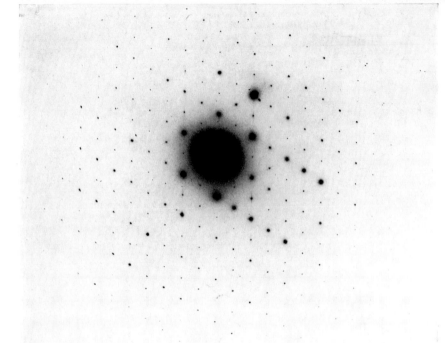

FIG. 15(a). Electron-diffraction photograph of low tridymite, which is orthorhombic. The photograph shows reflections of the type $hk0$ and is markedly pseudohexagonal.

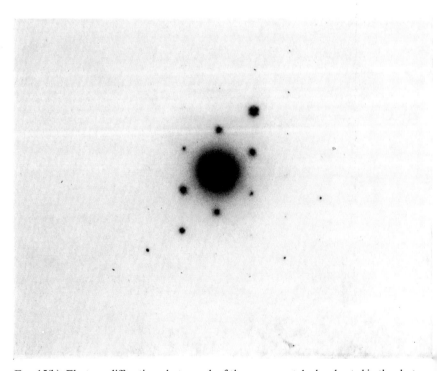

FIG. 15(b). Electron-diffraction photograph of the same crystal when heated in the electron beam. Note the loss of superlattice reflections due to inversion to high tridymite which is hexagonal.

particular those involving low-temperature dehydration. A good example of this technique is provided by the study of the dehydration of the polymorphs of Al(OH)$_3$ (Lippens and de Boer, 1964). The Al(OH)$_3$ polymorphs can only be examined in the electron microscope by using extensively defocused illumination. The corresponding change in the diffraction pattern associated with dehydration can be followed conveniently by gradually increasing the intensity of illumination on the specimen. Diffraction patterns obtained from norstrandite, Al(OH)$_3$, in the course of a heating programme of this kind are illustrated in Fig. 14. During the dehydration, the single crystal of norstrandite loses water with the development of a defect spinel structure, and becomes porous. The fine-pore structure gives intense phase-contrast effects even on thin-crystal specimens. The corresponding amplitude contrast can be observed in out-of-focus conditions.

A similar series of diffraction photographs taken during the heating of low-temperature tridymite in the electron beam are illustrated in Fig. 15. Tridymite is orthorhombic with a superlattice at room temperature. On being heated in the beam of the electron microscope, the superlattice reflections disappear, owing to transformation to the high-temperature structure.

### REFERENCES

Brindley, G. W. and Nakahira, M. (1959). The kaolinite–mullite reaction series: 1. A survey of outstanding problems. *J. Am. Ceram. Soc.* **42**, 311.
Chapman, J. A. and Zussman, J. (1959). Further electron optical observations on crystals of antigorite. *Acta cryst.* **12**, 550.
Comer, J. J. and Turley, J. W. (1955). Replica studies of bulk clays. *J. appl. Phys.* **26**, 346.
Doherty, P. E. and Leombruno, R. R. (1964). Transmission electron microscopy of glass-ceramics. *J. Am. Ceram. Soc.* **47**, 368.
Fleet, S. G. and Ribbe, P. H. (1963). An electron-microscope investigation of a moonstone. *Phil. Mag.* **8**, 1179.
Fleet, S. G. and Ribbe, P. H. (1965). An electron-microscope study of peristerite plagioclases. *Mineralog. Mag.* **35**, 165.
Gard, J. A. (1964). Electron Microscopy and Diffraction. *In* "The Chemistry of Cements," Vol. 2, p. 243. Academic Press, London.
Gard, J. A. and Taylor, H. F. W. (1957a). An investigation of two new minerals: rhodesite and mountainite. *Mineralog. Mag.* **31**, 611.
Gard, J. A. and Taylor, H. F. W. (1957b). A further investigation of tobermorite from Loch Eynort, Scotland. *Mineralog. Mag.* **31**, 361.
Glossop, A. B. and Pashley, D. W. (1959). The direct observation of anti-phase domain boundaries in ordered copper-gold (CuAu) alloy. *Proc. R. Soc.* **250A**, 132.
Heidenreich, R. D. (1964). "Fundamentals of Transmission Electron Microscopy." Wiley, New York.
Hirsch, P. B., Howie, A., Nicholson, R. B., Pashley, D. W. and Whelan, M. J. (1965). "Electron Microscopy of Thin Crystals." Butterworths, London.

Kay, D. (1965). "Techniques for Electron Microscopy." (2nd Ed.). Blackwell, Oxford.

Lippens, B. C. and de Boer, J. H. (1964). Study of phase transformations during calcination of aluminium hydroxides by selected area electron diffraction. *Acta cryst.* **17**, 1312.

McConnell, J. D. C. (1962). Electron-diffraction study of subsidiary maxima of scattered intensity in nepheline. *Mineralog. Mag.* **33**, 114.

McConnell, J. D. C. (1965). Electron optical study of effects associated with partial inversion in a silicate phase. *Phil. Mag.* **11**, 1289.

McConnell, J. D. C. and Fleet, S. G. (1963). Direct electron-optical resolution of anti-phase domains in a silicate. *Nature, Lond.* **199**, 586.

Menter, J. W. (1957). "Proceedings of the Stockholm Conference on Electron Microscopy," p. 88. Almqvist and Wiksell, Stockholm.

Nixon, H. L. and Weir, A. H. (1957). The morphology of the Unter-Rupsroth montmorillonite. *Mineralog. Mag.* **31**, 413.

Ross, M. (1959). Mineralogical applications of electron diffraction. II. Studies of some vanadium minerals of the Colorado Plateau. *Am. Miner.* **44**, 322.

Ross, M. and Christ, C. L. (1958). Mineralogical applications of electron diffraction. I. Theory and techniques. *Am. Miner.* **43**, 1157.

Smith, D. G. W. and McConnell, J. D. C. (1966). A comparative electron-diffraction study of sillimanite and some natural and artificial mullites. *Mineralog. Mag.* **35**, 810.

# CHAPTER 8

# Infrared Absorption Spectroscopy

## R. J. P. LYON

*Geophysics Department, Stanford University, California, U.S.A.*

## I. Introduction

The infrared spectrum of a mineral, obtained by absorption, reflection or emission techniques is uniquely characteristic of that mineral, and it can give in effect a "picture" of the structural formula without recourse to chemical investigation. Qualitative analyses of unknowns can generally be made without any detailed understanding of the mechanisms involved in the interaction between the infrared energy and the sample, and with carefully controlled experimental techniques it is possible to provide semi-quantitative evaluations for the constituent minerals of a mixture.

The character of each spectrum is determined by the vibrational modes of the atomic or molecular structure, be it solid, liquid or gas. These are a complex function of the inter-atomic distances, bond angles and bond forces, and the relative masses of the constituent atoms. The infrared analytical method is extremely sensitive to short-range (nearest-neighbour) ordering, and thus may be contrasted with X-ray diffraction, which relies mainly upon longer-range ordering and the periodic repetition of atoms. For example, infrared absorption is an extremely useful technique for the study of glasses, since the onset of crystallization can be simply revealed by the emergence of detail in a spectrum. A glass yields as strong a spectrum as does a crystalline material, but many more specific absorptions appear as secondary peaks when the crystal structure becomes more ordered.

Theoretical discussion on the origin of the infrared spectrum will be given only brief treatment here. This Chapter is designed rather as a guide to practical techniques and to the literature for further detailed reading and study. It is based upon nearly a decade of familiarity with the techniques used for mineralogical analysis. There are a great many excellent studies in the literature, and it is hoped that this compilation will show the breadth of the material available. As with many other sensitive techniques, only careful repetitious analytical practice will yield good results. Such a disciplined approach will reveal a wealth of chemical and structural information, a valuable complement to optical, X-ray and other methods.

## II. Theory of Infrared Spectroscopy

Electromagnetic radiation of different energy levels has markedly different effects upon atomic and molecular associations. The higher energy of ultra-violet and visible radiation causes transitions of energy levels of electrons in the outer orbitals of atoms, which appear as absorption bands in the resultant spectrum. When molecules or groups of atoms are excited by the lower-energy infrared radiation, however, then only vibrational and rotational modes can occur.

Molecules or groups of tightly bound atoms are in continuous motion, vibrating and rotating about their centre of gravity, and the frequencies of these periodic motions are specific for the particular atomic group and for no other.

Within a group of atoms any inter-atomic bond may be considered to be like a spring, and this can vibrate at a definite frequency. In addition to linear vibration there can also be bending and twisting kinds of motion, and in a complex molecule with many types of atoms with differing bond strengths and bond angles, the resultant motions can be compared to those of a complex spring system in continuous vibration. In addition to the fundamental vibrational frequencies, there are usually present in the spectrum weak harmonics (or overtones), and also "combination bands", which occur at the sum or difference of two or more fundamental or harmonic vibrations.

The stretching frequencies for pairs of bonded atoms make an important contribution to the infrared spectrum, and for the simple case of a diatomic molecule A-B the frequency is given theoretically by the expression† :

$$\nu = \frac{1}{2\pi c} \left(\frac{k}{\mu}\right)^{\frac{1}{2}}$$

where $\nu$ = wave number (cm$^{-1}$); $k$ = force constant (dynes/cm); $c$ = velocity of light; $\mu$ = reduced mass of the system A–B = $M_A \times M_B / M_A + M_B$.

If radiation of a given wave number impinges on a specimen containing a bond with corresponding vibration frequency, energy is absorbed. If radiation of some other wave number enters, not matched by that of an interatomic bond, then it passes through the specimen without change.

The assignment of absorptions to specific bond types in a solid is, however, an extremely difficult task. Although some degree of success can be had by an empirical approach to assignment, caution must be stressed.

Most metal–oxygen bonds have fundamental vibrations in the region between 2000 and 850 cm$^{-1}$ (5 and 12 $\mu$m). All silicates show marked absorption near 1000 cm$^{-1}$ (10 $\mu$m) caused by the Si–O stretching-vibration, with Si–O bending-vibrations at longer wavelengths 665 to 400 cm$^{-1}$ (15–25 $\mu$m). Considerable complexities appear in these longer wavelength spectra and these cause difficulties in their correct interpretation.

For simple molecules (usually gases), it is possible to predict from symmetry the fundamental vibrational frequencies to be expected. For most solids this is extremely difficult, and particularly so for complex minerals, such as the silicates. An empirical approach is therefore used. Groups of isostructural minerals are selected, and the patterns associated with characteristic Si–O configurations are determined for each major group (Adler, 1963). Launer (1952, p. 774) initiated an excellent study of this type, and there have been several more recent attempts to refine these procedures (see Nahin, 1955, p. 115; Stubican and Roy, 1961b).

Stubican and Roy (1961b) have successfully used *synthetic* clays of known composition to assign peak absorptions. Even when this degree of knowledge exists about a sample, however, assignments made for one structural group do not apply always to another. Stubican and Roy, elsewhere (1961a, p. 202)

† $\nu$ is defined as the reciprocal of the wavelength ($\nu = 1/\lambda$) of the radiation measured in cms. The product $\nu c$, where $c$ is the velocity of light ($3 \times 10^{10}$ cm/sec) gives the frequency in cycles/sec. The wavelength in microns is given by $10^4/\nu$.

P

emphasize this difficulty and describe the effect that atoms in the "secondary sphere of co-ordination" in an atomic configuration have upon the spectra. These two papers should be read carefully before advancing too far into the problems of assigning absorption peaks (see also Farmer, 1958, pp. 834–839; 1964, p. 301; Lyon and Tuddenham, 1960a; Tuddenham and Lyon, 1959).

The 1000–330 cm$^{-1}$ (10–30 $\mu$m) region is also the region of the fundamental vibrations for most of the inorganic salts (carbonates, nitrates, sulphates, phosphates, etc.) that occur in terrestrial rocks, and thus it is the region of paramount importance to the mineralogist.

Infrared analytical techniques are capable of distinguishing minerals within four major groups. (a) Minerals of relatively constant chemical composition (e.g., quartz). (b) Minerals that exhibit ranges of composition (e.g., between two end members as in the plagioclase feldspars, and olivines, or with more complex variations as in the pyroxenes and amphiboles). The infrared spectra clearly show small differences in the amounts of major elements, but not of trace elements or low-level impurities. (c) Minerals of constant chemical composition but with different crystal structures (e.g., $SiO_2$ as quartz, tridymite, cristobalite, stishovite or coesite; $CaCO_3$ as calcite or aragonite; potassium feldspar as microcline or orthoclase). (d) Minerals that vary both in structural modification and chemical composition (e.g. sodium-rich plagioclase with high- and low-temperature forms which may range from $An_0$ to $An_{20}$ in composition).

The inorganic anion groups have strong, usually simple, absorption spectra. Some of these are listed in Table 1 and shown in Fig. 1. A strong

TABLE 1

*Wavelengths and frequencies of the principal anion absorptions (see Fig. 1)*

| | Absorption peaks | |
| Group | Wavelength ($\mu$m) | Wave number (cm$^{-1}$) |
| --- | --- | --- |
| $CO_3^{2-}$ | 6·90– 7·09 | 1450–1410 |
| | 11·36–12·50 | 880–800 |
| $NO_2^-$ | 7·14– 7·70 | 1400–1300 |
| | 11·90–12·50 | 840–800 |
| $NO_3^-$ | 7·09– 7·46 | 1410–1340 |
| | 11·63–12·50 | 860–800 |
| $PO_4^{3-}$ | 9·09–10·50 | 1100–950 |
| $SO_4^{2-}$ | 8·85– 9·26 | 1130–1080 |
| | 14·71–16·40 | 680–610 |
| All silicates | 9·09–11·10 | 1100–900 |

absorption within one of these bands implies that a given functional anion group (e.g., $CO_3$) is present; the wave number of this strong absorption peak or that of smaller peaks elsewhere in the spectrum will indicate to which metal cation the group is bonded (such as to calcium in $CaCO_3$ or to magnesium in $MgCO_3$). Intermediate values for the principal absorptions may be considered diagnostic of solid solutions. Details for the carbonates appear in Table 4 (p. 385).

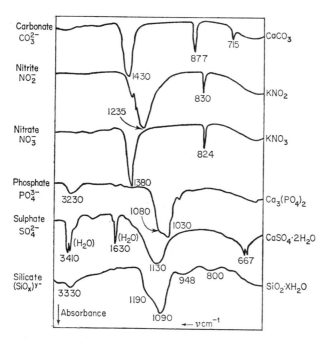

FIG. 1. Absorption spectra of the principal inorganic anions. This figure is redrawn from Miller and Wilkins (1952). Peak positions are given in wave numbers (reciprocals $\times$ $10^4$ of wavelengths in microns).

The following are useful references: Adler (1950), Hunt et al. (1950), Hunt and Turner (1953), Keller et al. (1952), Launer (1952), Miller and Wilkins (1952).

## III. Experimental Methods

The following discussion will be concerned with the techniques used to obtain the spectrum. Each method has its own specific operations, applications and merits and, wherever possible, drawbacks for mineralogical work will be noted and references cited for more detailed study.

This article is not concerned with the design and operating requirements of the recording equipment, as most laboratory texts in use today will give the reader adequate descriptions. A useful general handbook of this type is that of Willard *et al.* (1960), of which Chapter 6 is devoted to the practical methodology of infrared laboratory techniques and experimentation.

## A. CALIBRATION TECHNIQUES

To ensure that accuracy is maintained, "external" calibration techniques are used. These are so termed because calibration points are placed on each recording sheet *before* and *after* each analysis is made. Absorption peaks for water vapour ($2 \cdot 276$ $\mu$m, 3737 cm$^{-1}$ and $23 \cdot 87$ $\mu$m, 419 cm$^{-1}$); for carbon dioxide ($14 \cdot 99$ $\mu$m, 667 cm$^{-1}$); for ammonia vapour ($10 \cdot 07$ $\mu$m, 993 cm$^{-1}$ and $11 \cdot 01$ $\mu$m, 908 cm$^{-1}$) and for a polystyrene film ($6 \cdot 25$ $\mu$m, 1600 cm$^{-1}$) are often used for this purpose. Each point on the absorption curve can therefore be corrected to an absolute value based on the known wavelength or frequency values for these calibration points. For most curves it is not necessary to correct each peak by more than $\pm 2$ cm$^{-1}$.

If internal calibration is desired some material with known absorptions is introduced into the sample (Nujol, a common medium for specimen preparation can itself be used as an internal standard).

## B. ABSORPTION-ANALYSIS METHODS

Historically, spectral analysis commenced with transmission measurements (thin sheets or plates of minerals were placed in the infrared beam; see Coblentz 1908, part 5). With more interest being directed toward the absorption bands themselves (rather than the "transparent" regions), thinner and thinner plates were used. Crushed minerals were then mixed into slurries in infrared-transparent plates (rock salt). The compilations of Hunt *et al.* (1950) were the first major results to derive from this method. Other media (Nujol, mineral oils, etc.) were selected both because they possessed few absorption spectra, and for the ability of oils to decrease the scattering of the infrared beam.

Most standard techniques for infrared study involve analyses of samples in solvents, but these are of little use for minerals. The most popular method in use today for minerals is the alkali halide pelleting techniques, wherein the sample is mixed in powdered halide and pressed into a clear polycrystalline pellet. Scattering is reduced by the matching of refractive indices of the host and the powder. This method allows quantitative sample preparation.

## 1. Samples and Standards

So little sample (0·3 mg) is used with the potassium bromide pellet method in absorption analysis that purity is critical; hand-picking under a binocular microscope is often necessary to ensure homogeneity.

Standards are a vital necessity, and it should be emphasized that any analytical method is only as good as its standards. The small amount needed (1–10 mg) means that requests to other workers for material for standards are not unreasonable.

Infrared analysis is extremely sensitive to certain minerals, e.g., quartz, carbonates, etc., and 1 or 2% of these as impurities can give marked variations in the resultant spectrum. Also, one can often find variations in the spectra obtained from different parts of inhomogeneous crystals (e.g. adularia).

## 2. Sample Preparation

For optimum spectra the samples must be ground to a particle size below the wavelength of the radiation being used. Generally this is below 2 $\mu$m.

Duyckaerts (1959, p. 203) has drawn attention to several difficulties that are encountered when working with solid materials in making infrared-absorption studies. Control of particle size is of vital importance, and this control is necessary whether samples are handled as suspensions in potassium bromide, as mineral oil mulls or as deposited films. The grinding method described here ensures a good reproducibility of particle size.

## 3. Alkali Halide Pellets

Solid samples may be mixed at a low level of concentration with a powdered alkali halide, and then pressed into a clear disc or pellet. This technique is the most satisfactory method for handling rock and mineral samples for infrared analysis (Tuddenham and Lyon, 1960b; Farmer, 1955; Milkey, 1958), although with many minerals that contain OH groups, other techniques (such as mulling with Nujol) should be tried.

Most procedures use potassium bromide as the embedding medium, but other halides (potassium chloride or iodide) may be used. (For silver chloride techniques, see Pytlewski and Marchesain, 1965.)

Any concentration of sample in the KBr can be selected, but for silicates, from 0·15% to 0·25% is usually suitable. Preparation involves grinding 10 mg of the sample by hand with 10 drops of absolute alcohol in a 60-mm mullite mortar until the alcohol evaporates. This amount of grinding usually reduces the grains to a suitable size so that refractive index and scattering effects are minimized. A 1·5-mg portion of this pre-ground sample is added to 1·00 g of infrared-quality potassium bromide and blended in a dentist's amalgamator (Wig-L-Bug), by using a metal vial with a glass bead. A plastic vial or a plastic ball should not be used, or spurious absorptions due to abraded plastic will result (Lyon, 1963a). Enough of the blend (350 mg) is weighed to form a disc of the desired thickness and then pressed in a vacuum die. About 65 tons pressure per square inch is adequate to obtain permanently clear discs. These can be stored and used years later, as long as they are re-dried overnight in a vacuum oven.

It is necessary to grind the samples under alcohol because the structure of many minerals, particularly those containing OH groups, can be obliterated in a few minutes by vigorous dry grinding, and careful, consistent preparatory grinding, coupled with adequate blending, is quite critical. If these factors are controlled, then linear calibration curves for quantitative analysis can be obtained with mineral specimens (Nahin, 1955, p. 113; Lyon et al., 1959, p. 1048; Tuddenham and Lyon, 1959, p. 378; Tuddenham and Lyon, 1960a, p. 1631).

### 4. Deposited Films

Hunt et al. (1950) and Hunt and Turner (1953, p. 1170) describe the use of films of finely divided minerals deposited from alcohol dispersions on rock-salt windows. This was a good technique for rapid studies, but suffered from excessive scatter by the particles in the wavelength regions below 4 $\mu$m.† The window was weighed before and after depositing the film and so a rough quantitative measure could be obtained. This method gives preferred orientations in the specimen and is not recommended unless specifically desired, as for O–H orientation studies.

Ulrich (1961, p. 16) reviews methods for obtaining absorption spectra by reflection off an aluminium-mirror substrate. The infrared beam impinges at a 30° incident angle on to a very thin clay coating. It passes through to the mirror, is reflected back through the clay layer again, and emerges, producing an absorption spectra of the clay.

### 5. Nujol Mulls

To avoid the scattering effects given by finely divided materials, the mineral powders may be immersed or mulled in a suitable oil (Nujol). This oil is a fluor-hydrocarbon with a minimal number of absorptions. Few occur in the regions of interest for mineral analysis.

Miller and Wilkins (1952, p. 1254) compiled an excellent catalogue of inorganic and mineral spectra, in which these oil-mull techniques were used. The techniques suffer occasionally from interference with the absorption bands of the oil, but more from the lack of adequate specimen control, when quantitative estimations are required.

† These studies also suffered somewhat from the settling techniques used to collect the fine grain sizes. In a mixture of several minerals settling in the water column, an enhanced concentration of the heavier particles often resulted, and platy and micaceous minerals were left in suspension.

## 6. Oriented Clay "Papers"

Serratosa and Bradley (1958) and Serratosa et al. (1962) describe the use of a "paper" of oriented dried clay to secure spectral absorption data from the O–H vibrations of the clay particles (see also Wolfe, 1963).

## 7. Other Techniques—Differential Analysis

Modern spectrophotometers are of double-beam design. The signal is reflected along two parallel paths through the equipment and later integrated by a chopping mirror. If no sample is present, then the chopped signals are of equal strength and the pen traces a zero line. If the sample absorbs in one beam, the chopped signals differ and the pen records this absorption.

If both beams have samples of equal absorption at any given wavelength, then a matched signal will be sensed and a "zero" recorded. A given material (A) may be "subtracted" from a spectrum of a mixture (A and B): if it is placed at the correct concentration in the reference beam, only the spectrum of (B) will appear. If too much energy is removed from the signals by this process, however, the recording pen will become very sluggish. Scanning speeds must be kept very slow for this technique. The method requires two discs, one in each beam, and the difference spectrum between the two samples is automatically obtained. Minor components may be identified, or a contaminant "removed". Two potassium bromide discs are prepared: disc (a) containing the unknown sample and disc (b) containing the estimated concentration of the contaminant. If the differential absorption peaks now become slightly reversed into "reflection" peaks, then there is too much of the material in the reference beam (see also Clark, 1960, pp. 417–420).

Quartz is by far the principal contaminant with a strong absorption spectrum in most soil and mineral specimens. This technique is extremely useful for "removing" it from a sample. It is also applicable to cation- or anion-exchange studies (as with $PO_4^{3-}$, $NH_4^+$), by running discs of (a) clay + exchanged material, differentially against (b), the original clay before treatment (Eyring and Wadsworth, 1956).

# IV. General Analytical Results

## A. ADSORBED WATER, HYDROXYL AND WATER OF CRYSTALLIZATION

Mineralogists and chemists distinguish broadly two types of water by their behaviour upon heating—one that is lost at or below 110°C (called $H_2O^-$, and generally adsorbed on crystal surfaces), and one lost above 110°C (called $H_2O^+$ or "constitutional water").† In the potassium bromide pellet technique,

---

† Adsorbed water can, however, be held more tenaciously, particularly on materials of a very fine-grained nature.

"adsorbed water" can occur either on the sample or adhering to the potassium bromide itself, and the spectrum peaks for these are very nearly identical and cannot be distinguished.

In a mineral like gypsum there is "water of crystallization" that is essential to the preservation of the structure, and this can be present as "n" multiples of $H_2O$ per formula unit. In certain minerals there can be "zeolitic water", which can be driven in and out of the sample without affecting the structure in the temperature range of $100° - 250°C$. Minerals may also contain $(OH)^-$ groups of various types bonded in varying positions within the crystal lattice that give rise to the water driven off at higher temperatures. The infrared-absorption method can usually distinguish each type of water, adsorbed, water of crystallization, zeolitic water and $(OH)^-$, because they give rise to different peaks through differing degrees of hydrogen bonding in the various forms (see especially Wickersheim and Korpi, 1965, p. 582; Nakamoto et al., 1955). Much work remains to be done in this field, particularly with samples believed to contain "hydronium", $(H_3O)^+$, ions (White and Burns, 1963).

Figure 2 shows absorption spectra for the O–H stretching region in minerals, observed between 3700 and 3125 cm$^{-1}$ (2·7 and 3·2 μm). These

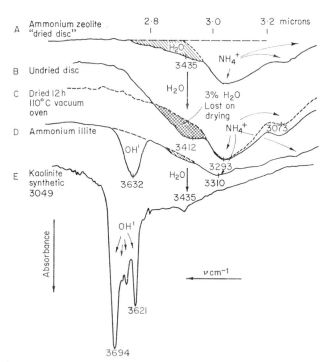

FIG. 2. Absorption spectra for $H_2O$, $NH_4^+$ and several types of $OH^-$.

spectra were prepared from samples at 0·15% concentration embedded in potassium bromide pellets and were examined with excellent resolution on a (Perkin–Elmer 221) grating infrared spectrophotometer.

Spectra A, B and C indicate the effects of drying overnight at 110°C in a vacuum oven, and it is clear that all of the water physically adsorbed on the potassium bromide pellet can be removed by this simple preliminary technique. Spectrum D indicates the presence of the $NH_4^+$ ion and shows the O–H stretching frequencies of an ammonium illite. Spectrum E indicates that with the degree of resolution available, *four* O–H stretching frequencies are observable in a kaolinite specimen (see Newnham, 1961). At the wavelength of maximum absorption for $H_2O^-$ (3435 cm$^{-1}$, approximately 2·91 $\mu$m) the spectrum indicates that little "water" is retained on the kaolinite specimen. This fact can be confirmed from the absence of an O–H bending absorption at 1610 cm$^{-1}$.

### B. SYNTHETIC VERSUS NATURAL MINERALS

Because of inherent difficulties in reaching equilibrium in times available in the laboratory, synthetic minerals often crystallize in a disordered state. When this involves aluminium/silicon diadochy as in the tetrahedral sites in silicate minerals, it is often undetectable by X-rays because of the similarities in scattering power for the two elements. Infrared analysis, however, is often sensitive to such substitutions, particularly where a charge unbalance arises ($Al^{3+}$ for $Si^{4+}$) (Laves and Hafner, 1956; Hafner and Laves, 1957; Laves, 1960).

Stubican and Roy (1961b) were unable to note the effects of the $Al^{3+}/Si^{4+}$ substitutions in a series of synthetic chlorites either by X-ray diffraction or by infrared-absorption analysis. By using natural chlorite samples, however, this substitution can be shown in the strong 9–10 $\mu$m region quite readily (Tuddenham and Lyon, 1959). Similar sensitivity to $Al^{3+}/Si^{4+}$ tetrahedral substitution was noted for lepidolite, muscovite, biotite and phlogopite micas (see Lyon and Tuddenham, 1960a) and is strongly suggested for the pyroxene and amphibole groups (Lyon, 1963b).

Excellent spectral resolution was obtained for samples of synthetic kaolinite (Fig. 3), and for a group of iron analogues of aluminous microcline and sanidine (Fig. 4).

In Fig. 3 the critical 1085 cm$^{-1}$ peak is not as well resolved for synthetic kaolinite as for natural kaolinite; this same poor resolution is often found in the disordered kaolinites called fire-clays. But the resolution of the four OH peaks in the stretching frequencies around 3700 cm$^{-1}$ and in the Al–O–OH bending frequency at 935–911 cm$^{-1}$ is excellent. (See also Wickersheim and Korpi, 1965.)

The differences between the two polymorphs of potassium feldspar, sanidine (high temperature, disordered) and microcline (low temperature,

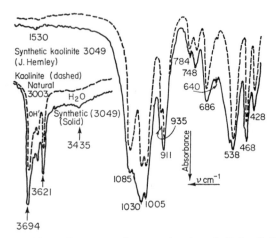

FIG. 3. Absorption spectra for natural and synthetic kaolinite.

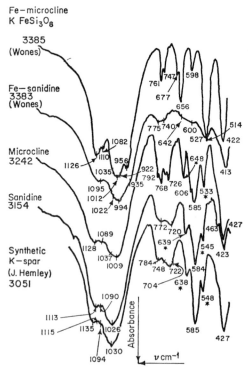

FIG. 4. Absorption spectra for natural and synthetic alkali feldspars. Synthetic iron feldspars are compared with the natural aluminium feldspars sanidine and microcline, and with a synthetic aluminium sanidine.

ordered), are clearly seen in Fig. 4. The region from 770–720 cm$^{-1}$ is the best for indicating disorder in the feldspar structure, and a definite peak shift (from 648–639 cm$^{-1}$) occurs between the two forms.

The ferric iron analogues are seen in the top two curves in Fig. 4. The iron microcline clearly shows a well resolved spectrum indicative of a high degree of order in the structure. The iron sanidine is quite disordered as deduced from the lack of resolution in its spectrum. The similarity between the two sets of spectra is striking in the 1100–900 cm$^{-1}$ (Si–O) region. The substitution of Fe$^{3+}$ for Al$^{3+}$ leads to the changes of peak location from the 650–530 cm$^{-1}$ region to about 514 cm$^{-1}$. The 422 cm$^{-1}$ peak is assigned to a Si–O bond because it does not change position.

### C. QUANTITATIVE ANALYSIS

Different analytical methods have different sensitivities for particular minerals in a mixture, and it is often advisable to use more than one method for this reason.

The infrared spectrometer can be used for quantitative analysis of mixtures, although sample preparation and dilution controls are critical. The alkali halide method gives the maximum accuracy, and statistical analyses can be made to assess this fact (Tuddenham and Lyon, 1960a).

Presuming that it has been possible to obtain a requisite high degree of precision in sample preparation, one can analyze the infrared spectrum for its components in the following manner. It is assumed that the sample in an alkali halide pellet is comparable to a true solution and that the Bouguer–Beer law is applicable. The absorbancy index (or molar absorbancy index) is then determined at a selected wavelength of strong absorbance for the sample.

The spectrum for a mixture in most cases is the sum of the spectra of the components, and it is possible to determine the concentration of each component at particular wavelengths, specific to each compound, where absorbancies do not overlap. A few such wavelengths for minerals in typical silicate matrices (soils, rocks and clays) are presented in Table 2.

TABLE 2

*Specific absorption wavelengths for common minerals*

| Mineral | Specific absorption wavelength ($\mu$m) |
|---|---|
| Quartz | 12·6 or 12·8 |
| Plagioclase | 13·48 (varies with An %) |
| Microcline | 15·45 |
| Orthoclase | 15·6 (broad) |
| Calcite | 7·0, 11·4, 14·02 |
| Kaolinite | 2·7, 2·75, 10·8, 11·0 |

The absorbancy index for each mineral in a mixture can be determined from a baseline drawn across the curve at one or more characteristic wavelengths. A series of linear equations can be set up and solved simultaneously for the desired concentrations. (Lyon et al., 1959, p. 1051; Willard et al.,1960, p. 146.)

## V. Infrared Absorption Spectra for Selected Mineral Groups

### A. CARBONATES

Commonly carbonates show strong spectral peaks at about 1400 cm⁻¹, medium strength at 860–880 cm⁻¹, and occasionally an extra band at 700–760 cm⁻¹ (see Moenke, 1962, p. 16–19).

### 1. Calcite Group

Magnesite, smithsonite, cobalt carbonate, siderite, rhodochrosite and calcite can be distinguished by band structures between 710 and 1820 cm⁻¹. Strong $CO_3^-$ group bands appear between 1425–1450 and 860–887 cm⁻¹. The diagnostic peak is weak and varies from 712 cm⁻¹ (calcite) to 748 cm⁻¹ (magnesite). (See Table 3.)

TABLE 3

*Carbonate group absorption peaks showing variation with bonded metal*

| Mineral | Wavelength (μm) | | | Wave number (cm⁻¹) | | |
|---|---|---|---|---|---|---|
| *Calcite sub-group* | | | | | | |
| Calcite ($CaCO_3$) | 6·97 | 11·45 | 14·04 | 1435 | 873 | 712 |
| Rhodochrosite ($MnCO_3$) | 6·98 | 11·53 | 13·76 | 1433 | 867 | 727 |
| Siderite ($FeCO_3$) | 7·03 | 11·55 | 13·57 | 1422 | 866 | 737 |
| Smithsonite ($ZnCO_3$) | 7·05 | 11·49 | 13·43 | 1418 | 871 | 744 |
| Magnesite ($MgCO_3$) | 6·90 | 11·27 | 13·37 | 1450 | 887 | 748 |
| *Dolomite sub-group* | | | | | | |
| Ankerite Ca [$Fe(CO_3)_2$] | 6·90 | 11·27 | 13·77 | 1450 | 877 | 726 |
| Dolomite Ca [$Mg(CO_3)_2$] | 6·97 | 11·35 | 13·70 | 1435 | 881 | 730 |
| *Aragonite sub-group* | | | | | | |
| Cerussite ($PbCO_3$) | 6·94 | 11·89 | 14·77 | 1440 | 841 | 677 |
| | (7·17) | | | (1395) | | |
| Witherite ($BaCO_3$) | 6·92 | 11·61 | 14·43 | 1445 | 860 | 693 |
| Strontianite ($SrCO_3$) | 6·80 | 11·61 | 14·14 | 1470 | 860 | 707 |
| Aragonite ($CaCO_3$) | 6·90 | 11·61 | 14·04 | 1450 | 860 | 712 |

Data from Moenke (1962) and Adler and Kerr (1963a, Table 1).

### 2. Dolomite Group

Dolomite, ankerite and kutnahorite [$CaMn(CO_3)_2$] may be distinguished by peaks at about 726–729 cm⁻¹.

### 3. Aragonite Group

The main peak at about 700 cm$^{-1}$ may be used to recognize members of this group except that for both aragonite and calcite it is at 712 cm$^{-1}$. In the detailed study of the aragonite–calcite problem by Adler and Kerr (1962), the authors suggest that a band at 860 cm$^{-1}$ is specific to aragonite, with one at 877 cm$^{-1}$ specific to calcite.

Adler and Kerr (1963a, b) discuss in detail the relationships between the spectra of the various carbonates and their dependence upon such factors as cation substitution and symmetry changes.

Azurite, malachite, hydrozincite, auricalcite and phosgenite may show a doublet- or triplet-splitting of the main 1425–1450 cm$^{-1}$ peak. The OH groups of the basic zinc and copper carbonates give peaks between 3250 and 3450 cm$^{-1}$ in the O–H stretching region. Spectra from other less common carbonates are given by Moenke (1962, p. 18).

### B. SULPHATES

The sulphate minerals have also received considerable attention, and among the better studies are those by Adler (1963) and Adler and Kerr (1965) whose paper on "Variations in Infrared Spectra, Molecular Symmetry, and Site Symmetry of Sulphate Minerals" is an excellent exposition of the relationship between decreasing symmetry and the details of an infrared spectrum.

A few typical sulphate mineral absorptions are tabulated below. The fundamental frequencies of the SO$_4^{2-}$ ion are given by Herzberg (1945) as 1104 ($\nu_3$), 981 ($\nu_1$), 613 ($\nu_4$) and 415 ($\nu_2$) cm$^{-1}$.

TABLE 4

*Absorption maxima for the sulphate ion in various sulphate minerals*

| Mineral | Site Symmetry | Absorptions (cm$^{-1}$) | | | | | |
|---------|---------------|------|------|------|------|------|------|
| | | $\nu_3$ | | | $\nu_1$ | $\nu_4$ * | |
| Potash alum | C$_3$ | 1093 | | | | — | |
| Thenardite | D$_2$ | 1116 | | | 991 | 635 | 615 |
| Anhydrite | C$_{2v}$ | 1153 | 1119 | | — | 675 | 616 | 597 |
| Gypsum | C$_2$ | 1142 | 1114 | | 1004 | 670 | 600 |
| Glauberite | C$_2$ or C$_1^4$ | 1135 | 1101 | | — | 633 | 608 |
| Blödite | C$_1^4$ | 1157 | 1124 | 1009 | 993 | 823 | 720 | 615 |
| Barytes | C$_s$ | 1179 | 1120 | 1083 | 985 | — | |
| Antlerite | C$_s^4$ | 1152 | 1111 | 1071 | 987 | — | |
| Jarosite | C$_{3v}$ | 1185 | 1087 | 1006 | ? | — | |
| Alunite | C$_{3v}$ | 1205 | 1087 | 1027 | 911 | — | |

(Table after Adler and Kerr, 1965, Table 5). * Data from Moenke (1959, Tables 4–6).

## C. BORATES

Borates may be formed of $BO_3$ (plane 3-fold) functional groups and also may contain $BO_4$ tetrahedra (see Moenke, 1959 and 1960).

TABLE 5

*Absorption maxima for borate minerals*

| | Absorptions expressed in $cm^{-1}$ | | |
|---|---|---|---|
| *Sodium borates and sassolite* | | | |
| Kernite | 832 | 1020 | 1375 |
| Borax | | 1002 | 1358 |
| Sassolite | 805 | 1198 | 1460 |
| *Magnesium borate* | | | |
| Pinnoite | 960 | 1165 | 1300 |
| Lüneburgite | 890 | 1022 | 1285 |
| Inderite | 822 | 1020 | 1415 |
| *Calcium borate* | | | |
| Colemanite | 940 | 1232 | 1340 |
| Pandermite | 902 | 1066 | 1378 |
| Ulexite | 1008 | 1065 | 1420 |
| Hydroboracite | 950 | | 1400 |
| *Borosilicates* | | | |
| Howlite | | 962 | |
| Danburite | 950 | 971 | 1142 |

After Moenke (1960, p. 191–215). See also Akmanova (1962).

## D. NITRATES

Nitrate minerals have a strong absorption around 1300–1400 $cm^{-1}$ with a subsidiary peak at 800–840 $cm^{-1}$. Ammonium nitrate is further differentiated by the N–H valence vibrations at 3020 and 3130 $cm^{-1}$. Occasionally the N–H vibration at 1400 $cm^{-1}$ is covered by a two-fold degeneracy of the $NO_3^-$ vibration. No far-infrared absorptions were found between 820 and 300 $cm^{-1}$ by Moenke (1962, p. 15).

## E. PHOSPHATES

A detailed study of the effect of symmetry and chemical substitution is again that of Adler (1964) in his study of the phosphate mineral series, the pyromorphites. In particular the effects of $PO_4^{3-}$, $AsO_4^{3-}$ and $VO_4^{3-}$ substitutions are discussed therein, as tabulated on p. 387.

Several studies of the calcium phosphate (apatite) series have been made in connection with dental and bone studies. The hydroxy and carbonate–apatite forms can be readily identified by peaks at 3540 and 1450 cm$^{-1}$, respectively (see Fischer and Ring, 1957; Romo, 1954; Underwood et al., 1955; Posner et al., 1960).

TABLE 6

Absorption maxima for phosphate, arsenate and vanadate minerals

| Mineral | $PO_4{}^{3-}$ | | $AsO_4{}^{3-}$ | | $VO_4{}^{3-}$ | |
|---|---|---|---|---|---|---|
| Apatite | 1092 | 966 | — | — | — | — |
| | 1040 | | | | | |
| Pyromorphite | 1022 | 926 | — | — | — | — |
| | 967 | | | | | |
| Mimetite | — | — | 816 | — | — | — |
| | | | 714 | | | |
| Vanadinite | — | — | — | — | 801 | 871 |
| | | | | | 739 | |

After Adler (1964, Table 2).

### F. HALIDES

The naturally occurring halides, halite, sylvine and fluorite, are transparent or show only weak absorptions from 5000–400 cm$^{-1}$, and therefore are often used for prisms, lenses and cells for other infrared studies (see especially Moenke, 1962, p. 11).

Ammonium chloride is an exception here, and the effect of the $NH_4^+$ ion can be seen by its absorption at 3100 and 1410 cm$^{-1}$. Moenke's assignments are: 1410 cm$^{-1}$, bending vibrations; 3100 cm$^{-1}$, NH valence vibrations.

Atacamite, $Cu_2(OH)_3Cl$, shows some spectral peaks below 600 cm$^{-1}$.

### G. OXIDES

Magnetite, gahnite, franklinite, chromite and Mg–Al spinel can be differentiated between 400 and 800 cm$^{-1}$. Hematite shows spectral peaks at beyond 600 cm$^{-1}$. Alumina shows marked differences between its α- and γ-forms by peaks shifting from 725 to 825 cm$^{-1}$, respectively, and a peak at 450 cm$^{-1}$ occurring in α-alumina.

Titanates show marked differences, as follows:

| | | | |
|---|---|---|---|
| Rutile $TiO_2$ | 600 | 535 cm$^{-1}$ | |
| $FeTiO_3$ (synthetic) | 674 | 533 | 439 cm$^{-1}$ |
| $MgTiO_3$ (synthetic) | 723 | 563 | 472 cm$^{-1}$ |
| $NiTiO_3$ (synthetic) | 692 | 615–565 | 455 cm$^{-1}$ |

The more important peaks for certain oxide minerals are listed in Table 7.

TABLE 7

*Peak absorptions ($cm^{-1}$) for certain oxides*

| Spinel | 685* | | | 515B |
|---|---|---|---|---|
| Magnetite | | | 575BW | |
| Hematite | | | 550* | 475 |
| Anatase | 1000W | | | 470BW |
| Brookite | | | 530B | 420 |
| Rutile | | | 530B | 415 |
| Ilmenite (natural) | 700W | | 540SB | 455 |

W – Weak; S – Strong; B – Broad.
* Diagnostic peak.

## H. HYDROXIDES

The ability of infrared absorption to distinguish $H_2O$ molecules from OH groups was recognized by Coblentz in his early (1905) studies. The O–H stretching vibrations occur between 3300 and 3700 $cm^{-1}$, and the bending vibration is located between 1640 $cm^{-1}$ ($H_2O$) and 980 $cm^{-1}$ (Al–O–H) in kaolinite group minerals. An excellent study of the bauxite group minerals is one by Frederickson (1954). A considerable amount of process control in aluminium production is performed with infrared-analysis techniques (see also Wickersheim and Korpi, 1965; Kolesova, 1959; Kolesova and Ryskin, 1959).

Care must be exercised that one examines hydrated minerals in some medium like Nujol, as well as in a potassium bromide pellet, and that extensive dehydration studies are made of the embedded sample. (See Farmer, 1958, p. 830, for results of dehydration of pellets to 200°C.)

Assignment of absorption peaks to single O–H bond types is probably an over-simplification of the complex interactions present (Farmer, 1964, p. 303).

## J. SILICATES

*1. Nesosilicates*

*Olivine.* The olivines show very simple spectra (comparable to many of the inorganic anions shown in Fig. 1), with peaks changing wave number in a predictable manner with cation substitution. The general pattern of the spectrum persists unchanged from sample to sample, but the whole pattern shows marked wave-number changes, as seen in Table 8.

This mineral group has been the subject of an exhaustive study by Duke and Stephens (1964). The slight departures from complete independence of the tetrahedra in the solid crystal were considered by Duke to be justifiable reasons for modifying theoretically derived absorption-peak positions to match observed values.

TABLE 8

*Absorption maxima for the olivine group*

| Sample | Absorption peaks (cm$^{-1}$) | | | | | | |
|--------|-----|-----|-----|-----|-----|-----|-----|
| Forsterite ($Mg_2SiO_4$) | 983 | 950 | 888 | 838 | 605 | 504 | 410 |
| Fayalite ($Fe_2SiO_4$) | | 947 | 873 | 829 | 558 | 482 | 410 |

*Garnet.* The garnet minerals may be divided into two main groups: the "pyralspites" with a general formula of $R_3^{2+}Al_2(SiO_4)_3$, and the "grandites" with a general formula of $Ca_3R_2^{3+}(SiO_4)_3$. Within each group intermediate compositions commonly occur, but intermediates between the two groups are comparatively rare.

The general appearance of the spectrum is again similar for all members of the groups, but peak positions vary in wave number with cation substitution, as shown in Table 9.

TABLE 9

*Peak positions in the garnet group*

| Mineral | Formula | Absorption peaks (cm$^{-1}$) | | | | | | |
|---------|---------|-----|-----|-----|-----|-----|-----|-----|
| *Pyralspite* | | | | | | | | |
| Pyrope | $Mg_3Al_2(SiO_4)_3$ | 970 | 900 | 872 | — | 575 | 482 | 460 |
| Almandite | $Fe_3Al_2(SiO_4)_3$ | 970 | 902 | 880 | 638 | 570 | 480 | 455 |
| Spessartite | $Mn_3Al_2(SiO_4)_3$ | 955 | 892 | 870 | 632 | 562 | 472 | 452 |
| *Grandites* | | | | | | | | |
| Grossularite | $Ca_3Al_2(SiO_4)_3$ | 915 | 860 | 840 | 618 | — | 540 | 470 | 450 |
| Andradite | $Ca_3Fe_2(SiO_4)_3$ | 895 | 840 | 820 | 592 | — | 512 | 482 | 438 |
| Uvarovite | $Ca_3Cr_2(SiO_4)_3$ | 900 | 840 | 825 | 609 | — | 540 | 455 | 425 |

(After Moenke, 1961b, Table 3).

## 2. Sorosilicates

The sorosilicate group is more complex structurally, and this is shown by the infrared spectra. Idocrase has both ($SiO_4$) and ($Si_2O_7$) groups and is half-way between the nesosilicates and the sorosilicates. Spectral maxima occur at 1012, 916 and 480 cm$^{-1}$. Lawsonite, $CaAl_2(OH)_2[Si_2O_7]H_2O$, has 6-co-ordinated Al–O and (OH$^-$) groups, linked by the $Si_2O_7$ groups. Spectral maxima occur at 995, 945, 880 and 580–480 cm$^{-1}$. The spectrum also shows evidence of the presence of weakly bonded $H_2O$.

## 3. Cyclosilicates

*Beryl*, $Be_3Al_2Si_6O_{18}$. Minerals of the beryl group are very similar to one another and have very strong, contrasting spectra. Compositional changes may be clearly defined in the 1015–1058 cm$^{-1}$ region with hydroxyl absorptions shown at 3700 and 3600 cm$^{-1}$. (For beryl studies and for Be–O assignments in other beryllium minerals, see Plyusnina, 1963.)

*Cordierite*, $Al_3(MgFe)_2(Si_5Al)O_{18}$. Cordierite spectra are broadly comparable to beryl spectra, but differ in the 800–700 cm$^{-1}$ region in having fewer peaks. Substitution of $Al^{3+}$ for $Si^{4+}$ takes place in parallel with the replacement of $Mg^{2+}$ for $Al^{3+}$ in the above structural formulae. Both substitutions tend to produce spectra characterized by fewer peaks. Hydroxyl absorptions can be defined at 3676 cm$^{-1}$.

### 4. Inosilicates

*Pyroxenes.* (a) Orthopyroxenes. An analyzed series of 21 orthopyroxenes has been studied, and their spectra clearly show the changing detail of an infrared spectrum with substitution in the Mg–Fe series (see Lyon, 1963b, Fig. 21). (b) Clinopyroxenes. Figure 5 shows the absorption spectra for a group of clinopyroxenes. The similarity of the infrared spectra of omphacite,

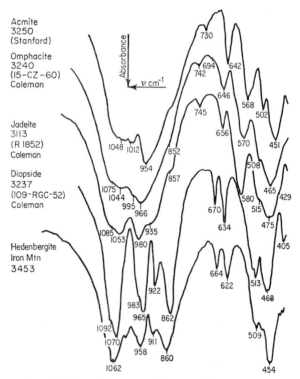

FIG. 5. Absorption spectra for some clinopyroxenes.

acmite and jadeite may be noted, and also the fact that although diopside and hedenbergite show comparable spectra, they differ considerably from the other three pyroxenes.

*Amphiboles.* The amphibole group is chemically very similar to the pyroxene group, but permits a much greater Al–Si substitution in the $Si_4O_{11}$ chains. Hydroxyl groups are an essential part of the structural formulae of the amphiboles, and evidence of these may be seen in the infrared spectra in the region around 3600 cm$^{-1}$.

Amphiboles that have been studied are the hornblendes and the alkali amphiboles, glaucophane and riebeckite. Because of almost overwhelming complexities of possible substitutions, spectra of these minerals are the most difficult to interpret. Examples are shown in Fig. 6.

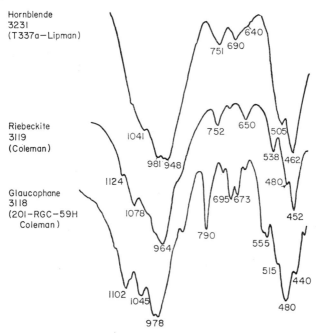

FIG. 6. Absorption spectra for some amphiboles. Hydroxyl absorptions often appear near 3600 cm$^{-1}$.

## 5. Phyllosilicates

*Tri-octahedral layered silicates.* Spectral assignments by Stubican and Roy (1961c, p. 35) for $Mg^{2+}$ and $Fe^{2+}$ tri-octahedral phyllosilicates are as follows:

| | |
|---|---|
| (Si–O) stretching | 1100–900 cm$^{-1}$ |
| (Si–O) "unassigned" | 668 cm$^{-1}$ |
| (Si–O–Si) bending | 460–430 cm$^{-1}$ |

Examples of spectra from the micas are shown in Fig. 7, and it is seen that the tri-octahedral mica biotite is readily distinguished from the di-octahedral muscovite and paragonite. Strong absorption bands appear near 3636 and 3610 cm$^{-1}$ for most micas, but are not shown in Fig. 7.

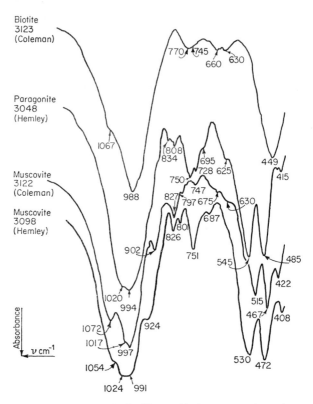

FIG. 7. Absorption spectra for the phyllosilicates, biotite, paragonite and muscovite. Note the effects of substitution in the micas. Hydroxyl region not shown.

The anisotropy of absorption in the infrared spectra of single crystals of natural micas has been investigated by Vedder (1964).

The changes in spectrum with Al–Si substitution in a series of lepidolites have been noted by Lyon and Tuddenham (1960a) and are illustrated in Fig. 8. Changes in the 1084–1130 cm$^{-1}$ peak and in the detail of the 992–962 cm$^{-1}$ peak should be noted.

The spectrum for talc has been given by Farmer (1958) and for chlorites by Tuddenham and Lyon (1959) and by Hayashi and Oinuma (1965). Spectra of

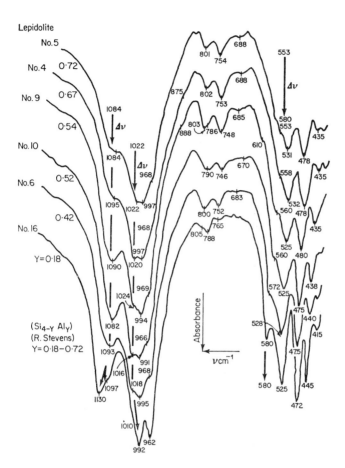

FIG. 8. Absorption spectra for a series of lepidolite micas showing the spectral variation with various amounts of tetrahedral aluminium.

the different varieties of serpentine mineral were published by Brindley and Zussman (1959).

*Di-octahedral phyllosilicates.* The spectra of muscovite and paragonite are shown in Fig. 7. Strong hydroxyl absorptions appear near 3636 and 3610 $cm^{-1}$ in most micas (and illites, montmorillonites).

For the kandite (kaolinite group) minerals, the principal peaks are shown in Table 10.

TABLE 10

*Absorption peaks for the kandite minerals*

| Kaolinite | 3705, 3670W; | 1105, | 1031*, 1006; | 935*, 909*, 545 |
| | | | | (and others) |
| Nacrite | 3712*, 3660*, 3630; | 1120, | 1035, 1003; | 940*, 918, 800 |
| | | | 700, | 545 |
| Halloysite | 3705*, 3630; | 1100W, 1040; | | 951, 545 |
| | | | | (and others) |

W = Weak      * = Diagnostic

See van der Marel (1961); Newnham (1961); Lyon (1963b).

## 6. Tectosilicates

*Polymorphs of SiO$_2$.* General assignments by Saksena (1940, 1958, 1961) for spectral peaks in quartz are:

Si–O stretching    9·10–12·5   $\mu$m   (1100–800 cm$^{-1}$)

Si–Si stretching   12·5 –16·67 $\mu$m   (800–600 cm$^{-1}$)

Si–O–Si bending   21·74–23·3  $\mu$m   (460–430 cm$^{-1}$)

Important peaks in the spectra from several forms of silica are given in Table 11 and are illustrated in Fig. 9.

TABLE 11

*Absorption peaks for forms of silica (cm$^{-1}$)*

| Fused quartz | 1098 | | 804* | | | 468 | | | |
| Quartz | 1078 | | 798 | 779* | | 455 | | | |
| Cristobalite | 1100 | | 800 | 730* | | 650 | 600* | 485 | 435 |
| Coesite | 1218 | 1077 | 837 | 814 | 794* | 683 | 600 | 561 | 445 |
| Stishovite | | | 885* | | | 672 | 560 | | |

* Diagnostic

Further calculations have been carried out for quartz by Kleinman and Spitzer (1962).

The stishovite structure has silicon in sixfold co-ordination, and so its spectrum is markedly different from those of the other polymorphs of SiO$_2$ in which Si has the more usual fourfold co-ordination. A parallel comparison is that between the "quartz" and "rutile" forms of GeO$_2$, which show absorption peaks at 870 cm$^{-1}$ for the former, and 715 cm$^{-1}$ for the latter (Dachille and Roy, 1959).

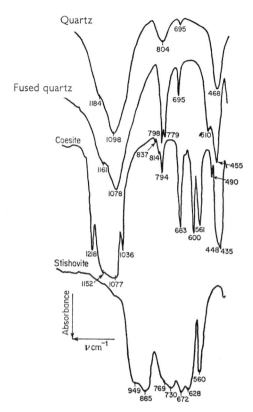

FIG. 9. Absorption spectra for polymorphs of $SiO_2$: quartz, fused quartz, coesite and stishovite.

*Feldspars.* The spectra of some potassium feldspars are illustrated in Fig. 4 (p. 382), and those for some members of the low temperature plagioclase series are shown in Fig. 10.

For the low-temperature plagioclases, the infrared spectra indicate some kind of structural discontinuity at about $An_{33}$ and $An_{67}$. The latter effect is more clearly shown; the peaks at 929 cm$^{-1}$ and 755 cm$^{-1}$ become significant above the break at $An_{67}$. (See Thompson and Wadsworth, 1957; Lyon, 1963b).

*Nepheline.* The absorption spectra of the main hexagonal phases of the nepheline-kalsilite system are given by Sahama (1965).

## 7. Glasses

One of the most significant uses of the infrared method is its ability to analyze for amorphous materials, such as glasses (Simon and McMahon, 1953), whether these are natural volcanic glasses like obsidian, or synthetic

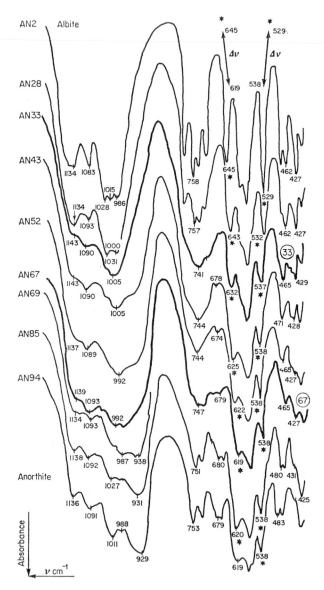

FIG. 10. Absorption spectra for plagioclase feldspars showing the spectral changes with increasing amounts of anorthite (An). Note the marked spectral changes at An$_{33}$ and An$_{67}$. Diagnostic peaks are marked with an asterisk.

glasses, like Pyrex. Glasses possess "molecular" structure and thus show infrared spectra whose patterns change with composition and progressive ordering (as in the beginning of crystallization). Glasses can often be treated, therefore, as separate phases whose infrared spectra appear in addition to those of any crystalline material present (Lyon, 1963b, Fig. 43, obsidian).

As the crystallites form in the earliest stages of crystallization of the glass, smaller peaks form on the flanks of the major absorptions. These modifications increase in size as the crystals grow and as the glass loses its content of disordered material. The silicate spectrum changes from a simple, single peaked bimodal curve to that of a multiply-peaked spectral curve typical of fully crystalline materials. In a devitrified obsidian glass, the feldspar peaks at 638, 582 and 542 $cm^{-1}$ are resolvable at an early stage.

### K. ORGANIC-SILICATE ABSORPTIONS

The following tabulation, taken from Cross (1960, p. 74), for organosilicon compounds is of some interest to mineralogists.

TABLE 12

*Organo-silicon absorption frequencies (all very strong)*

| | $cm^{-1}$ | Microns | Assignment |
|---|---|---|---|
| *Si–C vibrations* | | | |
| Si–CH₃ | 1260 | 7·94 | sym.CH₃def. |
| | *ca.* 800 | *ca.* 12·50 | Si–CH₃str. |
| Si(CH₃)₂ | 1260 | 7·94 | sym.CH₃def. |
| | 815–800 | 12·27–12·50 | Si–CH₃str. |
| Si(CH₃)₃ | 1250 | 8·00 | sym.CH₃def. |
| | 840 | 11·90 | Si–CH₃str. |
| | 755 | 13·25 | Si–CH₃str. |
| Si–phenyl | 1430–1425 | 6·99–7·02 | |
| | 1135–1090 | 8·81–9·17 | |
| *Stretching vibrations* | | | |
| Si–H | 2280–2080 | 4·39–4·81 | |
| *Si–O stretching vibrations* | | | |
| Si–O–Si and | 1090–1020 | 9·17–9·80 | Si–O str. |
| Si–O–C | | | |

*Organic compounds adsorbed on clays.* This relatively new field of study is concerned with the modifications of the infrared spectra of organic compounds induced by their adsorption on to clay and other mineral surfaces. Effects of changes in bond lengths, bond angles, and strengths are noticeable, and may be used diagnostically both for the organic material and for the clay. Differential analysis techniques are usually used and the clay matrix "cancelled out" (see p. 379).

Recent studies have shown that oriented clay (as "slips" or "papers") can induce orientations in the adsorbed organics, detectable in the same manner as orientation in the stretching of O–H bonds. See Brindley and Rustom (1958); Hoffman and Brindley (1961, p. 446); Serratosa *et al.* (1962).

### L. AVAILABLE SPECTRAL COMPILATIONS

A detailed compilation of spectra is that of Moenke (1962), containing over 350 photographic reproductions of spectra prepared for the Commission for Spectroscopy of the German Academy of Sciences in Berlin. This compilation contains infrared spectra of the most abundant and economically important halides, oxides, hydroxides, carbonates, nitrates, borates, sulphates, chromates, tungstates, molybdates, phosphates, arsenates, vanadates, and silicate minerals in the spectral region 4000–400 cm$^{-1}$ (2·5–25 $\mu$m). It also contains 40 pages of text describing the mineral groups investigated.

A secondary source (Lyon, 1962) is a bibliography with a detailed *spectral curve index*. This 76-page booklet embraces the fields of absorption, reflection, emission and transmission in minerals and related materials. More than 440 references are contained in this text. Annotations were not made, but Chemical Abstracts listings (volume, number and index pagination) are given for more than 95% of the references. The subject index contains approximately 1200 listings. Mineral spectra indicated include borates (29), carbonates (30), halides (3), phosphates and vanadates (30), sulphates, tungstates, and arsenates (30), sulphides (8), oxides and hydroxides (53). The silicates alone comprise 269 spectral listings.

Another source, now somewhat out of date, is contained in the "Infrared Spectra of Reference Clay Minerals" (Adler, 1950).

## VI. Conclusion

Mineralogists are interested in the chemical and structural composition of natural minerals and their synthetic equivalents. Infrared spectrophotometry represents a somewhat neglected, but highly promising, avenue for investigation. The study of crystal structure has been greatly advanced by the use of one particular instrument—the X-ray diffractometer—and not enough notice has been paid by mineralogists to other definitive tools. X-ray analysis is sensitive to long-range order and to the periodic arrangement of atoms in a crystalline structure. Infrared analysis, however, is much more sensitive to short-range ordering, or ordering on a nearest-atom basis. The two are essentially complementary. Infrared analysis has the ability to bridge the gaps between crystalline solid, liquid, and gas. It is an extremely valuable technique with which the mineralogist can study the effects of major-element substitutions so common in the mineralogical realm.

## BIBLIOGRAPHY

Bellamy, L. J. (1954). "The Infrared Spectra of Complex Molecules." Wiley, New York. (Principally IR theory and assignment of organic materials, some inorganic.)

Brugel, W. (1962). "An Introduction to Infrared Spectroscopy." (Translated by A. R. and A. J. D. Katritzky). Wiley, New York.

Clark, G. L. (editor) (1960). "The Encyclopedia of Spectroscopy," p. 784. Reinhold, New York. (Contains several articles, pages 450–453, 487–494, etc.)

Cross, A. D. (1960). "An Introduction to Practical Infrared Spectroscopy," p. 80. Butterworths, London. (Very useful small handbook – an excellent student first text.)

Grim, R. E. (1953). "Clay Mineralogy." McGraw-Hill, New York. (Chapter 12b is devoted to IR spectra of clays.)

Hertzberg, G. (1945). "Molecular Spectra and Molecular Structure." II, Infrared and Raman Spectra of Polyatomic Molecules, D. Van Nostrand Co., New York.

Nakamoto, K. (1963). "Infrared Spectra of Inorganic and Co-ordination Compounds." Wiley, New York.

Ulrich, W. F. (1961). "Bibliography of Infrared Applications." Bull. 754, Beckman, Inc. (collection of 678 references).

Vedder, W. and Hornig, D. F. (1961). Infrared spectra of crystals. *Advanc. Spectroscopy* **2**, 189.

Willard, H. H., Merritt, L. L., Jr. and Dean, J. A. (1960). "Instrumental Methods of Analysis" (3rd Ed.). Van Nostrand, New York. (Chapter 6 is devoted entirely to infrared techniques.)

## REFERENCES

Adler, H. H. (1950). Infrared investigations of clay and related minerals: Preliminary Report No. 8 (Infrared spectra of reference clay minerals). *Q. Am. Petrol. Inst.*, Project **49**, 1.

Adler, H. H. (1963). Some basic considerations of the application of infrared spectroscopy to mineral analysis. *Econ. Geol.* **58**, 558.

Adler, H. H. (1964). Infrared spectra of phosphate minerals: symmetry and substitutional effects in the pyromorphite series. *Am. Miner.* **49**, 1002.

Adler, H. H. and Kerr, P. F. (1962). Infrared study of aragonite and calcite. *Am. Miner.* **47**, 700.

Adler, H. H. and Kerr, P. F. (1963a). Infrared spectra, symmetry and structural relations of some carbonate minerals. *Am. Miner.* **48**, 839.

Adler, H. H. and Kerr, P. F. (1963b). Infrared absorption frequency trends for anhydrous normal carbonates. *Am. Miner.* **48**, 124.

Adler, H. H. and Kerr, P. F. (1965). Variations in infrared spectra molecular symmetry and site symmetry of sulfate minerals. *Am. Miner.* **50**, 132.

Akmanova, M. V. (1962). Investigation of the structure of native borates by infra-red spectroscopy. *J. struct. Chem.* (English Translation.) **3** (1), 24.

Brindley, G. W. and Rustom, M. (1958). Adsorption and retention of an organic material by montmorillonite in the presence of water. *Am. Miner.* **43**, 627.

Brindley, G. W. and Zussman, J. (1959). Infrared absorption data for serpentine minerals. *Am. Miner.* **44**, 185.

Coblentz, W. W. (1905, 1906, 1908). "Investigations of Infra-red Spectra." *Publs Carnegie Instn* **35, 65, 97**. Republished in a single volume with same title under the joint sponsorship of the Coblentz Society and the Perkin–Elmer Corporation, 1962.

Cross, A. D. (1960). "An Introduction to Practical Infrared Spectroscopy." Butterworths, London.

Dachille, F. and Roy, R. (1959). The use of infrared absorption and molar re-fractivities to check coordination. *Z. Kristallogr. Kristallgeom.* **11**, No. 6, 462.

Duke, D. A. and Stephens, J. D. (1964). IR Investigation of the olivine group minerals. *Am. Miner.* **49**, 1388.

Duyckaerts, G. (1959). The infrared analysis of solid substances. *Analyst, Lond.* **84**, 201.

Eyring, E. M. and Wadsworth, M. E. (1956). Differential spectra of adsorbed monolayers-n-hexanethiol on zinc minerals. *Min. Engng.* **8**, 531.

Farmer, V. C. (1955). Pressed disc technique in infrared spectroscopy. *Chemy Ind.* 586.

Farmer, V. C. (1958). The infrared spectra of talc, saponite and hectorite. *Mineralog. Mag.* **31**, 829.

Farmer, V. C. (1964). Infrared spectroscopy of silicates and related compounds. *In* "The Chemistry of Cements," Vol. II (Ed. H. F. W. Taylor), p. 289. Academic Press, London.

Fischer, R. B. and Ring, C. E. (1957). Quantitative IR analysis of apatite mixtures. *Analyt. Chem.* **29**, 431.

Frederickson, L. D. (1954). Characterization of hydrated aluminas by IR spectro-scopy. *Analyt. Chem.* **26**, 1883.

Hafner, St. and Laves, F. (1957). Order/disorder and infrared absorption. II: variation of the position and intensity of some absorptions of feldspars. The structure of orthoclase and adularia. *Z. Kristallogr. Kristallgeom.* **109**, 204.

Hayashi, H. and Oinuma, K. (1965). Relationship between infrared absorption spectra in the region of 450—900 $cm^{-1}$ and chemical composition of chlorite. *Am. Miner,* **50**, 476.

Hoffman, R. W. and Brindley, G. M. (1961). Infrared extinction coefficients of ketones adsorbed on Ca-montmorillonite in relation to surface coverage. Clay-organic studies. Part IV. *J. phys. Chem.* **65**, 443.

Hunt, J. M. and Turner, D. C. (1953). Determination of mineral constituents of rocks by infrared spectroscopy. *Analyt. Chem.* **25**, 1169.

Hunt, J. M., Wisherd, P. and Bonham, L. C. (1950). Infrared absorption spectra of minerals and other inorganic compounds. *Analyt. Chem.* **22**, 1478.

Keller, W. D., Spotts, J. H. and Biggs, D. L. (1952). Infrared spectra of some rock-forming minerals. *Am. J. Sci.* **250**, 453.

Kleinman, D. A. and Spitzer, W. G. (1962). Theory of the optical properties of quartz in the infrared. *Phys. Rev.* **125**, 16.

Kolesova, V. A. (1959). Infrared absorption spectra of the silicates containing aluminium and of certain crystalline aluminates. *Optika. Spectrosk.* **6**, 20.

Kolesova, V. A. and Ryskin, Ya. I. (1959). Infrared absorption spectrum of hydrar-gillite Al $(OH)_3$. *Optika. Spectrosk.* **7**, No. 2, 165.

Launer, P. J. (1952). Regularities in the infrared absorption spectra of silicate minerals. *Am. Miner.* 37, 764.

Laves, F. (1960). Al/Si distributions, phase transformations, and names of alkali feldspars. *Z. Kristallogr. Kristallgeom.* 113, 26.

Laves, F. and Hafner, St. (1956). Order/disorder and infrared absorption I. (Aluminium, Silicon) – Distribution in feldspars. *Z. Kristallogr. Kristallgeom.* 108, 52.

Lyon, R. J. P. (1962). "Minerals in the Infrared." Stanford Research Institute.

Lyon, R. J. P. (1963a). Sample container contamination in the infrared spectra of minerals. *Am. Miner.* 48, 1170.

Lyon, R. J. P. (1963b). "Evaluation of Infrared Spectrophotometry for Compositional Analysis of Lunar and Planetary Soils." Stanford Research Institute, Final Report under contract NASr – 49(04). Published by NASA as Technical Note D-1871.

Lyon, R. J. P. and Tuddenham, W. M. (1960a). Direct determination of aluminium in mica by infrared absorption analysis. *Nature, Lond.* 185, 374.

Lyon, R. J. P., Tuddenham, W. M. and Thompson, C. S. (1959). Quantitative Mineralogy in 30 minutes. *Econ. Geol.* 54, 1047.

van der Marel, H. W. (1961). Quantitative analysis of the clay separate of soils. *Acta Universitatis Carolinae, Geologica* Suppl. 1, 23.

Milkey, R. G. (1958). Potassium bromide method of infrared sampling. *Analyt. Chem.* 30, No. 12, 1931.

Miller, F. and Wilkins, C. H. (1952). Infrared spectra and characteristic frequencies of inorganic ions. *Analyt. Chem.* 24, 1253.

Moenke, H. (1959). Ultrarot-spektralphotometrische Bestimmung der Gesteinsbilden den Salzmineralien. *Jenaer Jb.* II, 361.

Moenke, H. (1960). Die Ultrarotabsorptionspektren Wasserhaltiger Bormineralien, des Howlithes und des Danburits im Bereich von 400 bis 1800 cm⁻¹. *Jenaer Jb.* I, 191.

Moenke, H. (1961b). Ultrarotspektralphotometrische Differenzierung von Mineralien der Granatgruppe in Spektrabereich 400 bis 650 cm⁻¹. *Veb. Carl. Zeiss, Jena, Nachrichten* 9, 82.

Moenke, H. (1962). "Mineral Specktren," p. 42. Deutsche Akademie der Wissenschaften zu Berlin, Akademie-Verlag, Berlin.

Nahin, P. G. (1955). Infrared analysis of clays and related minerals. *Bull. Div. Mines Calif.* 169, 112.

Nakamoto, K., Margoshes, M. and Rundle, R. E. (1955). Stretching frequencies as a function of distances in hydrogen bonds. *J. Am. chem. Soc.* 77, 6480.

Newnham, R. E. (1961). A refinement of the dickite structure and some remarks on polymorphism in kaolin minerals. *Mineralog. Mag.* 32, 683.

Plyusnina, I. I. (1963). IR absorption spectra of beryllium minerals. *Geochemistry, Wash.* 2, 174.

Posner, A. S., Stutman, J. M. and Lippincott, E. R. (1960). Hydrogen-bonding in calcium-deficient hydroxyapatites. *Nature, Lond.* 188, 486.

Pytlewski, L. L. and Marchesain, V. (1965). Silver chloride disk technique in infrared and visible spectrometry. *Analyt. Chem.* 37, 619.

Romo, L. A. (1954). Synthesis of carbonate-apatite. *J. Am. chem. Soc.* 76, 3924.

Sahama, Th. G. (1965). Infrared absorption of nepheline. *Bull. Commn. geol. Finl.* 218, 107.

Saksena, B. D. (1940). Analysis of the Raman and infrared spectra of α-quartz. *Proc. Indian Acad. Sci.*, Sec. A, 12, 93.

Saksena, B. D. (1958). Infrared absorption spectra of α-quartz between 4 and 15 microns. *Proc. phys. Soc. Lond.* **72**, 9.

Saksena, B. D. (1961). Infrared absorption studies of some silicate structures. *Trans. Faraday Soc.* **57** (2), 242.

Serratosa, J. M. and Bradley, W. F. (1958). Determination of the orientation of OH bond axes in layer silicates by infrared absorption. *J. phys. Chem.* **62**, No. 10, 1164.

Serratosa, J. M., Hidalgo, A. and Vinas, J. M. (1962). Orientation of OH bonds in kaolinite. *Nature, Lond.* **195**, 486.

Simon, J. and McMahon, H. O. (1953). Study of the structure of quartz, cristobalite, and vitreous silica by reflection in infrared. *J. chem. Phys.* **21**, 23.

Stubican, V. and Roy, R. (1961a). A new approach to assignment of infrared absorption bands in layer-lattice silicates. *Z. Kristallogr. Kristallgeom.* **115**, 200.

Stubican, V. and Roy, R. (1961b). Infrared spectra of layer-lattice silicates. *J. Am. Ceram. Soc.* **44** (12), 625.

Stubican, V. and Roy, R. (1961c). Isomorphous substitution and infrared spectra of the layer-lattice silicates. *Am. Miner.* **46**, 32.

Thompson, C. S. and Wadsworth, M. E. (1957). Determination of the composition of plagioclase feldspars by means of infrared spectroscopy. *Am. Miner.* **42**, 334.

Tuddenham, W. M. and Lyon, R. J. P. (1959). Relation of infrared spectra and chemical analysis for some chlorites and related minerals. *Analyt. Chem.* **31**, 377.

Tuddenham, W. M. and Lyon, R. J. P. (1960a). Infrared techniques in the identification and measurement of minerals. *Analyt. Chem.* **32**, 1630.

Tuddenham, W. M. and Lyon, R. J. P. (1960b). Infrared analysis of minerals and rocks. *In* "Encyclopedia of Spectroscopy," p. 491. Ed. G. L. Clark. Reinhold, New York.

Ulrich, W. F. (1961). Applications of specular infrared reflectance. *In* "The Analyzer," p. 14. (Beckman Inc. house journal.)

Underwood, A. L., Toribara, T. Y. and Neuman, W. F. (1955). An infrared study of the nature of bone carbonate. *J. Am. chem. Soc.* **77**, 317.

Vedder, W. (1964). Correlations between infrared spectrum and chemical composition of mica. *Am. Miner.* **49**, 736.

White, J. L. and Burns, A. F. (1963). Infrared spectra of hydronium ion in micaceous minerals. *Science* **141**, 800.

Wickersheim, K. A. and Korpi, G. K. (1965). Interpretation of the infrared spectrum of Boehmite. *J. chem. Phys.* **42** (2), 579.

Willard, H. H., Merritt, L. L., Jr. and Dean, J. A. (1960). "Instrumental Methods of Analysis" (3rd Ed.). Van Nostrand, New York.

Wolfe, R. G. (1963). Structural aspects of kaolinite using infrared absorption. *Am. Miner.* **48**, 390.

REFERENCES NOT CITED IN THE TEXT

Bassett, W. A. (1960). Role of hydroxyl orientation in mica alteration. *Bull. geol. Soc. Am.* **71**, 449.

Beutelspacher, H. (1956a). Infrared studies on soil colloids. *Sixième Congrès de la Science du Sol* **1**, 47, p. 329. Paris.

Beutelspacher, H. (1956b). Beiträge zur Ultrarotspektroskopie von Boden-Kolloiden. *Landw. Forsch.* **7**, 74.

Erd, R. C., Lyon, R. J. P. and Madsen, B. M. (1965). Infrared studies of saline sulfate minerals: discussion. *Bull. geol. Soc. Am.* **76**, 271.

Farmer, V. C. (1957). Effects of grinding during the preparation of alkali-halide discs on the infrared spectra of hydroxylic compounds. *Spectrochim. Acta* **8**, 374.

Huang, C. K. and Kerr, P. F. (1960). Infrared study of the carbonate minerals. *Am. Miner.* **45**, 311.

Kolesova, V. A. (1954). Problem of the interpretation of vibration spectra of the silicates and the silicate glasses. *Zh. eksp. teor. Fiz.* **26**, 124.

Lazarev, A. N. (1961). Spectroscopic identification of $Si_2O_7$ groups in silicates. *Kristallografiya* **6**, No. 1, 124. [Translation in *Crystallography*, 1961–62, **6**, 101.]

Lazarev, A. N. (1962). Vibrational spectra of silicates IV. Interpretation of the spectra of silicates and germanates with ring anions. *Optika Spectrosk.* **12**, 28.

Lazarev, A. N. and Tenisheva, T. F. (1961a). Vibrational spectra of the silicates. II. Infrared absorption spectra of silicates and germanates with chain anions. *Optika Spectrosk.* **10**, 79.

Lazarev, A. N. and Tenisheva, T. F. (1961b). Vibrational spectra of silicates. III. Infrared spectra of the pyroxenoids and other chain meta-silicates. *Optika Spectrosk.* **11**, 584.

Lippincott, E. R., van Valkenberg, A., Weir, C. R. and Bunting, E. N. (1958). Infrared studies on polymorphs of silicon dioxide and germanium dioxide. *J. Res. natn. Bur. Stand.* **61**, 61.

Lyon, R. J. P. and Tuddenham, W. M. (1959). Quantitative mineralogy as a guide in exploration. *Min. Engng.* **11**, 1233.

Lyon, R. J. P. and Tuddenham, W. M. (1960b). Infrared determination of the kaolin group minerals. *Nature, Lond.* **185**, 835.

Milkey, R. G. (1960a). Infrared spectra of some tectosilicates. *Am. Miner.* **45**, 990.

Milkey, R. G. (1960b). Mineral compositions and structures. *In* "Encyclopedia of Spectroscopy." Ed. G. L. Clark, p. 487. Reinhold, New York.

Moenke, H. (1961a). Ultrarot Absorption Spektralphotometrie und Silikatforschung. *Silikattechnik* **12**, 323.

Moenke, H. (1963). Entwicklung, Stand und Möglichkeiten der Ultrarotspektralphotometrie von Mineralien. *Fortschr. Miner.* **40**, 76.

Omori, K. and Kerr, P. F. (1963). Infrared study of sulfate minerals. *Bull. geol. Soc. Am.* **74**, 709.

Pfund, A. H. (1945). The identification of gems. *J. opt. Soc. Am.* **35**, 611.

Roy, D. and Roy, R. (1957). Hydrogen deuterium exchange in clays and problems in assignment of infrared frequencies in the hydroxyl region. *Geochim. cosmochim. Acta* **11**, 72.

Roy, R. (1960). High pressure – a new chemical tool. Min. Industries, Pennsylvania State College of Mining Industries, **29**, No. 5.

Tarte, P. (1960a). The infrared spectra of some silicates. Behaviour of the antisymmetrical valence vibrations of solid solutions of orthosilicates in orthogermanates. *Bull. Acad. ro. Belg. Cl. Sci.* **46**, 169.

Tarte, P. (1960b). Infrared spectrum of garnets. *Nature, Lond.* **186**, 234.

Tarte, P. (1960c). Infrared spectra of silicates. II. Determination of the structural role of Ti in certain silicates. *Silic. ind.* **25**, 171.

Tarte, P. and Ringwood, A. E. (1962). Infrared spectrum of the spinels $Ni_2SiO_4$, $Ni_2GeO_4$ and their solid solutions. *Nature, Lond.* **193**, 971.

White, J. L., Bailey, G. W., Brown, C. B. and Alrichs, J. L. (1961). Infrared investigation of the migration of lithium ions into empty octohedral sites in muscovite and montmorillonite. *Nature, Lond.* **190**, 342.

# CHAPTER 9

# Thermal Techniques

## R. J. W. McLAUGHLIN

*Department of Geology and Mineralogy, University of Melbourne, Australia*

## I. Introduction

The term "thermal techniques" has been used in this work in the wider sense to cover the investigation of all of the various measurable changes, either physical or chemical that occur with rise in temperature. There have, however, been certain exclusions, e.g. specific heat, and the chemistry of phase changes, since otherwise the list would be too lengthy. Because differential thermal analysis is such a well established technique it has been dealt with in considerably more detail than the others, viz. thermogravimetry, dilatometry, gas-evolution analysis, decrepitation, electrical conductivity and thermoluminescence. For the more common techniques abbreviations have been employed, and these are given in Table 1.

## II. Differential Thermal Analysis

### A. METHOD

The technique of differential thermal analysis is a simple one. It consists of heating simultaneously two separate samples, an unknown and an inert material, and recording the evolution or absorption of heat by the unknown

TABLE 1

*Thermal methods of analysis*

| Method of analysis | Physical parameter measured | Instrument used for detection | Abbreviated title | Applications |
|---|---|---|---|---|
| Differential thermal | Heat evolved or absorbed compared with an inert standard | Differential thermal analyser | d.t.a. | Crystallo-graphic inversions, melting, decomposition, dehydration |
| Thermogravi-metric | Weight change | Thermobalance | t.g.a. | Dehydration, decomposition |
| Dilatometric | Volume change | Dilatometer (or X-ray) | dil. | Phase changes |
| Gas evolution | Gas evolved identified | Thermal con-ductivity cell, gas chromato-graph etc. | g.e.a. | Pyrolysis (for carbon, sulphur etc.) |
| Decrepitation | Sound evolved | Amplifier | — | Geothermo-metry |
| Electrical conductivity | Resistance change | Potentiometer | — | Sintering, melting |
| Thermo-luminescence | Light evolved | Photo-multiplier | — | Geothermo-metry |

relative to the standard. The difference in temperature between the two is measured usually by a differential thermocouple. In this device, there are two junctions, one in the sample and one in the inert standard, linked to form a circuit in which the current flowing depends on the temperature difference between the two junctions. The samples are mounted symmetrically in a block of metal or ceramic, and they should have approximately the same dimensions and volume. The differences in temperature between sample and unknown as they are heated are recorded, either manually or automatically, and are plotted against actual temperature of the sample.

There are a number of changes that take place on heating which give rise to evolution or absorption of heat. Reactions involving crystallization, oxi-dation and some other chemical reactions are exothermic, whereas phase changes, dehydration, decomposition and crystalline inversions are generally endothermic. A substance undergoing any of these changes will become hotter or cooler than the inert material, and the current flowing in the differ-ential thermocouple is recorded. The more energetic the reaction the greater the differential. The differential temperature when plotted against temperature of the sample gives a curve that will be fairly reproducible for the mineral in question and for a particular apparatus. A typical curve is given in Fig. 1 for a specimen of anglesite, $PbSO_4$, contaminated with a very small amount of pyrite ($<5\%$). This curve shows three distinctly different types of reaction with increasing temperature. Peak A is exothermic and is due to the oxidation of the small quantity of pyrite; the reaction begins slowly and gains impetus as temperature rises. Peak B represents a crystalline inversion of orthorhombic

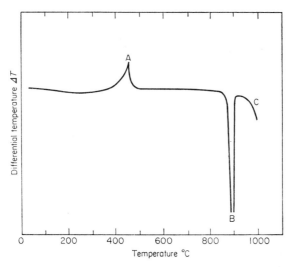

FIG. 1. Differential thermal curve for anglesite with approximately 5 % pyrite contamination. A, oxidation of pyrite; B, crystallographic inversion of anglesite; C, decomposition of anglesite.

to monoclinic anglesite and like many inversions the change is rapid. Peak C represents decomposition and, like peak A, initial reaction is slow, but impetus is gained quite quickly. In these curves it is relatively easy to recognize the points at which the curve leaves the baseline and later returns to it, as well as the point of maximum deflection (the thermal peak). Sluggish reactions are broader and sometimes extrapolation is necessary to decide the positions of departure from, and return to, the baseline. As the mass of reacting material is approximately proportional to the area under the curve, the recognition of departure from, and return to, the baseline is most important where quantitative estimations are desired.

### B. APPARATUS

Considerable variability exists in the component parts of the apparatus, and because of this and the dynamic nature of the method, results are variable from one apparatus to another. Excellent descriptions of different types of instruments have been given by Mackenzie (1957) and Wendlandt (1964). The essential constituents (see Fig. 2) are sample block, differential-temperature measurer, furnace, furnace controller, temperature and differential-temperature recorder. The technique lends itself quite well to multiple analysis (Kulp and Kerr, 1947).

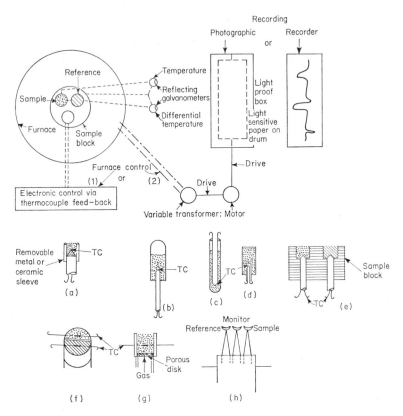

Fig. 2. Differential thermal analysis equipment. Block diagram of apparatus for differential thermal analysis. Two types of recording and furnace control are illustrated schematically. Various types of thermocouple assembly are shown. (a) Has a removable ceramic or metal sleeve about the sample, facilitating sample removal and cleaning; (b) shows the arrangement for volatile materials in which a sealed tube is used; (c) illustrates the thermojunction introduced from above the sample; (d) illustrates a thermocouple in a sheath, surrounded by reactive sample; (e) is a common type of sample block of either metal or ceramic and is very similar to the one shown in the block diagram above; (f) shows a ceramic crucible with a central division enabling both sample and reference to be contained in the same holder; (g) shows the arrangement used for dynamic d.t.a. permitting gas flow through a porous disc; (h) is an enlargement of the type of holder in which the thermocouple junctions have been drilled out to give cavities that hold small amounts of sample and reference material.

*1. Sample Block*

Sample blocks vary in their size, the choice depending somewhat on the weight of sample used. Controversy has arisen about the choice of material for the block, either metal or ceramic. Generally ceramic appears favourable at temperatures up to about 500°C and metal above this temperature. Ceramic holders give larger peaks, but being porous may influence reactions involving gases. Metal holders are much more easily machined.

Sample weights have varied from 5–6 g to 10 μg, with the most usual ranging from 100–400 mg. The sample well in the block is nearly always cylindrical and ca ¼ inch diameter by ⅜ inch deep. The geometry of the sample well will influence certain reactions, e.g. long narrow holders with impervious walls will inhibit oxidation (Arens, 1951). Similarly, the use of covering devices over the sample, unless a gap is introduced, will give variable results (e.g. decomposition of siderite; Rowland and Jonas, 1949). For special investigations tightly fitting screwed lids have been employed, e.g. for isobaric dehydration (Bohon, 1961). The use of small sample weights has much to recommend it and has been less explored than might be expected. In regimes in which such small samples are used, the temperature-sensing element itself serves as a sample holder (Mazières, 1959; Bogue, 1954; Welch, 1954).

*2. Temperature Measurement*

Although a thermometer, thermopile, thermistor or resistance element may be used as a temperature-measuring device, the general choice has been a thermocouple. Unless special applications are being investigated, the choice of thermocouple is either chromel/alumel or platinum/platinum–rhodium. The lower e.m.f. of the latter may be boosted either electronically or by the technique of multiple thermocouples (Lodding and Sturm, 1957). They are furthermore less liable to corrosion and will operate to higher temperatures. Where corrosion has become a problem, various preventative methods have been devised, and these usually involve shielding the thermocouple (Kopp and Kerr, 1957, 1959; Fitch and Hurd, 1959; Dunne and Kerr, 1960; Kupka, 1960). At high temperatures, the thermocouples must be shielded from inductive effects, by earthed metallic shielding (Pask and Warner, 1954). For operation at extremely elevated temperatures, a tantalum carbide/graphite junction has been suggested (Brewer and Zavitsanos, 1957).

*3. Furnace and Controller*

Heating is usually by an electrical furnace surrounding the specimens. [There have been only a few applications in which the thermocouples themselves constitute the heating elements (Bogue, 1954; Welch, 1954), but such methods offer considerable possibilities.] The winding of the electric furnace is normally of nichrome, but kanthal or platinum are used if temperatures in excess of 1000°C are required. The rate of temperature rise in the furnace must be controlled and reproducible, since otherwise results would not be comparable between experiments. There are many methods of accomplishing the smooth temperature rise required, and they generally fall into two groups; either motor-driven variable transformers are used, or the more refined technique of employing a feed-back circuit controlled by a thermocouple in the furnace. The variable transformer type of control is regulated by coupling a synchronous electric motor to the voltage control of the transformer. By using a specific gearing ratio, determined by experiment, fixed rates of voltage rise are obtained. The transformer output, fed to the electric furnace, gives regulated temperature rise. A still smoother effect may be obtained by using cams in place of gears, the cam shapes being pre-determined by experiment. The use of variable-speed motors, combined with cams, makes a very flexible arrangement, whereby the thermal regimes may be varied or kept constant as required. The electronic devices for furnace-temperature control, in which feed-back from a thermocouple or resistance element is used, although giving excellent stability, are generally much more expensive than the variable transformer method, and in many ways not as flexible. One important point that must always be considered is that on–off types of control, typical of some electronic circuitry, must be avoided, because of the danger of inducing spurious currents in the thermocouples. This danger becomes considerable at elevated temperatures unless earthed shielding is employed (Gérard-Hirne and Lamy, 1951; Peco, 1952).

*4. Recorders*

The recording of the differential temperature between sample and standard, as well as the actual sample temperature, may be manual, photographic (via a reflecting galvanometer) or potentiometric (XY recorder). Only comparatively recently has the sensitivity of potentiometric recorders equalled that of the simple reflecting galvanometer–photographic paper technique. It is recommended that the temperature of the sample rather than that of the inert material be used (Mackenzie, 1957). Calibration of temperature should present no problem, but it is useful to note the method of Barshad (1952) who introduced minute quantities of a reactive constituent with known reaction temperature into the sample. McLaughlin (1954) used the Curie point of nickel as a standard on each curve.

### C. VARIABLES IN APPARATUS AND IN SAMPLES

In interpretative work involving d.t.a. there has frequently been disagreement amongst workers regarding the shape of the curve for a particular mineral. In view of the large number of variables that may influence the result, and the differing types of instrument, the disagreement is not surprising. The variables may be divided broadly into two groups; those arising from the apparatus and those due to the sample.

## 1. Apparatus Variables

The effects of apparatus design have already been briefly mentioned. The choice of either metal or ceramic sample holder will influence results, largely owing to the greater porosity of the ceramic, and only slightly because of differing specific heats. The latter causes higher peak temperatures to be obtained when a ceramic block is used (Webb, 1954; Mackenzie, 1954). Conduction of heat along thermocouples, especially thick wires, has been shown to be considerable (Boersma, 1955; Sewell, 1957), and lower peak areas result (Vassallo and Harden, 1962). Geometrical asymmetry of thermocouples, as well as giving baseline drift (Smyth, 1951; Cole and Rowland, 1961; Barrall and Rogers, 1962), may give other spurious effects, and thus the size of the thermocouple junction will become important especially with small sample weights. It follows that the samples must be placed symmetrically in the furnace. Variation in the heating rate gives sharper peaks as the rate increases, for at rapid heating rates there is less chance of equilibrium between sample and standard (Speil et al., 1945). The most frequently adopted heating rate is 10°C/min, but it should be noted that for specific purposes a different heating rate is more desirable, e.g. Wittels (1951) with amphiboles, Keith and Tuttle (1952) with quartz. The heating rate must be linear with no rapid fluctuations, or spurious peaks may appear (Vold, 1949).

## 2. Sample Variables

It will be readily understood that many of the differences caused by the sample will be closely inter-related. Neglecting such effects as reaction of sample on container or thermocouple, and variability of a mineral species, there are still a large number of variables.

Particle-size effects have been investigated by many workers with conflicting results (Speil et al., 1945; Grimshaw et al., 1945; Gruver, 1948; Kulp and Trites, 1951; Chihari and Seki, 1953; Carthew, 1955; Takahashi, 1959; Martinez, 1961; Bayliss, 1964). Although the area under the peak should remain the same for similar weights of substances of different particle sizes, many other factors intrude to cause variation, such as degree of crystallinity, packing tightness, gaseous diffusivity and thermal diffusivity. It is therefore desirable to work with samples that lie within close particle-size limits. Grinding should be kept to a minimum [Dempster and Ritchie (1953) on quartz; McLaughlin (1955) on dickite; Mackenzie and Milne (1953) on micas; Takahashi (1959) on talc and kaolinite]. Packing tightness is dependent upon particle size and packing technique. Only in exceptional circumstances are pressure plungers employed (Whitehead and Breger, 1950; Webb and Heystek, 1957). The general conclusion is that packing to "finger tightness" gives fairly reproducible weights and thermal curves (McConnell and Earley, 1951). The greatest effects of packing density will be shown by reactions involving loss of a volatile, or oxidation by the atmosphere.

Sample weights are kept small, for preference 100–400 mg. Above and below certain critical masses there seems to be marked departure from linearity in the curve relating weight of reactive constituent to area under the thermal peak (Wittels, 1951; Barrall and Rogers, 1962). Sample dilution will also affect the curve. Although peaks are displaced to lower temperatures, many workers have reported that dilution in the region of 25% sample–75% inert diluent, gives superior quantitative data (Grimshaw and Roberts, 1953; Sabatier, 1954; Webb and Heystek, 1957; McLaughlin, 1961). Dilution of the specimen aids evolution of a volatile, especially if this reaction is of a violent nature. This, and the higher thermal conductivity of the diluents normally used, cause larger thermal peaks (Barrall and Rogers, 1962).

### D. QUALITATIVE AND QUANTITATIVE ESTIMATION

Mathematical relationships have in most cases been concerned with relating the area under the peak to the amount of reacting material and thus to the heat of reaction. The original work of Speil *et al.* (1945) has been greatly elaborated and for a fuller discussion the reader is referred to Sewell (1957) and Wendlandt (1964). Sewell (1957) and Boersma (1955) have proposed mathematical models in which the effect of heat loss along thermocouple wires has been considered and shown to be significant; in consequence fine thermocouple wires are recommended (ca 0·2 mm). Herald and Planje (1948) devised an apparatus in which the thermocouple junction was on the exterior, and thus the effects of differing thermal conductivities of samples and standard could be minimized.

One factor that is most difficult to evaluate is the effect of gaseous diffusion on thermal conductivity. Quite apart from the heat transferred by gaseous flow, conductivity of a powder varies considerably with the gas occupying the pores (McAdams, 1942; Fishenden and Saunders, 1950). Attempts to overcome this by the use of vacuum techniques have been claimed to give more reproducible results (Whitehead and Breger, 1950; Linseis, 1951; Wittels, 1951). Dilution of the sample, by mixture with inert material, lowers the amount of gas evolved in a reaction, and thus gives results more compatible with theory. The technique is strongly recommended.

Assuming that the variables referred to previously have been kept constant, d.t.a. records are reproducible from one apparatus to another in a qualitative way, but quantitative data require greater standardization. Excellent reproducibility is usual for a sharp crystallographic inversion, e.g. $\alpha \rightleftharpoons \beta$ quartz. Changes involving release of a volatile, or oxidation, are less reproducible for reasons given before.

Although extrapolation may sometimes be necessary in determining the position at which a thermal curve leaves and returns to the baseline, the thermal peak is usually somewhat easier to decide (see Fig. 3). A point that has been termed the "characteristic temperature", by Murray and Fischer (1951), is the temperature of the reference material at the point where a line drawn at 45° to the baseline is tangential to the commencement of the peak on the d.t.a. record. This characteristic temperature is of considerable use in qualitative work for it is claimed to be less subject to variation than peak temperature (Webb and Heystek, 1957) (see Fig. 3).

The d.t.a. technique as an analytical tool has certain limitations; for example it is difficult to take a completely unknown material, subject it to d.t.a. and identify the substance from the record. Not only will various curves overlap, but the chance of reaction between components in a mixture must be remembered. The usefulness of d.t.a. as a qualitative tool does however reside in its ability to determine fairly rapidly, whether or not specific substances are present, e.g. small quantities of carbonate, kaolinite in a clay, etc. The recent publication by Mackenzie (1962) of a wide variety of data in card-index form has considerably simplified both qualitative and quantitative estimation.

Quantitative estimations have nearly all been concerned with the area outlined by the peak hence the concern in deciding the points of departure and return to the baseline. Despite the voluminous literature claiming straight-line relationship between area of peak and mass of reactant, evidence for departure at high and low mass values is strong (Sementovskii, 1958). Here is another reason why the dilution technique already mentioned is favoured. With minerals of reasonable crystallinity (e.g. quartz, carbonates) it is possible to obtain accurate estimates of the amount present. With minerals where disorder of the structure is frequent, e.g. clays, great care must be exercised in quantitative work, but this is true for most physico-chemical methods of evaluation.

Other peak parameters that have been used for quantitative estimation have been shown in Fig. 3 and are as follows. Gallitelli et al. (1954) used peak height. Dean (1947) used the cosecant of the angle formed between two lines tangential to the sides of the curve. This relationship is now known to be of limited applicability, especially as the geometrical shape of a peak is a function of many variables. Bramão et al. (1952) devised a technique termed "slope ratio". In this the peak was divided into two angles by a vertical from the baseline to the peak, and straight lines from the peak and tangential to the sides of the curve. The difference between the Dean and Bramão technique

Q*

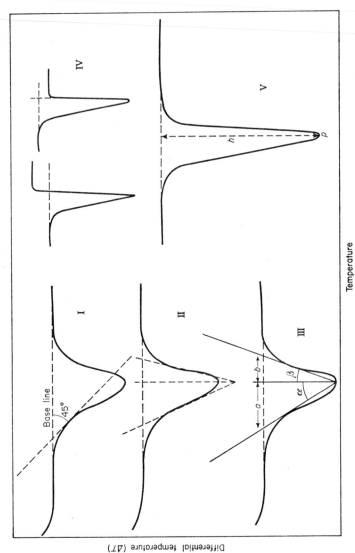

Fig. 3. Measurements on d.t.a. curves. I. Determination of "characteristic" temperature. II. Determination of peak angle, by the method of Dean (1947). III. Determination of slope ratio of the peak, by the method of Bramão *et al.* (1952). IV. Measurement of areas under the thermal peak when the baseline after the peak is above or below the original, by the method of Webb and Heystek (1957). V. Measurement of area under a peak in the normal case; peak temperature given at *p*; peak height by *h*.

is shown in Fig. 3. Dean extrapolated past the peak, Bramão measured at the peak. The slope ratio was tan $\alpha$/tan $\beta$, i.e. $a/b$ (see Fig. 3), and it was found to be related to degree of structure disorder. Carthew (1955) enlarged upon this technique and plotted curves relating slope ratio to peak area/peak width at half the amplitude. From such curves, quantitative estimations of minerals in the same group could be made and effects of particle size and lattice disorder neglected. Vold (1949) evolved a method in which the rate of change of the d.t.a. curve was measured, but difficulties arise in materials where the thermal peak does not indicate the finish of the reaction, e.g. kaolinite (Murray and White, 1949).

Double d.t.a. has been employed. In this method an estimated equivalent amount of reactant is mixed with the inert reference sample. Although in theory the two thermal effects should cancel if the amounts are the same, this is rarely true in practice. Somewhat similar is the method of Yagfarov (1961) who included a substance of known heat capacity as an internal standard. From his technique a wide variety of physical parameters may be calculated.

### E. TYPICAL CURVES

It would be impossible to include in a short space d.t.a. curves for all the various minerals that have been investigated by the method and in consequence only important species have been considered. Further data may be obtained from the index of McLaughlin and Mackenzie (1957) and the Scifax index of Mackenzie (1962). Földvàri-Vogl (1958) also gives many curves for minerals.

### 1. Elements

Unless they give a polymorphic inversion, oxidize or volatilize, elements do not give rise to d.t.a. records. Their simple compounds and silicates are much more widespread, and attention is focused upon these.

### 2. Halides

The most common, halite, shows no d.t.a. peak until melting at 801°C (Kopp and Kerr, 1959). Cryolite shows the monoclinic-to-cubic inversion. Atacamite shows loss of water from $Cu(OH)_2$ in the first two endotherms and the third represents melting of the $CuCl_2$ component (Frondel, 1950). Curves for these three minerals are given in Fig. 4.

### 3. Sulphides

Curves are complicated by the rate of oxidation, especially at elevated temperatures, where complexity is due to successive bursts of oxidation, as oxygen replaces a reducing, sulphur-rich environment (McLaughlin, 1957). With sulphides, considerable dilution is necessary, and thermocouple protection is desirable or attack will occur. Inert atmospheres have also been

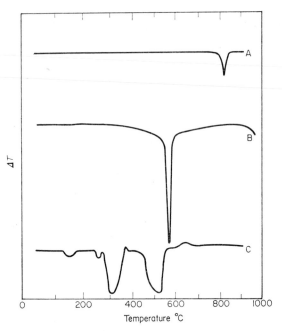

Fɪɢ. 4. Differential thermal curves for several halides. A, halite, Columbia, S. America; B, cryolite, Ivigtut, Greenland; C, atacamite. (A, B, after Kopp and Kerr, 1959; C after Frondel, 1950).

used in these investigations. Data are given in Kopp and Kerr (1957), Papailhau (1958), Kupka (1960) and Pickering (1963). A modified technique has been used by Bollin and Kerr (1961), termed pyrosynthesis, in which systems such as $CuS-Cu_2S$ and $FeS-FeS_2$ were investigated, and the d.t.a. curves showed quite clearly the development of various phases of fixed composition. Curves for the more important sulphides are given in Fig. 5.

*4. Silica Minerals*

Because of the considerable research on this group, they are considered separately and at some length. The first extensive investigation by d.t.a. was by Fenner (1913), and a second intensive research by Keith and Tuttle (1952) demonstrated by d.t.a. techniques that the $\alpha \rightleftharpoons \beta$ inversion of quartz was in fact very variable, an important result with applications to geothermometry. The main cause of the variability has been shown to be the inclusion of foreign ions in the structure, and this effect is also exhibited by the other forms of silica. A curve for quartz showing the $\alpha \rightleftharpoons \beta$ transformation is given in Fig. 6. The inversion has an energy change of 4·5 cal/g. Dempster and Ritchie (1952,

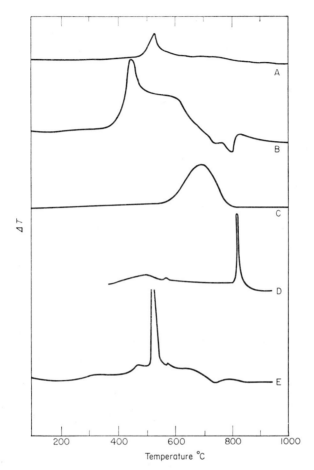

FIG. 5. Differential thermal curves for various sulphides. A, arsenopyrite (1:15), Del Oro, Canada; B, chalcopyrite (1:9), Sudbury, Canada; C, sphalerite (1:15), Ellenville, New York, U.S.A.; D, galena (1:9); E, chalcocite (1:9). Dilutions with alumina are given in brackets. (A, B, C after Kopp and Kerr, 1957; D, E after Pickering, 1963.)

1953) have demonstrated that finely ground quartz is difficult to estimate quantitatively by d.t.a., unless the distorted layer caused by grinding is removed by hydrofluoric acid or by sodium hydroxide (Boyer, 1954). The distorted layer apparently triggers the inversion, and it should be noted that this supports the conclusion reached by Perrier and de Mandrot (1922), who from a study of the elasticity moduli, claimed that the transition in quartz was a second-order reaction of the type $\alpha$ quartz $\rightleftharpoons$ disorder $\rightleftharpoons$ $\beta$ quartz (Jay, 1933; Megaw, 1946; White, 1953). Grimshaw (1953) and Grimshaw and

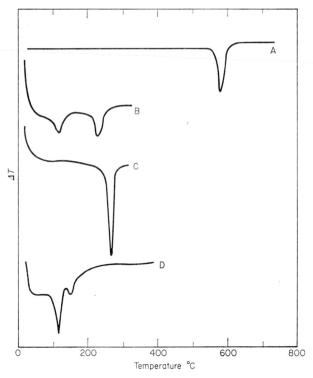

FIG. 6. Differential thermal curves of silica polymorphs. A, quartz, S. America; B, silica brick containing tridymite and cristobalite; C, silica brick containing mainly cristobalite; D, silica cement containing predominantly tridymite. (A, R. J. W. McLaughlin, unpublished work; B, C, D after Grimshaw and Roberts, 1957.)

Roberts (1957) have shown that d.t.a. of rocks, such as fine-grained quartzite give lower results for silica content than X-ray and chemical methods. They suggest that this is due to the occurrence of chalcedonic silica which does not give a reliable thermal effect at inversion. Considerable evidence from technological experience with ceramics supports their conclusions, but to the author's knowledge this finding has not been applied to geological investigations, where its importance is obvious.

The high-temperature modifications of silica considered here are tridymite and cristobalite. These have low temperature $\alpha \rightarrow \beta$ inversions that are irreversible, are extremely variable in temperature and therefore render quantitative determinations difficult (Gaskell et al., 1956). The variable temperature of these inversions has been ascribed generally to the inclusion of foreign ions, which are more readily accommodated in their more open structures. Thermal history also is important, and a longer period at high temperature favours a higher and less variable inversion temperature (Grim-shaw *et al.*, 1948). The cristobalite inversion temperature varies from 220–277°C. Tridymite has two inversions, the first at about 110°C, the second at about 163°C. In view of the findings of Keith and Tuttle (1952) on quartz, the variable inversions of tridymite and cristobalite are probably also due to in-clusion of foreign ions. It should be noted, however, that Mosesman and Pitzer (1941) and Rigby (1948) believe that rotational disorder of oxygen atoms on the Si—O tetrahedra is the cause. Curves for silica minerals are given in Fig. 6.

### 5. Iron Oxides

The two common oxides are hematite ($\alpha$-$Fe_2O_3$) and magnetite ($Fe_3O_4$). Hematite gives no observable change in d.t.a. Magnetite gives a small exo-therm at 350–400°C, due to surface oxidation (Schmidt and Vermaas, 1955) and this is greater therefore with decreasing particle size. Above the Curie point of 590°C, which may be observed in an inert atmosphere, the more open structure permits fuller oxidation, hence the broad exotherm in Fig. 7A, above 600°C.

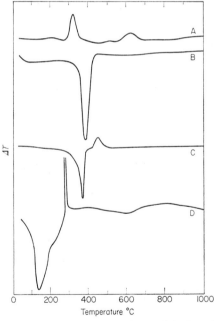

FIG. 7. Differential thermal curves for iron oxide and hydrated iron oxide minerals. A, magnetite, synthetic; B, goethite, Příbram, Czechoslovakia; C, lepidocrocite, North-ampton, Pennsylvania, U.S.A.; D, hydrated gel prepared by adding 9 M ($NH_4$)OH to 0·2 M $FeCl_3$ until the pH reached 5·0. (A after Gheith, 1952; B, D after Mackenzie, 1957; C after Kulp and Trites, 1951.)

Hydrated oxides contain two well defined species, goethite ($\alpha$-FeO.OH) and lepidocrocite ($\gamma$-FeO.OH). Goethite (Fig. 7B) shows a single endothermic peak, variable with particle size, from 385–405°C, and quantitative estimation is possible in some circumstances (Kulp and Trites, 1951). Lepidocrocite (Fig. 7C) is rarer in nature and its d.t.a. curve is more complex. The preliminary endotherm represents dehydration to maghemite ($\gamma$-Fe$_2$O$_3$) and the subsequent exotherm the transition to hematite. This latter transition is catalysed by water, and hence is variable and may render differentiation from goethite difficult (Kelly, 1956).

The truly amorphous material of nature is generally in dynamic equilibrium with natural processes; investigations on synthetic gels demonstrate rapid ageing especially at high pH or elevated temperatures. Synthetic gels give a strong exotherm (Fig. 7D), variable from 250–500°C, with position dependent upon conditions of formation. It is due to the formation of hematite (Mackenzie, 1957). Ageing of gels gives hematite or goethite (Weiser, 1935; Gheith, 1952). Most limonites of nature appear to be finely divided goethite, but lepidocrocite has also been recorded (Grim and Rowland, 1942; Kauffman and Dilling, 1950; Kulp and Trites, 1951).

### 6. Aluminium Oxides

The only natural oxide is corundum ($\alpha$-Al$_2$O$_3$), which is stable to heat. Hydrated oxides are of common occurrence and have been extensively investigated (Mackenzie, 1957). The natural minerals are gibbsite [$\gamma$-Al(OH)$_3$], diaspore ($\alpha$-AlO.OH) and boehmite ($\gamma$-AlO.OH). Gibbsite is the most common and forms the bulk of most bauxite deposits. The d.t.a. records show a strong fixed endotherm with peak at 320–330°C due to the formation of $\chi$-Al$_2$O$_3$ (Fig. 8A). There are two other endotherms, which are related to each other in size; the first at 250–300°C, the second at 525°C; they are related to the formation of boehmite which appears to be variable in amount. Many records show exotherms at 925°C and 1040°C, which represent $\chi \rightarrow \kappa \rightarrow \alpha$ transformations of Al$_2$O$_3$ (Bradley and Grim, 1951). Diaspore gives one strong endothermic peak at 540–585°C (Fig. 8B) which varies in position with particle size. Boehmite gives a peak in the same region as diaspore; (510–580°C; Fig. 8C), variable with particle size and other conditions: Karšulin et al. (1949) have suggested that its position may be related to conditions of formation. By contrast with the iron analogues, there is no evidence for hydrated alumina gels in nature; even fresh synthetic gels are crystalline boehmite, with bayerite [$\alpha$-Al(OH)$_3$] as a rarer constituent.

The determination of gibbsite and goethite, both singly and in admixture has been refined by Lodding and Hammell (1960). The sample is subjected to d.t.a. first in a hydrogen atmosphere, under which condition goethite dehydrates below 300°C and iron reduces giving amorphous Fe$_3$O$_4$, which subsequently crystallizes between 300–360°C. Replacement of the atmosphere first by nitrogen and then by air gives $Fe_3O_4 \rightarrow \gamma\text{-}Fe_2O_3 \rightarrow \alpha\text{-}Fe_2O_3$ reactions, from which the iron content can be evaluated. Allowance can then be made for the goethite contribution to the total dehydration curve (in air) for the mixture, the remainder being due to gibbsite.

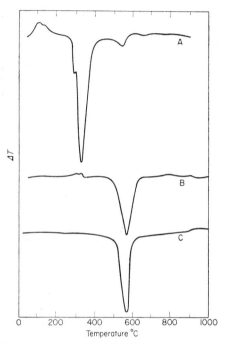

FIG. 8. Differential thermal curves for hydrated alumina minerals. A, gibbsite, New Caledonia; B, diaspore, Sverdlovsk, Urals, U.S.S.R.; C, boehmite, Swiss, Missouri, U.S.A. (After Mackenzie, 1957.)

*7. Manganese Oxides*

There is great variation in d.t.a. curves owing to the variability in composition and high sorptive properties of these minerals. On nearly all curves, there is evidence for the $\alpha \rightarrow \beta$ hausmannite transition with its peak at about 950°C and the $\beta \rightleftharpoons \gamma$ transition in the region of 1200°C (Rode, 1955). Hausmannite (Fig. 9A) naturally gives such curves, and pyrolusite (Fig. 9B) in addition to the above shows an endotherm at 650–700°C owing to bixbyite ($Mn_2O_3$) formation: this above 950°C gives hausmannite (Berg *et al.*, 1944; McMurdie and Golovato, 1948; Kulp and Perfetti, 1950). Hydrated oxides include manganite (MnO.OH) (Fig. 9C) which dehydrates with an endothermic peak at 370°C and subsequently behaves as above. Cryptomelane (Fig. 9D), on most curves, shows an endotherm in the 900–1000°C region (McMurdie and Golovato, 1948; Prasad and Patel, 1954; Rode, 1955), whereas psilomelane (Fig. 9E) shows an exotherm in this same temperature range (Kulp and Perfetti, 1950; Mackenzie, 1957).

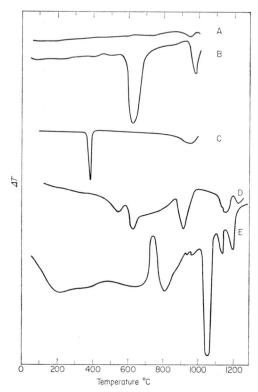

FIG. 9. Differential thermal curves for manganese oxide and hydrated manganese oxide minerals. A, hausmannite, Ilfeld, Harz, Germany; B, pyrolusite (polianite), Platten, Bohemia, Czechoslovakia; C, manganite, Ilfeld, Harz, Germany; D, cryptomelane, Tombstone, Arizona, U.S.A.; E, psilomelane (romanèchite), Schneeburg, Saxony, Germany, containing 10·45% BaO. (A, B after Mackenzie, 1957; C after Kulp and Perfetti, 1950; D after McMurdie and Golovato, 1948; E after Rode, 1955.)

## 8. Carbonates

During dissociation considerable heat is absorbed, and thus excellent d.t.a. curves are obtainable and even very small amounts (0·3%) may be estimated (Kulp et al., 1951; Rowland and Beck, 1952). There are numerous interferences that can cause variability. Decreasing particle size increases the rate of decomposition. Subsidiary oxidation phenomena as in iron and manganese carbonates, give differing curves. Increase of carbon dioxide pressure will inhibit breakdown of some carbonates more than others (Haul and Heystek, 1952), and the packing tightness and the porosity of the sample holder may cause a variable atmosphere about decomposing particles. The calcium component re-carbonates most readily, and this has been used as a method of differentiation (Rowland and Lewis, 1951). The influence of salts such as sodium chloride on the thermograms is very marked. Murray *et al.* (1951) record that the first endotherm of dolomite can be raised by 40°C or lowered by 255°C. Even 0·01% of sodium chloride will affect this reaction, and for magnesite 1% lowers the endotherm by 50°C and causes a 10% increase in peak area (Berg, 1943). Martin (1958) has demonstrated similar

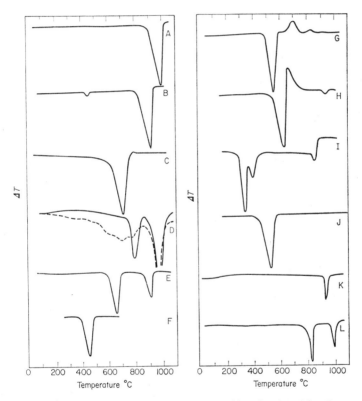

FIG. 10. Differential thermal curves for carbonates and brucite. A, calcite; B, aragonite; C, magnesite; D, dolomite, solid curve = untreated, dashed curve = treated with sodium chloride; E, huntite; F, brucite; G, chalybite (siderite); H, rhodochrosite; I, cerussite; J, smithsonite; K, strontianite; L, witherite. (A–K inclusive taken from Webb and Heystek, 1957; L, R. J. W. McLaughlin, unpublished work.)

anomalous results with soils containing soluble salts. The effects due to reaction with other oxides must also be considered (Hedvall and Heuberger, 1922).

Representative curves for various carbonates and the effects outlined above are illustrated in Fig. 10. The difference between aragonite and calcite is shown by the extra endotherm for the former. It is variable in position even under equilibrium conditions. Magnesite decomposes at a lower temperature than calcite, and in dolomite the decomposition temperature for the magnesium component is higher than for pure magnesite (Rowland, 1955). Prolonged grinding breaks down the dolomite structure and the d.t.a. record resembles a calcite and magnesite mixture (Bradley et al., 1953). Huntite behaves as a calcite and magnesite mixture. Magnesian limestones or calcareous magnesites almost invariably contain the minor component as dolomite. Chalybite ($FeCO_3$) gives exothermic changes due to the transition $\gamma$-$Fe_2O_3 \rightarrow \alpha$-$Fe_2O_3$, and magnetite formation (Rowland and Jonas, 1949), but technique plays a considerable part in determining the shape of the final record. Rhodochrosite is similar in having an exothermic peak due to conversion of MnO to $Mn_3O_4$; with increase in calcium content the peak temperature rises (Kulp et al., 1949). Stronianite shows an $\alpha \rightleftharpoons \beta$ transition, which being reversible permits estimation on cooling, even though a hysteresis of up to 80°C occurs. Witherite shows two such inversions $\alpha \rightleftharpoons \beta \rightleftharpoons \gamma$. Smithsonite and cerussite (Bayliss and Warne, 1962) are believed to form oxycarbonates as a partial break-down product. A curve for brucite has been included because of the frequent association of this mineral with carbonates. Water vapour and carbon dioxide modify the single large dehydration endotherm and shift the peak to higher temperatures. Grinding profoundly influences the bulk density and affects the area of the thermal peak. Webb and Heystek (1957) give an excellent account of the analysis of these and other carbonates.

## 9. Borates

Curves for various borates have been given by Berg et al. (1944), McLaughlin (1957), Allen (1957), and Kopp and Kerr (1959). In Fig. 11 curves are given for borax, colemanite and ulexite. Steps in the curves represent dehydration in stages, but frequently in d.t.a. the loss of water may be violent due to the rapid heating rate.

## 10. Nitrates

The common nitrates are those of potassium and sodium. They exhibit peaks due to crystallographic inversions as well as showing the melting point (Kopp and Kerr, 1959). Curves for these two nitrates and also a curve for crude "caliche" are given in Fig. 11. In this latter, arrests on the curves are due to impurities, e.g. iodates.

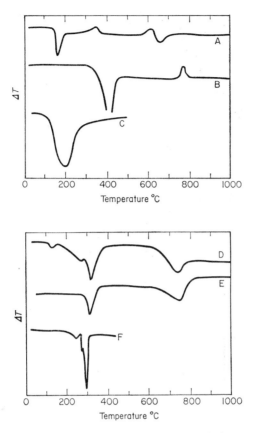

FIG. 11. Differential thermal curves for certain borates and nitrates. A, borax, Tibet;
B, colemanite, Mojave Area, California, U.S.A.; C, ulexite, Kramer District, California,
U.S.A.; D, nitre, Humboldt, Nevada, U.S.A.; E, soda nitre (nitratine), Tarapaca, Chile;
F, caliche, Chile. (A, B, D, E after Kopp and Kerr, 1959; C after Allen 1957; F after
McLaughlin, 1957.)

## 11. Sulphates

Although there are many occurrences of sulphates in nature, especially
hydrated ones, only four have been selected, and their curves are given in
Fig. 12 (and Fig. 1). For other less well known species, Cocco (1952) gives a
series of curves. The curve for gypsum shows its dehydration in two stages
and some curves show in addition a small exothermic reaction which has been
interpreted in various ways (Schedling and Wien, 1956). Anglesite (Fig. 1)
gives two endothermic effects; the first is a crystallographic inversion, and the
second is due to decomposition. The alunite-jarosite minerals being of somewhat similar structure have related d.t.a. curves. Their first endotherm repre-

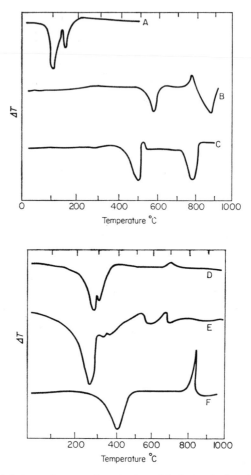

FIG. 12. Differential thermal curves for certain sulphates and phosphates. A, gypsum; B, alunite, Santa Rita, New Mexico, U.S.A.; C, jarosite, Darwin, California, U.S.A.; D, wavellite, Montgomery, Arkansas, U.S.A.; E, vivianite, Keystone, S. Dakota, U.S.A.; F, turquoise, New Mexico, U.S.A. (A after Barshad, 1952; B, C after Kulp and Adler, 1950; D, E, F after Manly, 1950.)

sents dehydration, the next exotherm the crystallization of $\alpha$-$Al_2O_3$ or $\alpha$-$Fe_2O_3$ and the final endotherm the decomposition of the double sulphate formed on dehydration.

## 12. Phosphates

Data have been given by Hummel (1949); Manly (1950); Sahores (1955); Flörke and Lachenmayr (1962). Curves are given in Fig. 12 for vivianite and wavellite. There is stepwise dehydration, and hence the d.t.a. record shows several endothermic peaks in the 300°C region (8 $H_2O \rightarrow 3\ H_2O \rightarrow H_2O \rightarrow$ anhydrous). The later portions of the record are complicated by the oxidation of ferrous iron. Turquoise gives two peaks, one endothermic and the other exothermic.

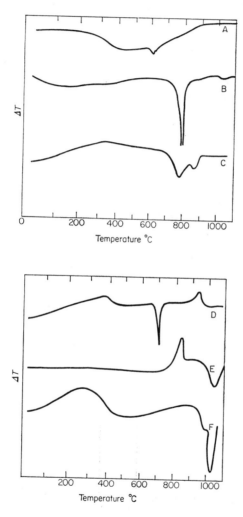

FIG. 13. Differential thermal curves for certain silicates. A, leucite, Vesuvius, Italy; B, pectolite, Bergen Hill, New Jersey, U.S.A.; C, prehnite, Cradock, Cape Province, S. Africa; D, hemimorphite, Franklin, New Jersey, U.S.A.; E, tremolite, Harvard Collection No. 97956; F, epidote, Zermatt, Switzerland. (A, B, C, D, F after McLaughlin, 1957; E after Wittels, 1951.)

## 13. Non-clay Silicates

In only a few cases do anhydrous species give any d.t.a. record, and these are due to crystallographic inversions, e.g. the leucite (Fig. 13A) endotherm commencing at 580°C. The hydrated silicates may dehydrate in one stage, as does pectolite (Fig. 13B), or in several stages, as do prehnite and hemimorphite (Fig. 13C and D). The exotherm in the 900°C region for the latter mineral is considered to be due to the formation of willemite (Faust, 1951). Amphiboles have been investigated by Wittels (1951, 1952). Water loss is at elevated temperatures and there are distinct differences between different species. The exotherm at about 820°C was related by Wittels to a crystallographic contraction, which was reversible for tremolite but irreversible for magnesian anthophyllite. A curve for epidote shows the high-temperature breakdown of this mineral.

## 14. Clay Minerals

Clay minerals have been very extensively investigated by d.t.a., and in fact it is with these minerals that the method has found greatest use. The literature is extensive, and for further information the reader is referred to the various chapters of Mackenzie (1957).

The kaolin minerals (kandites) (Fig. 14) give curves of similar type, the most highly ordered members of the group (nacrite, dickite) having higher dehydration temperatures. As disorder increases, there is more loosely held

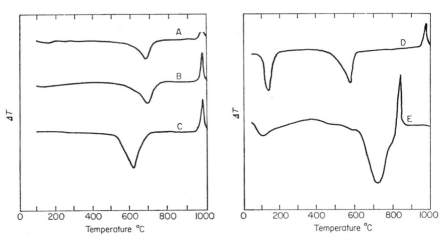

Fig. 14. Differential thermal curves for clay minerals. A, nacrite, Germany; B, dickite, Mexico; C, kaolinite (T), Utah, U.S.A.; D, halloysite, N Carolina, U.S.A.; E, serpentine (antigorite), Antigorio, Piedmont, Italy. (A, B, C, D after Kerr *et al.*, 1949 and Holdridge and Vaughan, 1957; E after Caillère and Hénin, 1957.)

inter-layer water, and hence larger endotherms in the 150°C region (halloysite). The high-temperature exotherm in the 980°C region is modified considerably by various factors, especially the presence of iron oxide and structural disorder. It is due to the nucleation of mullite and in some cases $\gamma$-$Al_2O_3$. There are additional exothermic effects above 1000°C due to further mullite development and crystallization of cristobalite (Glass, 1954). Members of the kandite group with little disorder may frequently give a small endotherm in the 900°C region, and this is considered to be due to structural collapse from the anhydride to an amorphous structure. Quantitative estimation is usually carried out by using the endotherm in the 500°C region (Holdridge and Vaughan, 1957), but attempts have also been made to utilize the 980°C exotherm (Pospíšil, 1959). Differentiation between halloysite and kaolinite may be effected by d.t.a. after treatment with ethylene glycol, which halloysite absorbs, so giving strong exothermic peaks during carbonization (Sand and Bates, 1953). Serpentine and related minerals may be regarded as the trioctahedral analogues of kandites and their d.t.a. curves are somewhat similar. The endotherm represents dehydroxylation, and the variable exotherm represents the crystallization of forsterite (Caillère and Hénin, 1957; Veniale, 1962).

The smectites (montmorillonite minerals) (Fig. 15), because of their expansile structures, show loss of inter-layer water below 200°C. This region is profoundly affected by the nature of the inter-layer cation, giving stepped curves of great diagnostic value (Greene-Kelly, 1957). The dehydroxylation process is variable, but above 500°C gives a broad endotherm. Some types give a weak endotherm in the 900°C region, and its origin is probably similar to that occurring for kandites (Grim and Bradley, 1940; Bradley and Grim, 1951; Earley et al., 1953; Brindley and Nakahira, 1959). Allaway (1949) has demonstrated the usefulness of piperidine treatment as a diagnostic aid. On subsequent heating, a coating of carbon appears to be left on the silicate layers and it does not disappear until full dehydroxylation occurs. Oades and Townsend (1963) have used this technique for determining the proportions and positions of exchangeable ions.

D.t.a. curves for minerals based on mica-like structures are given in Fig. 16. With increasing disorder there is an increasingly larger low-temperature endothermic effect. The lower the extent of $Al^{3+}$ for $Si^{4+}$ substitution in the tetrahedral layer, the lower the temperature of dehydroxylation. The endothermic and exothermic effects above 800°C are believed to originate in a fashion similar to those in smectites, but do not appear to occur in di-octahedral types. Dependent on type, the products of heating are iron-rich spinel, corundum, or $\gamma$-$Al_2O_3$. For vermiculite, the behaviour of the inter-layer water is reminiscent of that in smectites. It is influenced greatly by the nature of the exchangeable cation; thus magnesium varieties have a higher temperature of

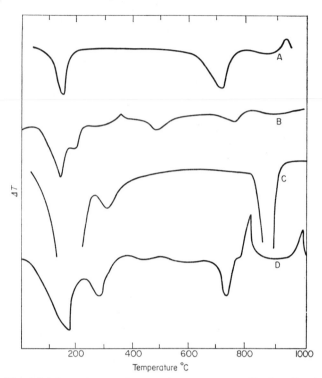

FIG. 15. Differential thermal curves for smectites. A, montmorillonite, Clay Spur, Wyoming, U.S.A.; B, nontronite, Hoher Hagen, Göttingen, Germany; C, saponite, Krugersdorp, Transvaal, S. Africa; D, sauconite, Coon Hollow Mine, Arkansas, U.S.A. (all taken from Greene-Kelly, 1957).

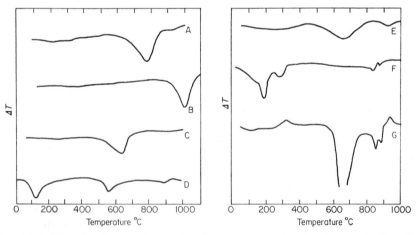

FIG. 16. Differential thermal curves for sheet-structure minerals. A, pyrophyllite, N Carolina, U.S.A.; B, talc, Vermont, U.S.A.; C, sericite, Ashio, Saisei, Japan (ca 1·3 μm); D, illite, Fithian, Illinois, U.S.A.; E, glauconite, Washington, U.S.A.; F, vermiculite (Mg-saturated) West Chester, Pennsylvania, U.S.A.; G, chlorite (clinochlore) Besafotra, Madagascar. (A, B after Grim and Rowland, 1942; C after Sudo et al., 1952; D after Mackenzie et al., 1949; E after Grim, 1953; F after Walker and Cole, 1957; G after Caillère and Hénin, 1957.)

breakdown than sodium, and large cations, e.g. rubidium, exclude much of the water and hence cause much smaller endothermic peaks (Walker and Cole, 1957). The higher-temperature region of the d.t.a. curves is variable; endotherms generally represent dehydroxylation, and an exotherm, perhaps, the development of enstatite. Chloritic types give two endothermic reactions but not until above 600°C. The first represents breakdown of the "brucite" layer and is lowered by iron content, the second represents dehydration of the structure and varies with aluminium-for-silicon substitution. The anhydrous, disordered material crystallizes to olivine or spinel, giving an exotherm on the d.t.a. record (Caillère and Hénin, 1957).

### F. OTHER APPLICATIONS

The technique of d.t.a. has been applied increasingly in various fields. In technological applications, d.t.a. has proven of value, not only in mineralogical fields, such as the estimation of bauxite deposits or the evaluation of cements (Bláha, 1960), but also in the investigation of glass sintering (Wilburn and Thomasson, 1958; Kühne, 1960; Thomasson and Wilburn, 1960, 1961). Of more direct geological application has been the investigation of recalescence phenomena in metamict minerals (Kerr and Holland, 1951), very high-temperature phase changes (Newkirk, 1958), quantitative estimation of minerals in fine-grained rocks (Black, 1959) and volcanic glasses (Ewart and Fieldes, 1962), and of organic materials in carbonaceous rocks (Radchenko and Koperina, 1961).

The use of controlled atmospheres has already been mentioned, e.g. in goethite and gibbsite estimation by Lodding and Hammell (1960) or in studying the effect of carbon dioxide on re-carbonation (Rowland and Jonas, 1949; Rowland and Lewis, 1951; Webb and Heystek, 1957; Papailhau, 1958). The technique has been extended from a static to a dynamic one by streaming a controlled atmosphere at varying pressures through the sample and the inert material while d.t.a. is taking place. The general scheme employs porous tops and bottoms to the sample containers (Stone, 1951, 1952, 1954). It is surprising that the extension to experiments conducted at elevated pressures (e.g. Harker, 1964) has received scant attention, in view of the possible applications to the investigation of geological phenomena.

## III. Thermogravimetric Analysis

This technique consists of studying the loss in weight of a substance as it is being heated. Like d.t.a., it is a dynamic method, and older methods where constant weight was attained before the temperature was raised, have been replaced by thermobalances where heating is at a constant rate, as with d.t.a.

Although a differential type of curve is generally more revealing than an ordinary t.g.a. curve, this method has not been extensively used (DeKeyser, 1953; Paulik *et al.*, 1958; Paulik *et al.*, 1963). Similar remarks apply to the differential-thermobalance technique (Paulik *et al.*, 1958; Waters, 1958), in which two identical samples are heated at the same rate but one is kept approximately 5°C below the other. Variability of results is common for t.g.a., and for the same reasons as for d.t.a. (see, for example, the effects of carbon dioxide and particle size on carbonate decomposition [Garn and Kessler, 1960; Rabatin and Card, 1959; Richer and Vallet, 1953; Martinez, 1961].)

Obviously the combination of d.t.a. with t.g.a. offers considerable possibilities, and various types of apparatus for simultaneous determinations have been described (Papailhau, 1959; Markowitz and Boryta, 1960). A thermal change occurring with heating may immediately be designated as associated with either weight loss or crystallographic inversion. A refinement by Waters (1960) is to differentiate the liberated gases by fractionally absorbing a portion (e.g. absorb water but allow carbon dioxide to be liberated). Quantitative data are fairly readily obtainable from t.g.a. curves. Whereas d.t.a. records usually cease at liquefaction, the t.g.a. procedure permits study to proceed above these temperatures, and thus is useful for studying materials with low melting points.

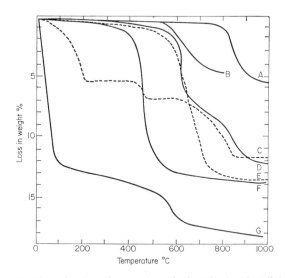

FIG. 17. Thermogravimetric curves for a variety of minerals. A, talc, Tijola, Spain; B, pyrophyllite, North Carolina, U.S.A.; C, carbonate precipitate of calcium and magnesium; D, chlorite (clinochlore), Besafotra, Madagascar (–60 mesh); E, serpentine, Ambindavato, Madagascar (–60 mesh); F, kaolinite, Cornwall; G, montmorillonite, Mississippi, U.S.A. (A, B taken from Muñoz Taboadela and Ferrandis, 1957; C after Dupuis and Dupuis, 1958; D, F after Caillère and Hénin, 1957; E. G after Mackenzie, 1957.)

The applications of the t.g.a. method are various. They include the investigation of solid-state reactions, the derivation of various physical constants, and applications to technology, such as the roasting of minerals, and investigations of ceramics and glass (Lukaszewski and Redfern, 1961; Duval, 1963). In mineralogical investigations, most attention has been focused on the examination of clay minerals. Curves are given for these and several other minerals in Fig. 17 and they may be related to the d.t.a. curves given in previous figures. Carbonates have been investigated, and dolomite has been quantitatively evaluated (Dupuis and Dupuis, 1958; Biffen, 1956). Data may be obtained for various minerals in Nutting (1943) and in Schnitzer *et al.* (1959), and an excellent review with numerous references has recently been published by Coats and Redfern (1963).

## IV. Subsidiary Techniques

### A. DILATOMETRY

The control of material in the ceramics industry is closely regulated by the relation between the thermal expansions of body and glaze. Similarly in metallurgy, behaviour with temperature rise is of vital importance. In mineralogical investigations there has been little application of dilatometric procedures, which is surprising, in view of their ability to detect solid → liquid and other phase changes with great accuracy. The application to polycrystalline unconsolidated materials (e.g. Holdridge, 1959) and to orientation phenomena, demonstrate distinct possibilities for this method.

The normal apparatus consists of a furnace that has a controlled linear heating rate. The sample is placed in the furnace and held rigidly between two rods, which are usually of vitreous silica. One of the rods transmits the expansion or contraction of the test piece as the temperature is raised or lowered. Recording of the expansion may be carried out manually (Pearce and Mardon, 1959), photographically by using an optical lever principle (Chevenard, 1917; Jetter and Mehl, 1943), or electronically (Turnbull, 1950; Vanderman, 1951; Barford, 1963). Interferometric techniques (Saunders, 1939) are superior, since smaller samples may be used and they permit comparison with a standard. A high-temperature inert-atmosphere apparatus has been described by Schaffer and Mark (1963). The most serious complication in the dilatometric method is the effect of orientation phenomena when polycrystalline materials are investigated, for most investigations have used linear thermal expansion. The volumetric dilatometer (Beals and Zerfoss, 1944) does much to eliminate orientation defects. Curves have been given for clays by Geller and Bunting (1940), cristobalite by Beals and Zerfoss (1944), berlinite etc. by Beck (1949), tridymite by Austin (1954) and garnets by Skinner (1956).

It should be pointed out that normally X-ray techniques are employed for the study of thermal-expansion phenomena. Such techniques are less frequently applied to polycrystalline aggregates, especially as in diffractometer measurements a large volume change may so affect the specimen geometry as to give misleading data.

### B. GAS-EVOLUTION ANALYSIS

This analytical method is an extension of the dynamic gas method of Stone (1960) combined with the t.g.a. technique of Waters (1960). The method uses a very small sample, and as the temperature rises in the furnace, an inert carrier gas is streamed through the material. The effluent is analysed by various techniques, e.g. thermal conductivity (Rogers *et al.*, 1960) or gas phase chromatography. The term pyrolysis involves a specific application of g.e.a. to organic materials, especially polymers, that degrade to smaller molecules on rapid heating and these latter are identified by techniques outlined above. Application of this in the field of organic geochemistry is obvious, e.g. for carbonaceous shales. For a purely volatile compound, g.e.a. curves show a gradual rise with increasing temperature and then a sudden cessation, whereas a specific reaction involving a decomposition with gas evolution commences sharply. A comparison between d.t.a. and g.e.a. curves for chalcanthite is given in Fig. 18.

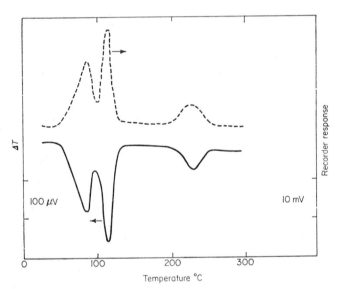

FIG. 18. Simultaneous d.t.a. (solid) and g.e.a. (dashed) curves of chalcanthite (35·7 mg of sample). (After Wendlandt, 1962.)

The Hahn emanation technique is appropriately considered here (Hahn, 1929; Zimens, 1937). A radioactive element is introduced into the substance; when a change such as crystallographic inversion occurs on heating, there is a striking increase in the evolution of radioactive emanation. It should be noted that the observed emanation capacity depends quite considerably on the ageing, recrystallization etc. of reacting powders, and thus the technique is of variable success and may be used purely as an indicator rather than as a precise means of measuring temperatures.

Decrepitation analysis consists of heating the mineral and observing the temperature at which the small fluid and/or gas inclusions burst. This represents a critical temperature, and it can be inferred that the mineral crystallized below this limit. The bursting of the inclusions is measured by a sensitive amplifier and curves of the number of decrepitations per second or minute against temperature generally give a marked peak, which is taken as the critical temperature (Peach, 1949; Smith, 1953).

## C. ELECTRICAL CONDUCTIVITY

Because of the high resistance of nearly all minerals at normal temperatures, investigations of electrical conductivity are largely confined to elevated temperatures. The equipment necessary is quite simple; a potentiometer, electrodes and a furnace. The method is excellent for following phase changes, for there is a sharp increase in conductivity at the onset of a solid-state reaction, and even sintering effects are shown by aggregates (Berg and Burmistrova, 1960; Reisman et al., 1955; Glaser and Moskowitz, 1953; Garn and Flaschen, 1957). Data for kaolinite and serpentine, relating conductivity to dilatometric measurements, have been given by Mchedlov-Petrosyan (1953).

## D. THERMOLUMINESCENCE

When a mineral containing filled electron traps (F centres) is heated, there is a release of energy in the form of light, and a lower energy state is reached by the mineral. The sample is heated on a block or in a furnace at a slow rate, e.g. 1°C/min, and the total evolution of light is measured by a photo-multiplier. When the amount of light is plotted against temperature, a series of peaks results and the curve is called a glow curve. Discussion of the development of electron traps is outside the scope of this text. They are generally produced during crystallization, by imperfections, distortions or vacancies in the structure, and may also arise by later mechanical, thermal or radiation treatment. Various instruments and applications have been described by the following: Urbach (1930); Randall and Wilkins (1945); Kroger (1948); Boyd (1949); Garlick (1949); Pringsheim (1949); Zeller (1952); Daniells et al. (1953); Lewis (1956); Ingerson (1956); Ashby and Kellagher (1958).

Limestones appear to form electron traps with considerable facility, and it has been possible to relate thermoluminescence to depth of burial, erosion intervals, and experimental effects of pressure (Saunders, 1953; Parkes, 1953; Angino, 1959; Medlin, 1961). Roach (1960) has suggested that thermoluminescence and porosity may be related to ore distance and deposition. Curves for quartz have been given by Medlin (1963), and for various other minerals by Angino (1964).

The spectra involved in the thermoluminescence of minerals have not been extensively investigated. Variation does occur; thus calcium and magnesium carbonates give light-yellow to orange emission, and sodium and potassium feldspars give white to blue-violet emission. Reflectance spectroscopy, as applied to organic materials, co-ordination complexes etc. (Wendlandt, 1964), would possibly yield interesting results.

## REFERENCES

Allaway, W. H. (1949). D.t.a. of clays treated with organic cations. *Proc. Soil. Sci. Soc. Am.* **13**, 183,

Allen, R. D. (1957). D.t.a. of selected borate minerals. *Bull. U.S. geol. Surv.* **1036K**, 193.

Angino, E. E. (1959). Pressure effects on thermoluminescence of limestones relative to geologic age. *J. geophys. Res.* **64**, 569, 1638.

Angino, E. E. (1964). Some effects of pressure on the thermoluminescence of certain minerals. *Am. Miner.* **49**, 387.

Arens, P. L. (1951). Study of d.t.a. of clays and clay minerals. Dissertation, Wageningen.

Ashby, G. E. and Kellagher, R. C. (1958). An apparatus for the study of thermoluminescence from minerals. *Am. Miner.* **43**, 695.

Austin, J. B. (1954). The coefficient of linear thermal expansion of tridymite. *J. Am. chem. Soc.* **76**, 6019.

Barford, J. J. (1963). An automatic recording differential dilatometer. *J. scient. Instrum.* **40**, 444.

Barrall, E. M. and Rogers, L. B. (1962). D.t.a. of organic compounds; effects of dilution and geometry. *Analyt. Chem.* **34**, 1101.

Barshad, I. (1952). Calibration of d.t.a. apparatus. *Am. Miner.* **37**, 667.

Bayliss, P. (1964). Effect of particle size on d.t.a. *Nature, Lond.* **201**, 1019.

Bayliss, P. and Warne, S. St. J. (1962). The d.t.a. of cerussite. *Am. Miner.* **47**, 1011.

Beals, M. D. and Zerfoss, S. (1944). Volume change on inversion of cristobalite. *J. Am. Ceram. Soc.* **27**, 285.

Beck, W. R. (1949). Inversions of aluminium phosphate polymorphs. *J. Am. Ceram. Soc.* **32**, 147.

Berg, L. G. (1943). Influence of salt on dissociation of dolomite. *C.r. Acad. Sci. U.R.S.S.* **49**, 648.

Berg, L. G. and Burmistrova, N. P. (1960). Thermographic analysis of salts with simultaneous determination of temperature effects and electric conductivity. *Zh. neorg. Khim.* **5**, 676.

Berg, L. G., Nikolaev, A. V. and Rode, E. J. (1944). Thermography. *Dokl. Acad. Nauk. SSSR.*

Biffen, F. M. (1956). Determination of free lime and carbonate in calcium silicate hydrates by thermobalance. *Analyt. Chem.* **28**, 1133.

Black, W. W. (1959). D.t.a. applied to some Lower Carboniferous sedimentary rocks. *Trans. Leeds geol. Ass.* **7**, 111.

Bláha, J. (1960). High temperature d.t.a. of the raw materials in cement production. *Silikáty* **4**, 36.

Boersma, S. L. (1955). Theory of d.t.a.; new methods of measuring and interpretation. *J. Am. Ceram. Soc.* **38**, 281.

Bogue, R. H. (1954). Constitution of cement clinker. Proceedings of the 3rd International Symposium on *The Chemistry of Cement*, p. 184. Cement and Concrete Association, London.

Bohon, R. L. (1961). D.t.a. of explosives and propellants under controlled atmospheres. *Analyt. Chem.* **33**, 1451.

Bollin, E. M. and Kerr, P. F. (1961). Differential thermal pyrosynthesis. *Am. Miner.* **46**, 823.

Boyd, C. A. (1949). A kinetic study of the thermoluminescence of lithium fluoride. *J. chem. Phys.* **17**, 1221.

Boyer, A. F. (1954). Influence of mechanical treatment on quartz. *Bull. Soc. fr. Minér. Cristallogr.* **77**, 1116.

Bradley, W. F. and Grim, R. E. (1951). High temperature thermal effects of clay and related materials. *Am. Miner.* **36**, 182.

Bradley, W. F., Burst, J. F. and Graf, D. L. (1953). D.t.a. of dolomite. *Am. Miner.* **38**, 207.

Bramão, L., Cady, J. G., Hendricks, S. B. and Swerdlow, M. (1952). Characterization of kaolin minerals. *Soil Sci.* **73**, 273.

Brewer, L. and Zavitsanos, P. (1957). Study of Ge–$GeO_2$ system by an inductively heated d.t.a. apparatus. *Physics Chem. Solids* **2**, 284.

Brindley, G. W. and Nakahira, M. (1959). The kaolinite mullite reaction series. *J. Am. Ceram. Soc.* **42**, 311.

Caillère, S. and Hénin, S. (1957). *In* "The Differential Thermal Investigation of Clays," p. 207. Ed. R. C. Mackenzie. Mineralogical Society, London.

Carthew, A. R. (1955). Quantitative estimation of kaolinite by d.t.a. *Am. Miner.* **40**, 107.

Chevenard, P. (1917). Recording differential dilatometer. *Revue Métall., Paris* **14**, 610.

Chihari, H. and Seki, S. (1953). Studies of crystalline hydrates. II. Thermal transition and dehydration of some sulphates. *Bull. chem. Soc. Japan* **26**, 88.

Coats, A. W. and Redfern, J. P. (1963). Thermogravimetric analysis. *Analyst, Lond.* **88**, 906.

Cocco, C. (1952). D.t.a. of various sulphates. *Periodico Miner.* **21**, 103.

Cole, W. F. and Rowland, N. M. (1961). An abnormal effect in d.t.a. of clay minerals. *Am. Miner.* **46**, 304.

Daniells, F., Boyd, C. A. and Saunders, D. F. (1953). Thermoluminescence as a research tool. *Science, N.Y.* **117**, 343.

Dean, L. A. (1947). D.t.a. of Hawaiian soils. *Soil. Sci.* **63**, 95.

DeKeyser, W. L. (1953). A differential thermobalance. *Nature, Lond.* **172**, 364.

Dempster, P. B. and Ritchie, P. D. (1952). The surface of finely ground silica. *Nature, Lond.* **169**, 538.

R

Dempster, P. B. and Ritchie, P. D. (1953). Physico-chemical studies on dusts V. *J. appl. Chem., Lond.* **3**, 182.

Dunne, J. A. and Kerr, P. F. (1960). An improved thermal head for d.t.a. of corrosive material. *Am. Miner.* **45**, 881.

Dupuis, T. and Dupuis, J. (1958). Estimation of Ca and Mg by t.g.a. of "dolomites." *Microchim. Acta* p. 186.

Duval, C. (1963). "Inorganic Thermogravimetric Analysis." Elsevier, Amsterdam.

Earley, J. W., Milne, I. H. and McVeagh, W. J. (1953). Dehydration of montmorillonite. *Am. Miner.* **38**, 770.

Ewart, A. and Fieldes, M. (1962). Low temperature thermal effects in volcanic glasses due to strain. *Mineralog. Mag.* **33**, 237.

Faust, G. T. (1951). D.t.a. and X-ray study of sauconite and related minerals. *Am. Miner.* **36**, 795.

Fenner, C. N. (1913). Stability relations of the silica minerals. *Am. J. Sci.* **36**, 331.

Fishenden, M. and Saunders, O. A. (1950). "An Introduction to Heat Transfer." Oxford University Press.

Fitch, J. C. and Hurd, B. G. (1959). A sample holder for d.t.a. of fusible or reactive samples. *Am. Miner.* **44**, 431.

Flörke, O. W. and Lachenmayr, H. (1962). D.t.a. and X-ray study of AlPO$_4$. *Ber. dt. keram. Ges.* **39**, 55.

Földvári-Vogl, M. (1958). The role of d.t.a. in mineralogy and geological prospecting. *Acta geol. hung.* **5**, 1.

Frondel, C. (1950). Paratacamite and some related copper chlorides. *Mineralog. Mag.* **29**, 34.

Gallitelli, P., Cola, M. and Alietti, A. (1954). D.t.a. technique. *Atti. Acad. naz. Lincei Memorie* **4**, 49.

Garlick, G. F. (1949). "Luminescent Materials." Oxford University Press.

Garn, P. D. and Flaschen, S. S. (1957). Analytical applications of a d.t.a. apparatus. *Analyt. Chem.* **29**, 271.

Garn, P. D. and Kessler, J. E. (1960). Thermogravimetry in self-generated atmospheres. *Analyt. Chem.* **32**, 1563.

Gaskell, J. A., Grimshaw, R. W. and Roberts, A. L. (1956). 47th report of the refractory materials joint committee. *Communs. Gas Res. Bd.* No. 30.

Geller, R. F. and Bunting, E. N. (1940). Length changes of whiteware clays and bodies during initial heating, with supplementary data on mica. *J. Res. natn. Bur. Stand.* **25**, 15.

Gérard-Hirne and Lamy, C. (1951). D.t.a. identification of clays. *Bull. Soc. fr. Céram.* **10**, 26.

Gheith, M. (1952). D.t.a. of certain iron oxide and iron oxide hydrates. *Am. J. Sci.* **250**, 677.

Glaser, F. W. and Moskowitz, D. (1953). Electrical measurements at high temperatures as an efficient tool for thermal analysis. *Powder. Metall. Bull.* **6**, 178.

Glass, H. D. (1954). High temperature phases from kaolinite and halloysite. *Am. Miner.* **39**, 193.

Greene-Kelly, R. (1957). *In* "The Differential Thermal Investigation of Clays," p. 140. Ed. R. C. Mackenzie. Mineralogical Society, London.

Grim, R. E. (1953). "Clay Mineralogy." McGraw-Hill, London.

Grim, R. E. and Bradley, W. F. (1940). Effect of heat on illite and montmorillonite. *J. Am. Ceram. Soc.* **23**, 242.

Grim, R. E. and Rowland, R. A. (1942). D.t.a. of clay minerals etc. *Am. Miner.* **27**, 746, 801.

Grimshaw, R. W. (1953). Quantitative estimation of silica minerals. *Clay Miner. Bull.* **2**, 2.

Grimshaw, R. W. and Roberts, A. L. (1953). Quantitative d.t.a. of some minerals in ceramic materials. *Trans. Br. Ceram. Soc.* **52**, 50.

Grimshaw, R. W. and Roberts, A. L. (1957). *In* "The Differential Thermal Investigation of Clays," p. 275. Ed. R. C. Mackenzie. Mineralogical Society, London.

Grimshaw, R. W., Heaton, E. and Roberts, A. L. (1945). Constitution of refractory clays. II. *Trans. Br. Ceram. Soc.* **44**, 76.

Grimshaw, R. W., Westerman, A. and Roberts, A. L. (1948). Thermal effects accompanying silica inversion. *Trans. Br. Ceram. Soc.* **47**, 269.

Gruver, R. M. (1948). Precision method of thermal analysis. *J. Am. Ceram. Soc.* **31**, 323.

Hahn, O. (1929). The emanation method for chemical and physicochemical research. *Naturwissenschaften* **17**, 295.

Harker, R. I. (1964). D.t.a. in closed systems at high hydrostatic pressures. *Am. Miner.* **49**, 1741.

Haul, R. A. W. and Heystek, H. (1952). D.t.a. of dolomite. *Am. Miner.* **37**, 166.

Hedvall, J. A. and Heuberger, J. (1922). Solid phase reactions. I. *Z. anorg. allg. Chem.* **122**, 181.

Herald, P. G. and Planje, T. J. (1948). Modified d.t.a. apparatus. *J. Am. Ceram. Soc.* **31**, 20.

Holdridge, D. A. (1959). Thermal expansion as a method for checking the composition of ceramic clays and of studying mineralogical changes during firing. *Clay. Miner. Bull.* **4**, 94.

Holdridge, D. A. and Vaughan, F. (1957). *In* "The Differential Thermal Investigation of Clays," p. 98. Ed. R. C. Mackenzie. Mineralogical Society, London.

Hummel, F. A. (1949). Properties of some substances isostructural with silica. *J. Am. Ceram. Soc.* **32**, 320.

Ingerson, E. (1956). Methods and problems of geologic thermometry. *Econ. Geol.* **50** (Anniversary Vol.), 341.

Jay, A. H. (1933). Thermal expansion of quartz. *Proc. R. Soc.* **A142**, 237.

Jetter, L. K. and Mehl, R. F. (1943). Rate of precipitation of silicon from solid solution in aluminium. *Trans. Am. Inst. Min. metall. Engrs.* **152**, 166.

Karšulin, M., Tomić, A. and Lahodny, A. (1949). Studies on bauxites. I. *Rad. jugosl. Akad. Znan. Umjetn.* **276**, 125.

Kauffman, A. J., Jr. and Dilling, E. D. (1950). D.t.a. curves of certain minerals. *Econ. Geol.* **45**, 222.

Keith, M. L. and Tuttle, O. (1952). High-low inversion of quartz. *Am. J. Sci.* Bowen Vol. Part 1, 203.

Kelly, W. C. (1956). Application of d.t.a. to identification of natural hydrous ferric oxides. *Am. Miner.* **41**, 353.

Kerr, P. F. and Holland, H. D. (1951). D.t.a. of davidite. *Am. Miner.* **36**, 563.

Kerr, P. F., Kulp, J. L. and Hamilton, P. K. (1949). Reference clay minerals. Report No. 3. American Petroleum Institute, Project 49.

Kopp, O. C. and Kerr, P. F. (1957). D.t.a. of sulphides and arsenides. *Am. Miner.* **42**, 445.

Kopp, O. C. and Kerr, P. F. (1959). D.t.a. of evaporites. *Am. Miner.* **44**, 674.

440     R. J. W. MCLAUGHLIN

Kroger, F. A. (1948). "Some Aspects of the Luminescence of Solids." Elsevier, Amsterdam.

Kühne, K. (1960). Thermal effects of vitreous materials. *Silikattechnik* **11**, 106.

Kulp, J. L. and Adler, H. H. (1950). Thermal study of jarosite. *Am. J. Sci.* **248**, 475.

Kulp, J. L. and Kerr, P. F. (1947). Multiple thermal analyses. *Science, N. Y.* **105**, 413.

Kulp, J. L. and Perfetti, J. N. (1950). Thermal study of manganese oxides. *Mineralog. Mag.* **29**, 239.

Kulp, J. L. and Trites, A. F. (1951). D.t.a. of natural hydrous ferric oxides. *Am. Miner.* **36**, 23.

Kulp, J. L., Kent, P. and Kerr, P. F. (1951). D.t.a. of Ca – Mg – Fe carbonate minerals. *Am. Miner.* **36**, 643.

Kulp, J. L., Wright, H. D. and Holmes, R. J. (1949). D.t.a. of rhodochrosite. *Am. Miner.* **34**, 195.

Kupka, F. (1960). A contribution to the d.t.a. of corrosive materials. *Silikáty.* **4**, 176.

Lewis, D. R. (1956). The thermoluminescence of dolomite and calcite. *J. phys. Chem.* **60**, 698.

Linseis, M. (1951). D.t.a. *Tonind.-Ztg. Keram. Rdsh.* **75**, 243.

Lodding, W. and Hammell, L. (1960). D.t.a. of hydroxides in reducing atmosphere. *Analyt. Chem.* **32**, 657.

Lodding, W. and Sturm, E. (1957). A new method of d.t.a. employing multiple thermocouples. *Am. Miner.* **42**, 78.

Lukaszewski, G. M. and Redfern, J. P. (1961). T.g.a. *Lab. Pract.* **10**, 469, 552, 630, 721.

McAdams, W. H. (1942). "Heat Transmission." McGraw-Hill, New York.

McConnell, D. and Earley, J. W. (1951). D.t.a. apparatus. *J. Am. Ceram. Soc.* **34**, 183.

Mackenzie, R. C. (1954). Nickel and ceramic specimen holders. *Nature, Lond.* **174**, 688.

Mackenzie, R. C. (1957). *In* "Differential Thermal Investigation of Clays," pp. 1, 299. Ed. R. C. Mackenzie. Mineralogical Society, London.

Mackenzie, R. C. (1962). "Scifax, d.t.a. Data Index, with Mineral, Inorganic and Organic Sections." Clever-Hume, London.

Mackenzie, R. C. and Milne, A. A. (1953). Effect of grinding on muscovite. *Mineralog. Mag.* **30**, 178.

Mackenzie, R. C., Walker, G. F. and Hart, R. (1949). Illite from Ballater, *Mineralog. Mag.* **28**, 704.

McLaughlin, R. J. W. (1954). Quantitative d.t.a. of soil clays and silts. *Am. J. Sci.* **252**, 555.

McLaughlin, R. J. W. (1955). Effects of grinding on dickite. *Clay. Miner. Bull.* **2**, 309.

McLaughlin, R. J. W. (1957). *In* "The Differential Thermal Investigation of Clays," p. 364. Ed. R. C. Mackenzie. Mineralogical Society, London.

McLaughlin, R. J. W. (1961). Effect of dilution on the shape of d.t.a. curves of kaolinite. *Trans. Br. Ceram. Soc.* **60**, 177.

McLaughlin, R. J. W. and Mackenzie, R. C. (1957). *In* "The Differential Thermal Investigation of Clays," p. 379. Ed. R. C. Mackenzie. Mineralogical Society, London.

McMurdie, H. F. and Golovato, E. (1948). Manganese dioxide variations. *J. Res. natn. Bur. Stand.* **41**, 589.

Manly, R. L. (1950). D.t.a. of certain phosphates. *Am. Miner.* **35**, 108.

Markowitz, M. M. and Boryta, D. A. (1960). A convenient system of t.g.a. and of d.t.a. *Analyt. Chem.* **32**, 1588.

Martin, R. T. (1958). Clay-carbonate soluble salt interaction during d.t.a. *Am. Miner.* **43**, 649.

Martinez, E. (1961). Effect of particle size on thermal properties of serpentine minerals. *Am. Miner.* **46**, 901.

Mazières, C. (1959). Arrangement for micro d.t.a. *C.r. hebd. Séanc. Acad. Sci., Paris* **248**, 2990.

Mchedlov-Petrosyan, O. P. (1953). On the nature of the exothermal effect in hydrous silicates with layer structures. Transactions of the 1st Congress on Thermography (Kazan), p. 272. Academy of Sciences U.S.S.R., 1955.

Medlin, W. L. (1961). Thermoluminescence in dolomite. *J. chem. Phys.* **34**, 672.

Medlin, W. L. (1963). Thermoluminescence in quartz. *J. chem. Phys.* **38**, 1132.

Megaw, H. D. (1946). Structure of oxides of the perovskite type. *Trans. Faraday Soc.* **42A**, 224.

Mosesman, M. A. and Pitzer, K. S. (1941). Thermodynamic properties of the crystalline forms of silica. *J. Am. chem. Soc.* **63**, 2348.

Muñoz Taboadela, M. and Ferrandis, V. A. (1957). *In* "The Differential Thermal Investigation of Clays," p.165. Ed. R. C. Mackenzie. Mineralogical Society, London.

Murray, J. A. and Fischer, H. C. (1951). D.t.a. of white coat plaster. *Proc. Am. Soc. Test. Mater.* **51**, 1197.

Murray, J. A., Fischer, H. C. and Shade, R. W. (1951). M.I.T. fellowship report. *Proc. natn. Lime Ass.* **49**, 95.

Murray, P. and White, J. (1949). Kinetics of the thermal dehydration of clays. *Trans. Br. Ceram. Soc.* **48**, 187.

Newkirk, T. F. (1958). D.t.a. above 1200°C. *J. Am. Ceram. Soc.* **41**, 409.

Nutting, P. G. (1943). Standard dehydration curves. *Prof. Pap. U.S. geol. Surv.* No. **197E**, 197.

Oades, J. M. and Townsend, W. N. (1963). The use of piperidine as an aid to clay mineral identification. *Clay Miner. Bull.* **5**, 177.

Papailhau, J. (1958). A new arrangement for d.t.a. *Bull. Soc. fr. Minér. Cristallogr.* **81**, 142.

Papailhau, J. (1959). Apparatus for simultaneous t.g.a. and d.t.a. *Bull. Soc. fr. Minér. Cristallogr.* **82**, 367.

Parkes, J. M. (1953). Use of thermoluminescence of limestones in sub-surface stratigraphy. *Bull. Am. Ass. Petrol. Geol.* **37**, 125.

Pask, J. A. and Warner, M. F. (1954). D.t.a. methods and techniques. *Bull. Am. Ceram. Soc.* **33**, 168.

Paulik, F., Liptay, G. and Gal, S. (1963). Derivatographic determination of the calcite content of bauxites. *Talanta* **10**, 551.

Paulik, F., Paulik, J. and Erdey, L. (1958). The derivatograph; apparatus for simultaneous d.t.a., t.g.a. and d.t.g. *Z. Analyt. Chem.* **160**, 241, 321.

Peach, P. A. (1949). A decrepitation geothermometer. *Am. Miner.* **34**, 413.

Pearce, J. H. and Mardon, P. G. (1959). Apparatus for combined d.t.a. and dilatometry. *J. scient. Instrum.* **36**, 457.

Peco, G. (1952). D.t.a. in the study of clays. *Ceramica, Roma* **7**, 48.

Perrier, A. and de Mandrot, B. (1922). Elasticity and symmetry of quartz at high temperatures. *C.r. hebd. Séanc. Acad. Sci., Paris* **175**, 622.

Pickering, R. J. (1963). Apparatus for controlled atmosphere d.t.a. of corrosive materials. *Am. Miner.* **48**, 1383.

R*

Pospíšil, Z. (1959). Application of the first exothermic peak for the quantitative determination of kaolinite. *Silikáty* **3**, 36.

Prasad, N. S. and Patel, C. C. (1954). D.t.a. of synthetic manganese dioxides. *J. Indian Inst. Sci.* **36**, 23.

Pringsheim, P. (1949). "Fluorescence and Phosphorescence." Interscience, New York.

Rabatin, J. G. and Card, C. S. (1959). Simple recording thermobalance for vacuum and pressure studies. *Analyt. Chem.* **31**, 1689.

Radchenko, O. A. and Koperina, V. V. (1961). The applications of thermal analysis to the study of disseminated organic matter in rocks. *Dokl. (Proc.) Akad. Sci. U.S.S.R.* (Earth Sciences section) **135**, 1329.

Randall, J. T. and Wilkins, M. H. F. (1945). The phosphorescence of various solids. *Proc. R. Soc.* **A184**, 347.

Reisman, A., Triebwasser, S. and Holtzberg, F. (1955). Phase diagram of the system KNbO₃ – KTaO₃ by d.t.a. and resistance analysis. *J. Am. chem. Soc.* **77**, 4228.

Richer, A. and Vallet, P. (1953). T.g.a. of CaCO₃ in various atmospheres. *Bull. Soc. chim. Fr.* 148.

Rigby, G. R. (1948). Crystal chemistry of various silica minerals. *Trans. Br. Ceram. Soc.* **47**, 284.

Roach, C. H. (1960). Thermoluminescence and porosity of host rocks at the Eagle Mine, Gilman, Colorado. *Prof. Pap. U.S. geol. Surv.* **400B**, 107.

Rode, E. J. (1955). Thermography of manganese oxides. Transactions of the 1st Congress on Thermography, Kazan, 1953, p. 219.

Rogers, R. N., Yasuda, S. K. and Zinn, J. (1960). Pyrolysis as an analytical tool. *Analyt. Chem.* **32**, 672.

Rowland, R. A. (1955). D.t.a. of clays and carbonates. Proc. 1st Natl. Conf. Clays and Clay Mins. *Bull. Div. Mines Calif.* **169**, 151.

Rowland, R. A. and Beck, C. W. (1952). Determination of small quantities of dolomite by d.t.a. *Am. Miner.* **37**, 76.

Rowland, R. A. and Jonas, E. C. (1949). D.t.a. curves of siderite. *Am. Miner.* **34**, 550.

Rowland, R. A. and Lewis, D. R. (1951). Furnace atmosphere control in d.t.a. *Am. Miner.* **36**, 80.

Sabatier, G. (1954). Determination of heats of transformation by d.t.a. *Bull. Soc. fr. Minér. Cristallogr.* **77**, 953, 1077.

Sahores, J. (1955). Thermal study of phosphates of the vivianite group. *C.r. hebd. Séanc. Acad. Sci., Paris* **241**, 221.

Sand, L. B. and Bates, T. F. (1953). Quantitative analysis of some kandites by d.t.a. *Am. Miner.* **38**, 271.

Saunders, D. F. (1953). Thermoluminescence and surface correlation of limestones. *Bull. Am. Ass. Petrol. Geol.* **37**, 114.

Saunders, J. B. (1939). Improved interferometric procedure with application to expansion measurements. *J. Res. natn. Bur. Stand.* **23**, 179.

Schaffer, P. T. B. and Mark, S. D. (1963). Inert atmosphere dilatometer for use to 2000°C. *J. Am. Ceram. Soc.* **46**, 104.

Schedling, J. A. and Wien, J. (1956). D.t.a. of the exothermic reaction of gypsum. *Acta phys. austriaca.* **10**, 247.

Schmidt, E. R. and Vermaas, F. H. S. (1955). D.t.a. and cell dimensions of some natural magnetites. *Am. Miner.* **40**, 422.

Schnitzer, M., Wright, J. R. and Hoffman, I. (1959). Use of the thermobalance in analysis of soils and clays. *Analyt. Chem.* **31**, 440.

Sementovskii, Yu.V. (1958). Relation between mass of reactant and area on d.t.a. records. Transactions of the 5th Conference on Experimental Techniques in Mineralogy and Petrology, p. 79. Academy of Sciences, U.S.S.R.

Sewell, E. C. (1957). *In* "The Differential Thermal Investigation of Clays," p. 65. Ed. R. C. Mackenzie. Mineralogical Society, London.

Skinner, B. J. (1956). Physical properties of the garnet group of minerals. *Am. Miner.* **41**, 428.

Smith, F. G. (1953). "Historical Development of Inclusion Thermometry." University of Toronto, Canada.

Smyth, H. T. (1951). Temperature distribution during mineral inversion. *J. Am. Ceram. Soc.* **34**, 221.

Speil, S., Berkelhamer, L. H., Pask, J. A. and Davies, B. (1945). D.t.a. of clays and aluminous materials. *Tech. Pap. Bur. Mines, Wash.* No. **664**.

Stone, R. L. (1951). D.t.a. under controlled thermodynamic conditions. *Bull. Ohio Engng. Exp. Sta.* No. **146**.

Stone, R. L. (1952). Apparatus for d.t.a. in controlled atmospheres. *J. Am. Ceram. Soc.* **35**, 75.

Stone, R. L. (1954). Thermal analysis of magnesite in $CO_2$ atmospheres. *J. Am. Ceram. Soc.* **37**, 46.

Stone, R. L. (1960). D.t.a. by the dynamic gas technique. *Analyt. Chem.* **32**, 1582.

Sudo, T., Nagasawa, K., Amafuji, M., Kimura, M., Honda, S., Muto, T. and Tanemura, M. (1952). D.t.a. of Japanese clay minerals. *J. geol. Soc. Japan* **58**, 115.

Takahashi, H. (1959). Effects of dry grinding on talc; wet grinding on kaolinite. *Bull. chem. Soc. Japan* **32**, 374.

Thomasson, C. V. and Wilburn, F. W. (1960). *In* "Physical Chemistry of Glasses." Vol. I, p. 52. Interceram, Lübeck.

Thomasson, C. V. and Wilburn, F. W. (1961). *In* "Physical Chemistry of Glasses." Vol. 2, p. 126. Interceram, Lübeck.

Turnbull, J. C. (1950). Recording dilatometer for measuring thermal expansion of solids. *J. Am. Ceram. Soc.* **33**, 54.

Urbach, F. (1930). Theory, measurement and results of thermoluminescence. *Sber. Akad. Wiss. Wien KL.Abt. IIA*, **139**, 353.

Vanderman, E. J. (1951). Microformer dilatometer. *Rev. scient. Instrum.* **22**, 757.

Vassallo, D. A. and Harden, J. C. (1962). Precise phase transition measurements of organic materials by d.t.a. *Analyt. Chem.* **34**, 132.

Veniale, F. (1962). Effect of fibre dimensions on d.t.a. of the serpentine group. *Rc. Soc. miner. ital. P.* **18**, 277.

Vold, M. J. (1949). D.t.a. *Analyt. Chem.* **21**, 683.

Walker, G. F. and Cole, W. F. (1957). *In* "The Differential Thermal Investigation of Clays." p. 191. Ed. R. C. Mackenzie, Mineralogical Society, London.

Waters, P. L. (1958). New types of recording differential thermobalances. *J. scient. Instrum.* **35**, 41.

Waters, P. L. (1960). Fractional t.g.a. *Analyt. Chem.* **32**, 852.

Webb, T. L. (1954). Nickel and ceramic specimen holders. *Nature, Lond.* **174**, 686.

Webb, T. L. and Heystek, H. (1957). *In* "The Differential Thermal Investigation of Clays," p. 329. Ed. R. C. Mackenzie. Mineralogical Society, London.

Weiser, H. B. (1935). "Inorganic Colloid Chemistry." Vol. 2. Wiley, New York.

Welch, J. H. (1954). Discussion on: Constitution of cement clinker. Proceedings of the 3rd International Symposium on the Chemistry of Cement. London, 1952, p. 214. Cement and Concrete Association, London.

Wendlandt, W. W. (1962). A new apparatus for simultaneous d.t.a. and g.e.a. *Analytica chim. Acta* **27**, 309.

Wendlandt, W. W. (1964). "Thermal Methods of Analysis." Interscience, New York.

White, J. (1953). Discussion on: Quantitative estimation of silica minerals. *Clay. Miner. Bull.* **2**, 5.

Whitehead, W. L. and Breger, I. A. (1950). Vacuum d.t.a. *Science, N.Y.* **111**, 279.

Wilburn, F. W. and Thomasson, C. V. (1958). The applications of d.t.a. and t.g.a. to the study of reactions between glass making materials. I. The $Na_2CO_3–SiO_2$ system. *J. Soc. Glass Technol.* **42**, 158.

Wittels, M. (1951). Structural transformations in amphiboles at elevated temperatures. *Am. Miner.* **36**, 851.

Wittels, M. (1952). Structural disintegration of some amphiboles. *Am. Miner.* **37**, 28.

Yagfarov, M.Sh. (1961). A new method of quantitative thermography. *Zh. neorg. Khim.* **6**, 2440.

Zeller, E. J. (1952). Thermoluminescence as a radiation damage method of geologic age determination in carbonate sediments. *C.r. Int. Geol. Congr., Algiers* **12**, 365.

Zimens, K. E. (1937). Application of emanation technique to alkaline earth carbonates. *Z. phys. Chem.* **37B**, 231.

# CHAPTER 10

# Emission Spectrography [with Appendix on Spark Source Mass Spectrography]

## G. D. NICHOLLS

*Department of Geology, University of Manchester, England*

## I. Introduction

The principal use of emission spectrography in geology is, without doubt, in the study of trace-element distribution in rocks and minerals. Other methods lend themselves more readily to the determination of minerals by the estimation of major-element concentrations, and it is relatively rarely that the characterization of a mineral depends upon the content of trace elements. It is also true that the details of emission-spectrographic techniques are rather well established, and have already been described in published works (Nachtrieb, 1950; Ahrens, 1950; Taylor and Ahrens, 1960). There remain, nevertheless, certain aspects of emission spectrography that are validly dealt with in the present context of determinative mineralogy. Thus the following pages deal with: (a) the general strategy of choosing from a variety of emission-spectrographic techniques; (b) some examples of specific applications to problems of mineral determination; and (c) some industrial applications.

## II. Selection of Appropriate Technique

In order that the selection of working conditions can be appreciated, the essential features of emission spectrography and the apparatus used in this technique will be outlined first.

Flame tests have been used in simple determinative mineralogy for many years. The use of emission-spectrographic techniques in mineralogical studies is merely a more sophisticated modern development of the same approach. If atoms are sufficiently excited in an excitation source, whether it be a flame, an electric arc or a spark discharge, they emit energy of distinct wavelengths. Some elements excited in a flame source emit energy of only a few wavelengths within the visible spectrum, and the flame assumes a colouration distinctive for that element. For these elements a simple flame test is possible. Many elements, however, emit energy of a large number of discrete wavelengths, so that no distinctive colouration of a flame is obtained. Nevertheless, the wavelengths of the emitted energy are characteristic of these elements. Emission spectrographs are devices for dispersing and recording such energy emitted from excited material in an excitation source producing wavelengths falling within the range of about 2300–9000 Å.

Basically, emission spectrography is a method of analysis for the chemical elements present in the material under investigation. In qualitative studies its advantages are the extreme rapidity with which the presence of various elements can be certainly established, the large number of elements (between 30 and 40) that can be simultaneously sought for, and the small amounts of sample required for the analysis. In quantitative work its advantages again include speed and number of elements, together with considerable sensitivity for many elements of interest in determinative mineralogy and geochemistry.

Consideration of the apparatus used must begin with the recognition that there are three basic sections of an emission-spectrographic assembly—the excitation source, the dispersing medium and the recording device.

### A. THE EXCITATION SOURCE

The excitation source may be a flame, a d.c. electric arc, a spark discharge or a laser source accompanied by an auxiliary spark. Each of these has advantages and disadvantages. A flame excitation source is a low-energy source in which approximately 40 elements can be excited. The spectra produced are often simpler and easier to interpret than those produced by arc or spark sources. This source is extremely steady and the intensity of the emitted light is constant with time. Generally, samples must be brought into solution if this source is to be used, and with silicate materials this necessitates preexcitation chemical work.

The d.c. arc source is one yielding high sensitivity, though, unless special precautions are taken, the precision and accuracy of determination obtained by the use of this source are less than those from a spark discharge. This source tends to be less stable than a simple flame source, though stability can be improved by the use of such devices as the Stallwood air jet. Material need not be brought into solution for analysis in an arc source—solids are usually mixed with carbon and the mixture packed into a cavity in the end of a carbon rod used as the sample electrode. The sample-electrode may be made either the cathode or the anode of the arc, the counter-electrode being a carbon rod carrying no sample. With cathode excitation, a zone of enhanced emission occurs near the cathode (the cathode-layer effect), and if light from this part of the arc is directed into the dispersing medium, the sensitivity obtained by the use of this source is further increased. The development of a strong cathode-layer effect is most marked when the highest possible arc temperature is achieved, such as is favoured by introducing only small amounts of material (less than 10 mg) into the arc. The difficulty of ensuring that such a small sample is truly representative of the analysed material may result in sensitivity being gained at the expense of reproducibility. Anode excitation gives less sensitivity but more reproducible conditions, and is preferred by many analysts. Elements are not all volatilized and excited with equal ease, and variation in volatilization rates is particularly marked in the d.c. arc when anode excitation is used. For quantitative work it is necessary to adopt slightly different excitation conditions for the volatile elements, e.g. the alkali metals, zinc and gallium, the involatile elements, e.g. cobalt, nickel and vanadium, and the refractory elements, e.g. zirconium.

A high-frequency, high-voltage a.c. spark provides an excitation source of greater flexibility than the arc. It is somewhat less sensitive for many elements but more reproducible. It gives higher excitation energies than the arc and tends to excite ion rather than atom lines. Solids can be sparked, though frequently silicate samples are brought into solution first. Laser excitation permits a minute sample to be examined, but, at present, has only been used for qualitative and semi-quantitative determinations. It can be used to examine minerals still contained in the host rock and is likely to be much more widely used in future.

### B. THE DISPERSING MEDIUM

The dispersing medium may be a prism or a grating. Under most circumstances the prism has a smaller dispersion and, where photographic methods of recording are used, permits a wide range of wavelengths to be recorded on a 10-in. plate. However, resolution of beams of closely similar wavelength may be difficult with a prism, and line interference on the plate may result. The larger dispersion possible with a grating may be useful to resolve complex spectra (e.g. rare-earth spectra), but when a grating is used in this way the wavelength range normally recorded is restricted.

### C. THE RECORDING DEVICE

The most commonly used recording device is the photographic plate, though in simple, rapid, qualitative analysis the human eye may be used. However, direct-reading spectrographs, fitted with slits and phototubes, are available if large numbers of routine determinations have to be made and justify the high cost involved. The advantages of direct-reading devices lie in their greater speed and potentially greater accuracy. Electron-multiplier phototubes compare favourably with photographic emulsions in respect of the lower limit of radiant intensity to which they are responsive and can be made to integrate light intensity by using their output to charge a condenser, the discharge current of which is used as an index of the integrated light intensity. Direct-reading spectrographs are particularly valuable in some industrial applications of emission spectrography where ratios of elements may be as informative as knowledge of their absolute amounts. The main disadvantage of direct-reading devices, other than their high cost, is the rigid control that must be exercised over the atmospheric conditions (temperature, humidity, etc.) in which they are operated. Photographic plates are much cheaper and more versatile. Their use necessitates the adoption of some method of relating the density of the photographic image to the intensity of incident light. Various methods of plate calibration exist and analysts have personal preferences for different ways of interpreting their plates.

With choices available in each of the three basic sections of the assembly, it is not surprising that there is no typical emission-spectrographic arrangement. An operator having direct-reading facilities may choose to use a spark-excitation source because of its greater reproducibility, relying on his detectors to compensate for the lack of sensitivity of this source compared with the arc. One using photographic methods of recording may well choose arc excitation to obtain good sensitivity, even though selective-volatilization effects will then force him to adopt multiple arc-ing of samples. For maximum sensitivity, cathode excitation would be adopted, though if good reproducibility is also desired anode excitation is preferable. The nature of the investigation will, to a large degree, dictate the best assembly for the specific purpose. Skill in this field of analysis involves the careful selection of optimum working conditions in the light of objectives.

Spectrographic analysis of minerals may be qualitative or quantitative. If the latter is attempted, standardization with standards of known composition is a pre-requisite to determinations on unknown samples. Standards should have a similar major-element composition to the unknowns, as significant matrix effects are encountered in this type of analysis.

## III. Spectrographic Identification of Mineral Grains

Emission-spectrographic methods can be used for the characterization of minerals available only in milligram amounts or less, such as might represent minor components of a rock or even impurity inclusions in other minerals. The small size of these samples generally limits the period during which they emit spectra to a few seconds or less, so the excitation source chosen must be one of high sensitivity, and it is desirable that the dispersing medium and recording device shall cover a wide wavelength range. Stich (1953) used the d.c. arc with a medium quartz-prism spectrograph for semi-quantitative examination of more than 300 mineral samples in the laboratories of the U.S. Geological Survey. In many cases he arc-ed ethyl cellulose X-ray powder spindles containing the grains, thus enabling direct reference of the spectrographic data to X-ray diffraction patterns. The spectrographic data were used to confirm the chemical compositions of minerals tentatively identified from their X-ray diffraction patterns, to determine impurities in minerals and to differentiate between minerals producing similar X-ray patterns, such as metatorbenite and metazeunerite.

More recently a ruby laser has been used to excite vaporization of very small quantities of minerals, the vapour then being further excited in an auxiliary spark source (Maxwell, 1963). Because the sample need not be electrically conducting for laser excitation, separation of the mineral from its matrix is not necessary. Areas of 50–100 $\mu$m in diameter can be examined by using polished sections or rock slices, somewhat thicker than normal, without destruction of the specimen. The technique has been applied to the qualitative identification of ore minerals and is of obvious application to the determination of small inclusions in host crystals. The sample to be examined is placed on the stage of a microscope and the crystal to be analysed brought to the centre of the field of view and focused. The laser is fired ("lased") and the brief burst of intense light is reflected down the axis of the microscope by a prism or mirror. The burst lasts about 1 $\mu$sec and the power is of the order of 2–3 MW. The intense energy, concentrated on an area 100 $\mu$m or less in diameter, vaporizes a few micrograms of material leaving a hemispherical pit or crater marking the analysis area. Thus, after the analysis, it is possible to check that the target grain was actually "hit". The time involved is so short and the temperature reached (estimated as equivalent to over 10,000°K) so high, that selective volatilization is most unlikely. The vapour rising from the sample passes between two auxiliary charged electrodes, ionizes the gap between them and a spark discharge results. In this way the constituent elements in the vapour are excited and emit their characteristic spectra. The light is focused by a lens on to the slit of a spectrograph and the spectra recorded photographically. Maxwell (1963) gives further operational details and presents qualitative data on various ore minerals he examined. It has been claimed that the routine lower limit of sensitivity for certain elements is 0·001%, with a reproducibility of ±5%, but Maxwell was unable to check this. However it appears probable that not only major but also minor and trace components of minerals might be identified by this method in extremely small grains. Although still only in the pioneer stages of development, the technique of microemission spectrography with a laser microprobe shows promise of being an exceedingly valuable method in the determination of extremely small mineral grains.

Allied to this application of spectrography to mineralogy is the use of the emission spectrograph to detect and determine elements present in minor amounts in minerals when these elements are difficult to determine by other methods. The outstanding example is fluorine. In minerals such as amphiboles, micas, apatite and in other less common minerals, structural (OH) groups may be replaced by fluorine to give fluor-amphiboles, etc. It is sometimes difficult to distinguish the fluorine-bearing representatives from the hydroxy ones without determining the amount of fluorine present. This is tedious by wet chemical procedures. Fluorine can, and has, been determined spectrographically. Atomic fluorine is difficult to excite and it is customary to attempt excitation of one of the molecular emitters CaF or SrF and record the emission photographically. The CaF emitter is easily excited, but the molecule breaks down in a high-energy source. For this reason the d.c. arc has been preferred to a spark source for fluorine detection and determination. However, this does introduce a difficulty, in that selective volatilization tends to be more marked in the d.c. arc; fluorine volatilizes much earlier than calcium unless precautions are taken, and it is obviously necessary to have both elements in the arc atmosphere simultaneously. If the mineral under investigation contains little or no calcium (or strontium) this element must be added to the charges before excitation, and Ahrens (1950) points out that, for optimum sensitivity, calcium should be present in the arc atmosphere in excess. The difference in volatilization rates of the two elements means that, in order to achieve this condition, the amount of calcium in the charge must be considerably in excess of that of fluorine.

During work in the Department of Geology, University of Manchester, on samples with up to 3% by weight fluorine, it was found that optimum sensitivity was achieved when 3 parts by weight of $CaCO_3$ were added to 4 parts by weight Ca-free standard or sample. Excess of Ca over this ratio does not appear to have any deleterious effects. Selective volatilization can be minimized by adding carbon powder to the charges and by avoiding low-amperage arc conditions. In the Manchester work, 9 parts by weight of carbon powder were added to 14 parts by weight of sample–calcium carbonate mixture, and the arc current was maintained at 8 amps. Molecular emitters tend to produce band spectra and CaF is no exception. It is customary to use the CaF band with a head at 5291 Å, the components of which are not very sharp or narrow, so optimum sensitivity is attained with a prism dispersing medium. CaF 6064 is claimed to be much more sensitive than CaF 5291 (Ahrens, 1950), but tends to be masked by intense CaO bands. For determination rather than detection of fluorine, an internal-standard element may be introduced into the charges and the photographic plates are measured as described in the next Section. Copper added as CuO has been found to be a suitable internal-standard element, one part by weight of the oxide being added to 14 parts by weight of sample–calcium carbonate mixture before the addition of the carbon. Cu 5015 or Cu 5153 may be used as the internal-standard line.

Magnesium in magnesian calcites may also be determined by emission spectrographic methods. For such a determination a spark source is preferable, there being no difficulty in bringing the material into solution. Introduction of the internal standard uniformly through the charge excited, always a possible source of error, is far easier if only mixing of solutions is involved. Strontium (Ahrens, 1950) and copper (Jaycox, 1947) have been used as internal standards. Potassium in highly sodic feldspars has also been determined spectrographically with rubidium as an internal standard in a d.c. arc of 7 amps with anode excitation (Ahrens, 1950). A standard deviation of $\pm 3 \cdot 1\%$ was quoted for this method. In work such as this, the highest sensitivity is rarely required, and spark excitation would often be preferable by reason of its greater reproducibility.

## IV. Trace Element Determination to Characterize Mineral Paragenesis

Trace element contents of minerals may sometimes be useful indicators of their paragenesis. Faust and his co-workers have shown that serpentines derived from the alteration of ultrabasic rocks can be distinguished from those formed in metamorphosed dolomites and limestones and in hydrothermal deposits by their trace-element contents (Faust et al., 1956; Faust and Fahey, 1962; Faust, 1963). The former group are characterized by the presence of significant amounts of nickel, chromium, cobalt and scandium, whereas these elements are absent or present in very small amounts in serpentines of the second group. In studies on dolomite, Weber (1964) showed that "primary" dolomites contain higher contents of aluminium, barium, iron, lithium, potassium, sodium, and zinc than "secondary" dolomites, whereas strontium is significantly concentrated in the latter.

For such investigations both sensitivity and precision are required in the analytical technique, and methods very similar to those used in geochemical investigations (see Taylor and Ahrens, 1960) are adopted. The d.c. arc is commonly used as an excitation source, and to obtain maximum precision and accuracy an internal-standard element is introduced into both samples and standards in the same amount before arc-ing. Photographic recording has usually been used, though direct-reading methods would be applicable in such studies if the equipment was available. When photographic recording is used, care must be exercised to ensure standardized development of all plates. It is customary to interpolate a rapidly rotating seven-step sector between the excitation source and the entry slit of the spectrograph. In consequence the spectral lines on the developed photographic plate are stepped and, since adjacent steps on the rotating sector have a known log intensity ratio (e.g. 0·2), effectively seven stepped exposures of known relative intensity are produced from one arc-ing. The density of a line in each step is measured on a microdensitometer or similar instrument and for each line a plot of density against log intensity is prepared. For some emulsions it is advisable to convert density readings to the Seidel transformation of density before plotting. From these graphs the intensities required to give lines of a chosen density for both sample and internal-standard elements can be obtained and expressed as a ratio. By using the data obtained from the standards, working curves are prepared relating line-intensity ratios for two elements to concentration ratios for these two elements. Determination of the intensity ratio of the same two lines in the spectrum of a sample then permits the element-concentration ratio in the sample to be deduced. Since the content of the internal-standard element is known, the concentration of the element sought can be calculated. For the most precise work, graphs are made of background density against log intensity for each line, and the intensity required to give a chosen line density is corrected for the background contribution.

For accurate emission-spectrographic work the major-element composition of the standards must compare closely with that of the samples being analysed. Furthermore, selective volatilization within the d.c. arc necessitates the use of different excitation conditions and internal standards for determination of different elements. It is often necessary to arc a sample under two or three different sets of excitation conditions to cover a full range of element determinations.

Many similar applications of emission-spectrographic methods to determinative mineralogy can be envisaged. As an example, the so-called "heavy" minerals of clastic sedimentary rocks might be made to yield much more information about the source rocks from which they were derived than is usually obtained. These heavy minerals have been used to deduce that the source rocks were metamorphic or acid igneous or basic igneous, but rarely has the nature of these rocks been more precisely defined. Since trace-element contents of similar minerals from different rocks often show distinctive differences, trace-element studies on the heavy minerals of sediments should enable the source rock to be determined much more closely. In the sediments of the deep equatorial Atlantic, pyroxene is quite frequent as a "heavy" mineral, raising the question whether it has been derived from basalts exposed on the ocean floor, from ultrabasic rocks exposed on sea floor scarps and the islets of St. Paul's Rocks, or from the continents. Trace-element studies on the pyroxene might well clarify this uncertainty.

This approach of characterizing mineral paragenesis by trace element contents is promising and emission spectrography is a useful technique in such work. However the detection limits of emission-spectrographic analysis for most elements in silicate matrices are above 1 p.p.m. and in some cases over 100 p.p.m. This limits the effectiveness of the technique to about 25 elements. The new technique, spark-source mass spectrography (p. 456), is more versatile, and, perhaps, more suited to this type of work.

## V. Ratio Work and Industrial Applications

Another application of emission spectrography to mineralogical investigations may be called "ratio work". In such work there is no need to introduce internal standards into the excited material. Preliminary surveys of the Rb/Sr ratio in micas, feldspars, etc. intended for age-determination work by mass spectrometry afford an example. Speed of analysis is the advantage possessed by emission spectrography in this context. The absolute amounts of rubidium and strontium are not determined—only the relative amounts of the two, since the only interest is in establishing whether the relative amount of common strontium is sufficiently low to prevent radiogenic $^{87}$Sr being swamped by non-radiogenic $^{87}$Sr. Simple visual inspection of the plates suffices for discriminatory work of this nature.

In industrial use of minerals this kind of investigation is much more important. Mineral source materials used in the metallurgical industry frequently contain some impurities, often such as are difficult to remove by mineral dressing procedures. Zircon, derived from beach sand deposits, may be partially coated with finely divided titanium dioxide. Removal of this impurity before, or during, extraction of the zirconium is difficult, though the tolerable titanium content in the final metal must often be below certain specified amounts. From knowledge of the extraction efficiencies, the tolerable Ti/Zr ratio in submitted ore can be calculated. It is desirable to have a rapid method of estimating the relative amounts of titanium and zirconium in ore samples, as submitted, with the object of eliminating unsatisfactory supplies. But, since other factors such as cost and reliability of supply also influence industrial decision, it is necessary to know the Ti/Zr ratios more accurately than is required in the case of rubidium and strontium mentioned above. Since sensitivity is no longer a limiting factor, spark excitation is preferable for this work to ensure maximum accuracy. The higher energy source also ensures more complete excitation of these refractory elements. Direct-reading spectrographs are used in a number of industrial laboratories and can be adjusted to give a direct reading of the intensity ratio of selected Ti and Zr lines. The relationship of the intensity ratio to concentration ratio is determined from zircons of pre-determined Ti/Zr ratios and then the spectrographic technique permits large numbers of samples to be evaluated rapidly and the exercise of quality control on the raw material. Other examples could be given where ratios are important, in that knowledge of them permits estimation of the likely concentration of impurity in the extracted metal.

Modern industrial technology sometimes demands rare metals that do not form minerals in which they are a major constituent, e.g. hafnium. Such metals must be won from sources in which they are a minor, or even trace, constituent in a host mineral (zircon in the case of hafnium). Here again spectrographic techniques can be employed to distinguish the relative merits of different samples as sources of the desirable minor constituent. Usually there will be no shortage of sample to be analysed, and if a direct-reading spectrograph is employed the problem of sensitivity limits can be overcome by exciting abnormally large samples and integrating the light intensities over the period of excitation. Since only relative hafnium contents of samples from different sources of supply are required for management decision and all samples are virtually identical with respect to major-element composition, control of amount of sample excited and excitation conditions permits the required information to be acquired very rapidly in terms of cumulative intensities produced from the samples.

With the increasing demand for metals of very high purity with very tight specifications of tolerable amounts of impurity elements, spectrographic techniques have another role in industrial determinative mineralogy. Although a range of methods exists for the progressive purification of metals, there remain some purification problems which are almost impossible to surmount economically, e.g. extraction of zinc from magnesium. In such cases, high-purity metals can be produced by ensuring that the primary ore material contains such small quantities of the undesirable impurity elements that even after concentration in the metal during smelting their content is still below specification. Where a number of impurity elements are involved spectrographic techniques are especially valuable in exercising quality control on the raw material fed to the smelting plant.

An extension of this approach arises from the recognition that the trace-element contents of a mineral ore deposit are rarely constant throughout its whole extent. This is especially so in the case of sedimentary ore deposits. Some of the metal produced from deposits of variable geochemistry may have to meet a rigid specification, whereas the remainder, destined for alloy production, need not meet such stringent purity controls. Once the distribution of trace elements in the mineral deposit is known, different parts of the deposit may be worked at different times to produce raw material for the extraction process according to the impurity specifications to be met at any one time. Market demand for metal of different specifications can be allowed to determine the working of the deposit. Such "differential working" of a deposit raises interesting points about the economics of exploiting natural mineral deposits as sources of metals. It may emerge, in particular instances, that it is less costly in the long run to investigate the distribution of trace elements through a mineral deposit and then work it differentially than to install and operate post-extraction purification processes. The speed, range of elements determinable and sensitivity of spectrographic procedures make them particularly applicable in such work, though, here too, the greater range and sensitivity of spark-source mass spectrometry may result in emission spectrography being replaced by the more modern technique in work of this nature. With the probable extension of the "differential working" approach to the exploitation of natural deposits, it may be expected that the emission spectrograph and spark-source mass spectrograph will become increasingly useful in industrial determinative mineralogy.

Possible future exploitation of manganese nodules from the ocean floor raises similar questions. Considerable amounts of metallic elements other than iron and manganese are often present in these nodules, the extraction of which would be an important feature of the economic exploitation of these deposits. Mero (1964) has shown that there are considerable regional variations in the composition of the nodules and recognizes four regional types. In type A the average composition is: iron, 28·3%; manganese, 21·7%; cobalt, 0·35%; nickel, 0·46%; copper, 0·32%; and lead, 0·21%. Type C is quite different with: iron, 17·7%; manganese, 33·3%; cobalt, 0·39%; nickel, 1·52%; copper, 1·13%; and lead, 0·18%. In discussing the exploitation of these deposits Mero (*op. cit.*) suggests that "mining" sites may be chosen to yield nodules of such composition that all the extracted metals will be disposable without seriously disrupting world markets. In evaluating the suitability of nodule deposits for exploitation, emission spectrography and, even more so, solid-source spark mass spectrography should prove invaluable techniques.

## VI. Conclusion

In determinative mineralogy, the real value of emission-spectrographic methods lies in their use in distinguishing between mineral samples that are similar or almost identical in respect to their major-element composition, structure, optical properties, etc. The distinction achieved may be of academic interest, as in the case of serpentines, or of considerable economic value, as in the case of mineral ore deposits. Though spectrographic methods can be used for mineral identification, they are of most value in determination of minor, though significant, differences in minerals. For most effective determination, the assembly used and the working conditions adopted must be carefully selected, and this technique has suffered in reputation more than most through inadequate attention to the optimum working conditions for the solution of particular problems. It can be a most effective tool in the hands of a mineralogist.

## VII. Appendix. Spark Source Mass Spectrography

Although this Chapter is principally concerned with emission spectrography, mention should be made of a different technique, which has recently been applied to the determination of trace elements in geological samples. This is solid-source spark mass spectrography (Brown and Wolstenholme, 1964; Taylor, 1965a,b; Nicholls et al., 1967). By this technique some 70 elements may be determined from a single excitation event. It is much more sensitive than emission spectrography, sensitivity limits being 0·01 p.p.m. or below. Far smaller amounts of sample are consumed in each analysis, 1 mg frequently being adequate for a full analysis. Provided that careful attention is paid to electrode preparation, a precision of better than $\pm5\%$ can be obtained (Nicholls et al., 1967), which is better than that achieved in most emission-spectrographic analyses. At the present time the major drawback to the more extensive use of solid-source spark mass spectrography is the cost. Initial capital outlay on the equipment is substantially higher than that on conventional emission-spectrographic assemblies and running costs are also greater. Nevertheless, the advantages gained by the use of this technique are so pronounced that it may be expected to replace conventional emission spectrography for trace-element determination in rocks and minerals in the not-too-distant future.

In a general sense, three basic sections can also be recognized in a solid-source spark mass spectrograph. The excitation source is a spark discharge between two electrodes under vacuum (approximately $5 \times 10^{-6}$ torr). The electrodes are prepared by mixing the analysis sample (pre-treated or otherwise as occasion requires) with very pure graphite and compressing the resulting powder into bars of cross-sectional area 0·001 sq. in. under a pressure of 7500 lb/sq. in. A spark voltage of 25 kV is used with a pulse length of 200 $\mu$sec and a pulse repetition rate of 300/sec. This treatment liberates ions from the electrodes which are accelerated through an ion gun by an accelerating voltage of 19·6 kV. On leaving the ion gun, the beam of ions enters the dispersing section, an electrostatic analyser followed by a magnetic analyser. The original ion beam is dispersed into a large number of separate beams according to the mass/charge ratio of the ions. At the exit from the magnetic analyser, the separate beams meet the recording device, a photographic plate with an emulsion sensitive to charged particles. Up to 15 separate exposures can be accommodated on one photographic plate by translating the plate across the exit slit of the magnetic analyser. Usually all 15 positions are assigned to different exposures of the same sample arranged in a graduated series. The method of measuring exposures is quite different from that adopted in emission spectrography. Between the electrostatic analyser and the magnetic analyser the ion beam passes through a monitor slit, the jaws of which intercept half the beam. The charge accumulating on the monitor slit jaws is measured and the extent of an exposure determined by the amount of charge collected. Since this is related to the total number of ions that have passed on into the magnetic analyser and been recorded, a semi-quantitative estimation of element concentrations in the analysed material can be made by observation of the extent of exposure required to produce a detectable line on the photographic plate for chosen isotopes of the elements being sought. For more accurate work, the technique must be standardized against known standards and the use of an internal-standard element is necessary in the most precise work.

## REFERENCES

Ahrens, L. H. (1950). "Spectrochemical Analysis." Addison-Wesley Press, Cambridge, Mass.

Brown, R. and Wolstenholme, W. A. (1964). Analysis of geological samples by spark-source mass spectrography. *Nature, Lond.* **201**, 598.

Faust, G. T. (1963). Minor elements in serpentine – additional data. *Geochim. cosmochim. Acta* **27**, 665.

Faust, G. T. and Fahey, J. J. (1962). The serpentine-group minerals. *Prof. Pap. U.S. geol. Surv.* **384-A**.

Faust, G. T., Murata, K. J. and Fahey, J. J. (1956). Relation of minor-element content of serpentines to their geological origin. *Geochim. cosmochim. Acta* **10**. 316.

Jaycox, E. K. (1947). Spectrochemical analysis of ceramics and other non-metallic materials. *J. opt. Soc. Am.* **37**, 162.

Maxwell, J. A. (1963). The laser as a tool in mineral identification. *Can. Mineralogist* **7**, 727.

Mero, J. L. (1964). "The Mineral Resources of The Sea." Elsevier, Amsterdam.

Nachtrieb, N. H. (1950). "Principles and Practice of Spectrochemical Analysis." McGraw-Hill, New York.

Nicholls, G. D., Graham, A. L., Williams, E. and Wood, M. (1967). Precision and accuracy in trace-element analysis of geological materials using solid-source spark mass spectrography. *Analyt. Chem.* (In press).

Stich, J. N. (1953). Spectrographic identification of mineral grains. *Circ. U.S. geol. Surv.* **234**.

Taylor, S. R. (1965a). Geochemical application of spark-source mass spectrography. *Nature, Lond.* **205**, 34.

Taylor, S. R. (1965b). Geochemical analysis by spark-source mass spectrography. *Geochim. cosmochim. Acta* **29**, 1243.

Taylor, S. R. and Ahrens, L. H. (1960). Spectrochemical analysis. *In* "Methods in Geochemistry", (Ed. Smales, A. A. and Wager, L. R.), p. 81. Interscience, New York and London.

Weber, J. N. (1964). Trace element composition of dolostones and dolomites and its bearing on the dolomite problem. *Geochim. cosmochim. Acta* **28**, 1817.

# CHAPTER 11

# Density Determination

## L. D. MULLER

*Warren Spring Laboratory, Stevenage, Hertfordshire, England*

## I. Introduction

Density is a fundamental and characteristic property of a crystalline substance or solid, and as such is an important determinative property.* Though the terms "specific gravity" and "density" are frequently used synonymously, a distinction exists between them. Specific gravity is a number, being the ratio of the mass of any quantity of a substance to the mass of an equal volume of a standard substance (normally water at 4°C). Density is defined as the ratio of the mass of any quantity of a substance to its volume; it is numerically equivalent to "mass per unit volume" and may be expressed in either c.g.s. or f.p.s. units. In c.g.s. units this becomes the mass in grams of one cubic centimetre of the substance, and as that volume of water at 4°C weighs one gram, density, when so expressed, is numerically equal to specific gravity.

## II. Methods of Determining Density

Many methods for determining the density of solids on macro-, semi-micro- and microscales have been proposed, though as Bannister and Hey (1938) have pointed out, no one method is universally applicable. Reilly and Rae (1954)

* The publication by Mursky and Thompson (1958) of a specific gravity index giving data for more than 1400 minerals provides a valuable reference when this property is used for determinative purposes.

describe the classical techniques of density determination and Mason (1944) has critically reviewed many of the standard and also the less commonly used methods. A theoretical value of density can be calculated for a mineral if the parameters and chemical content of the unit cell are known (see Chapter 6, p. 294).

The choice of method is dictated primarily by the nature and amount of material available and the accuracy required. When the specific gravity of rocks is determined, relatively large samples are usually available and any simple method can yield acceptable results. With minerals, however, the need to ensure rigorously that the material is homogeneous often entails working with sample weights of between about 5 and 200 mg. Such amounts of material require a semi-micro- or micromethod, with consequent difficulty in many instances of obtaining a high level of accuracy.

## A. GENERAL PRECAUTIONS

Many methods depend on immersion of the solid in a liquid and so have a common major source of error due to the possible adherence of small air-bubbles to the surface of the solid. Liquids with a low surface tension, that is, with a high wetting power, should therefore be used whenever possible in preference to water. Organic liquids such as benzene, carbon tetrachloride, toluene and ethanol are suitable. If water is used, the addition of small amounts of a wetting agent is beneficial and care must be taken, particularly when powdered solids are being investigated, to eliminate all adhering or entrapped air by shaking or by application of a vacuum. With these methods it is assumed that no interaction takes place between the solid and selected liquid.

For maximum precision, meticulous control of temperature is essential, and great care must be exercised to ensure that the material is homogeneous and free from inclusions.

## B. DISPLACEMENT METHODS

These methods are based on the direct weighing of the solid, the volume being determined by measurement of the volume of the liquid displaced when the solid is immersed in it. In its simplest form, the weighed solid is immersed in a liquid contained in a measuring jar and the volume increase is read off.

With this method le Chatelier and Bogitch (1916) obtained an accuracy of about $\pm 0 \cdot 1\%$ with approximately 2 ml of solid. Their apparatus consisted of a 5 mm i.d. graduated glass tube closed at its lower end and mounted vertically. The tube was half-filled with a suitable liquid and a cathetometer could be employed accurately to measure the rise in liquid level due to the solid.

Caley (1930) adapted le Chatelier and Bogitch's method to a semi-micro-scale. He used a 2 mm diameter calibrated tube and diethyl ether as the immersion liquid. With weights of material of 30–100 mg an accuracy of $\pm 3 \cdot 0\%$ was obtained, the accuracy increasing with increasing weight of material. Rudenko and Vassilifsky (1957), using similar apparatus and boiled ethanol, quote accuracies of about $\pm 0 \cdot 1\%$ with weights of solids up to 70 mg.

Blank (1931) avoided the need to calibrate the graduated tube by using a weight burette. The tube used had a side capillary tube and the apparatus was filled with liquid to a lower mark on this tube. After insertion of the solid, liquid from the weight burette was added to the main tube to bring the liquid level to an upper mark on the capillary tube. Knowing the weight of liquid added, its density and the weight of liquid required to fill the apparatus from the lower to the upper mark, the volume of the added solid could be determined. With comparable weights of material, Blank obtained accuracies similar to those obtained by Caley.

Muller (1960) replaced the vertically mounted graduated tube by a precision-bore glass U tube of 1 mm diameter. This tube was mounted on the graduated stage of a microscope so arranged that the stage was exactly vertical. One meniscus of the toluene-filled U tube was accurately centred on the microscope crosswires, and the tube limb was aligned vertically. The rise in liquid level due to the added solid, and therefore to its volume, may be calculated from the measured angle of tilt necessary to return the meniscus to its datum point on the crosswires. With 2–30 mg weights of sample, accuracies of $\pm 0 \cdot 5$ to $\pm 1 \cdot 0\%$ were achieved.

## C. PYKNOMETRY

The principle underlying this familiar method of density determination requires that the weight of solid is measured directly and the volume obtained by weighing the volume of a suitable liquid displaced from the pyknometer by the solid.

In general all pyknometric methods entail the successive weighing, on some form of analytical balance, of the empty pyknometer ($W_1$), the pyknometer plus sample ($W_2$), the pyknometer plus sample plus liquid ($W_3$) and the pyknometer plus liquid ($W_4$). The specific gravity of the solid may then be calculated, as:

$$\text{sp.gr. of solid} = \rho \times \frac{(W_2 - W_1)}{(W_4 - W_1) - (W_3 - W_2)}$$

where $\rho$ = density of the liquid used. Sufficient material at least to quarter-fill the pyknometer is needed, and precautions must be taken to remove all entrained air.

With relatively large weights of material, the pyknometer, or specific-gravity bottle, can yield accurate results, but the method is subject to several possible sources of error. These include evaporation losses of liquid, and variation in

s

the pyknometer volume due to temperature variations and to the variable seating of the stopper. These sources of error have been examined in detail by Mason (1944).

Syromyatnikov (1935), Bannister and Hey (1938) and Winchell (1938) have all modified the pyknometric method to a semi-micro- or microscale. By use of these techniques, accuracies of $\pm 0.5$ to $\pm 1.0\%$ have been obtained with weights of solid from about 5–100 mg. Ksanda and Merwin (1939) improved the technique of micropyknometry by a more precise control of the volume of the pyknometer.

Straumanis (1953) used a micropyknometer accurately to measure the density of a liquid in which a few milligrams of solid wire were in hydrostatic balance with the liquid; the required density of the solid thus equalled that of the determined liquid. By this means Straumanis was able to measure the density of pure sublimed salt with an accuracy of $\pm 0.014\%$.

### D. METHODS OF HYDROSTATIC WEIGHING

These methods are based on the principle of Archimedes and require that the solid is first weighed in air ($W_1$) and then reweighed ($W_2$) in a liquid of accurately known density ($\rho$). From the apparent loss in weight, the volume of the solid may be computed and hence its density. Thus:

$$\text{Density of solid} = \rho \times \frac{W_1}{W_1 - W_2}$$

The major source of error is due to the effect of the surface tension of the liquid on the wire normally used for suspension of the sample during the second weighing. The wire should therefore be as fine as practicable and an organic liquid of low surface tension employed as the displacement liquid. With water, a few drops of a wetting agent should be added. With large enough specimens, reasonably accurate results are obtained with any simple weighing device and, as Mason (1944) has shown, the relative accuracy of the measured density is greater, the greater the weight of material taken, the greater the density of the displacement liquid and the smaller the density of the material.

For routine density determination of large specimens, such as rocks, any suitable laboratory balance or a Walker steelyard balance, which obviates the need for a set of weights, may be used. For smaller samples (1–50 g) the well known Jolly Spring balance is suitable. With this balance an accuracy of $\pm 0.2\%$ is possible. These three methods are fully described by Holmes (1930). For high precision measurement of density (of large single crystals of 1–20 c.c.), Smakula and Sils (1955) developed a refined technique of hydrostatic weighing, with which they obtained an accuracy of $\pm 0.001$ to $0.002\%$.

When only very small amounts of material are available, the microtorsion balance described by Berman (1939) may be used. This yields an accuracy of $\pm 0.2\%$ on 25 mg of solid with a density of 5. Toluene is used as the displacement liquid, and the technique is both very simple and rapid. It is usable for single grains or coarse powders.

Penfield (1895) and more recently Fahey (1961) have both adapted the method of hydrostatic weighing for use with fine powders. A small glass tube is first weighed in air ($W_1$) and then in the displacement liquid ($W_2$). After careful drying and addition of the powdered solid, the tube is reweighed in air ($W_3$) and finally in the liquid ($W_4$), ensuring that the tube is completely filled by the liquid and the powdered solid thoroughly wetted. The required density may then be calculated, as:

$$\text{Density of solid} = \rho \times \frac{(W_3 - W_1)}{(W_3 - W_1) - (W_4 - W_2)}$$

where $\rho$ = density of displacement liquid.

With from $0.5$–$1.5$ g of $-100$ $+200$ mesh quartz, Fahey demonstrated an accuracy of about $\pm 0.1\%$ for this technique.

### E. METHODS BASED ON COMPARISON WITH HEAVY LIQUIDS

These methods are among the most accurate available for the measurement of the density of solids and are all based on the direct comparison of the density of the solid with that of a suitable heavy liquid. The density of the liquid is subsequently accurately determined, in most instances by the standard methods of pyknometry, the Westphal balance (capable of an accuracy of at least $\pm 0.0005$) or refractive-index measurement (Midgeley, 1951). The methods are in general applicable to powders or to single grains as small as $0.001$ cu. mm, and are the only techniques that give an indication of the homogeneity of the material.

As the solid is immersed in a liquid, precautions must be taken to ensure complete wetting, and for very precise results rigorous temperature control must be applied and care taken to eliminate convection currents in the liquid by use of a thermostat.

The methods may be sub-divided into two classes. In the first, the solid is immersed in a suitable heavy liquid and its behaviour noted. If the solid floats, a miscible diluent of lower density is added until the specimen neither sinks nor floats. Conversely, if the solid sinks, a heavier density miscible liquid is added. The density of the liquid is then determined by one of the methods already referred to. Wulff and Heigl (1931) have described the technique of this method and state that an average accuracy of $\pm 0.02\%$ may be attained.

Reilly and Rae (1954) and Wunderlich (1957) both describe a modification of this method, by which final adjustment of the density of the liquid is achieved by varying the temperature.

With very small particles, their rate of movement in the liquid becomes extremely slow. Bernal and Crowfoot (1934) overcame this problem by using a centrifuge to increase the rate of rise or fall of the particles in the liquid. By this means they obtained acceptable results with particles as small as 0·001 cu. mm.

With these methods, if several separate grains of the same mineral are measured together, any unsuspected differences in their homogeneity are revealed by the nature of their behaviour in the liquid.

Methods belonging to the second class depend on the use of a heavy-liquid density gradient column. In such a column the density of the liquid increases *uniformly* from top to bottom; a suitably wetted mineral grain when added to the column settles to a density level exactly matching its own. As the change in density in a column can be made extremely gradual the technique is one capable of great sensitivity and can permit a density determination to $\pm 0·0002$ g/ml. The simultaneous addition of a number of grains of the same mineral will give an indication of their homogeneity by the absence of scattering of individual grains in the column. When very small grains are handled, the column should be centrifuged to ensure that the grains settle to the appropriate density level (Muller and Burton, 1965).

Many techniques of setting up a required density gradient have been described (Holmes, 1930; Jones, 1961b; Beevers, 1961; British Standards, 1964; Muller and Burton, 1965). These techniques all require the controlled mixing of two miscible liquids with densities equal to the maximum and minimum of the gradient sought. An alternative technique is to apply a *thermal gradient* to a column of *homogeneous* liquid (Moret, 1961; Pelsmaekers and Amelinckx, 1961 and Kats, 1962).

The density at any specified level in a column, that is the level at which the solid is in equilibrium with the liquid, may be determined by one of several methods. Beevers (1962), Pelsmaekers and Amelinckx (1961) and Moret (1961) have used small, accurately calibrated floats to delineate the gradient; the required density being derived by interpolation, where necessary. Kats (1962) developed a direct-reading density scale by measuring the thermal coefficient of cubical expansion of the liquid, its density for one value of temperature and the temperatures along the column when thermal equilibrium had been established. The use of a hollow prism by Jones (1961a) to contain the liquid density gradient column enables the refractive index of the liquid at any level to be determined directly by the method of minimum deviation. The required density is then obtained from a graph of refractive index v. density (normally linear), previously established for the liquids used in the gradient. Muller and Burton (1965) have described a thermostatted density-gradient-column separator that enables the liquid column to be precisely cut at any specific level in order to extract a small sample of liquid (and the specimen, if required) for refractive-index determination. Techniques of establishing the appropriate calibration curves are also given.

For the density determination of minerals, any of the commonly used heavy liquids, including Clerici solution, as discussed in Chapter 1, are suitable. A further list of liquids, covering a density range of 0·69–4·3 g/ml are given by Richards and Berger (1962).

## F. GAS VOLUMENOMETRY

This method is based on Boyles Law for perfect gases. A calibrated vessel is used and the change in gas pressure is measured when the volume of the vessel is altered by a known amount. A weighed sample of solid is then introduced into the vessel and the procedure repeated. From the measurements obtained, the volume of the solid may be calculated. This method is of value when it is undesirable to immerse the solid in a liquid or when the solid is in the form of a very fine powder or is porous and so not easily wetted.

Reilly and Rae (1954) give details of the method and Hauptmann and Schulze (1934) have adapted the method to a microscale. With relatively large amounts of material, a probable accuracy of $\pm 0 \cdot 1\%$ may be attained; with volumes of the order of 0·01 ml, the accuracy is about 1·0%.

### REFERENCES

Bannister, F. A. and Hey, M. H. (1938). A new micropyknometric method for the specific gravity of solids; with a note on the accuracy of specific gravity determinations. *Mineralog. Mag.* **25**, 30–34.

Beevers, A. H. (1961). Preparation of sensitive linear density gradients. *Soil. Sci.* **25**, 357.

Beevers, R. B. (1962). The use of density-gradient columns. *Lab. Pract.* 287.

Berman, H. (1939). A torsion microbalance for the determination of specific gravity of minerals. *Am. Miner.* **24**, 434.

Bernal, J. D. and Crowfoot, D. (1934). Use of the centrifuge in determining the density of small crystals. *Nature, Lond.* **134**, 809.

Blank, E. W. (1931). Device for estimation of density of gems and small amounts of solids. *Ind. Engng Chem. analyt. Edn.* **3**, 9.

British Standards Institution (1964). Specification for concentration density gradient columns. B. S. 3715. The Institution, London.

Caley, E. R. (1930). Device for the rapid estimation of the density of small amounts of solids. *Ind. Engng Chem. analyt. Edn.* **2**, 177.

le Chatelier, H. and Bogitch, F. (1916). Sur la détermination de la densité des corps solides. *C.r. hebd. Séanc. Acad. Sci., Paris* **163**, 459.

Fahey, J. J. (1961). A method for determining the specific gravity of sand and ground rocks or minerals. *Prof. Pap. U.S. geol. Surv.* **424**-C, C372.

S*

466 L. D. MULLER

Hauptmann, H. and Schulze, G. E. R. (1934). Über ein neues Mikrovolumeno-meter. *Z. phys. Chem.* **A171**, 36.
Holmes, A. (1930). "Petrographic Methods and Calculations." Thomas Murby, London.
Jones, J. M. (1961a). Method of density determination on a micro-scale. *J. scient. Instrum.* **38**, 303.
Jones, J. M. (1961b). Method of establishing a liquid column of graded density. *J. scient. Instrum.* **38**, 367.
Kats, M. Ya. (1962). Measurement of the density of solid bodies with a gradient tube. *Pribory Tekh. Eksp.* **No. 1**, 178.
Ksanda, C. J. and Merwin, H. E. (1939). Improved technique in micropyknometric density determination. *Am. Miner.* **24**, 482.
Mason, B. (1944). The determination of the density of solids. *Geol. För. Stokh. Förh.* **66**, H.1, 27.
Midgeley, H. G. (1951). A quick method of determining the density of liquid mix-tures. *Acta crystallogr.* **4**, 565.
Moret, H. (1961). Density gradient column. *Rev. scient. Instrum.* **32**, 1157.
Muller, L. D. (1960). Some laboratory techniques developed for ore dressing mineralogy. Proceedings of the International Mineral Processing Congress, London, paper No. 52. Institution of Mining and Metallurgy, London.
Muller, L. D. and Burton, C. J. (1965). "The heavy liquid density gradient and its applications in ore-dressing mineralogy." Paper No. 49, presented at the VIII Commonwealth Mining and Metallurgy Congress, Australia and New Zealand.
Mursky, G. A. and Thompson, R. M. (1958). A specific gravity index for minerals. *Can. Mineralogist* **6**, 273.
Pelsmaekers, J. and Amelinckx, S. (1961). Simple apparatus for comparative density measurements. *Rev. scient. Instrum.* **32**, 828.
Penfield, S. L. (1895). On some devices for the separation of minerals of high specific gravity. *Am. J. Sci.* (3), **50**, 446.
Reilly, J. and Rae, W. N. (1954). "Physico-Chemical Methods," 5th Edn., Vol. 1, p. 577. Van Nostrand, New York.
Richards, F. M. and Berger, J. E. (1962). Determination of the density of solids. *In* "International Tables for X-ray Crystallography," (Ed. Lonsdale, K.), Vol. III, p. 17. The Kynoch Press, Birmingham.
Rudenko, N. I. and Vassilifsky, M. M. (1957). A simplified method of determination of the S.G. of minerals. *Zap. vses. miner. Obshch.* **86**, 131.
Smakula, A. and Sils, V. (1955). Precision density determination of large single crystals by hydrostatic weighing. *Phys. Rev.* **99**, 1744.
Straumanis, M. E. (1953). Density determination by a modified suspension method; X-ray molecular weight, and soundness of sodium chloride. *Am. Miner.* **38**, 662.
Syromyatnikov, F. V. (1935). The micropyknometric method for the determination of specific gravity of minerals. *Am. Miner.* **20**, 364.
Winchell, H. (1938). A new pyknometer for the determination of densities of heavy solids. *Am. Miner.* **23**, 805.
Wulff, P. and Heigl, A. (1931). Methodisches zur Dichtebestimmung fester Stoffe, inbesondere anorganischer Salze. *Z. phys. Chem.* **A152**, 187.
Wunderlich, J. A. (1957). Une méthode rapide pour mesurer la densité d'un cristal. *Acta crystallogr.* **10**, 433.

# CHAPTER 12

# Autoradiography

## S. H. U. BOWIE

*Atomic Energy Division, Institute of Geological Sciences, London, England*

## I. Introduction

Autoradiographic techniques were first applied to the study of ore minerals by Stěp and Becke (1904) who used normal photographic emulsions to show the distribution of pitchblende in uranium ore from Joachimsthal. Five years later, Mügge (1909) demonstrated that when mineral grains were scattered on a photographic plate, radioactive minerals produced a blackening of the emulsion on development. Thick nuclear-type emulsions, however, were not used until Alexandrov (1927) made detailed autoradiographs showing the location of tyuyamunite in limestone from the Tyuya Muyen district of Russian Turkestan.

Modern nuclear emulsions are extremely fine grained and are relatively insensitive to visible light, but are activated by particles such as mesons, protons and alpha particles. Dye-sensitized emulsions capable of recording all charged particles are also available. These have a somewhat larger grain size and, as they are sensitive to beta particles, have a lower resolving power than alpha-sensitive emulsions. Most emulsion types suitable for mineralogical purposes are available on glass plates about 1·25–1·5 mm thick and in a variety of different sizes. The most common sizes are 3 × 1 in and $4\frac{1}{4} \times 3\frac{1}{4}$ in. Emulsions are also supplied on a backing that can readily be removed (stripping emulsions), as a hardened emulsion film without support (pellicles) or as gels (liquid emulsions).

## II. Alpha Particle Autoradiography

Alpha particles of the uranium-decay series have a mean range of about 20–25 $\mu$m in nuclear emulsion. Hence the resolving power, or ability to differentiate between two adjacent radioactive grains, is about 30 $\mu$m. Resolution is greatest when the emulsion is in direct contact with a flat surface (Fig. 1). Thus, uncovered thin sections produce good results in emul-

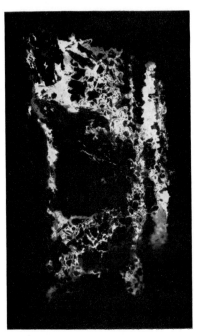

FIG. 1. Smooth-ground surface of quartz vein showing interstitial distribution of pitch-blende (white). Autoradiographic positive. Magnification, × 2.

sions backed by glass, as do polished sections, provided they are relatively free from cavities. It may be necessary to impregnate the surface of friable specimens with a fairly fluid plastic or resin before the surface is prepared for exposure (see Chapter 3, p. 103). If the material to be studied is fine grained and maximum resolution is required, this can be achieved by using emulsions not more than 5–10 $\mu$m thick.

The technique of mounting mineral grains directly on the moistened surface of a nuclear plate is particularly applicable to the study of sands, silts, and pulped rocks. The choice of emulsion thickness depends mainly on the size of the grains. If the sample is composed of grains differing widely in size, then it is advisable to prepare two or three sized fractions and to mount them on separate plates of appropriate emulsion thickness. Equal-sized grains can readily be pressed into the emulsion with a glass plate so that they remain in position throughout the process of developing, fixing and washing. Emulsions 200 μm thick are suitable for grains coarser than 30 B.S. ( >0·5 mm); 100 μm emulsions for grains 30–72 B.S. (0·5–0·2 mm); and those of 25 or 50 μm are of adequate thickness for material finer than 72 B.S.

Most emulsion types are available in thicknesses from 5–1000 μm, but layers thicker than 200 μm are rarely required for mineralogical studies. Suitable emulsions for alpha autoradiography are given in Table 1.

TABLE 1

| Emulsion type | Characteristics | Grain diameter |
|---|---|---|
| Ilford K.2 | Gives clearly defined alpha tracks | 0·20 μm |
| Ilford L.2 | Fine-grained emulsion suitable for track-length measurements | 0·14 μm |
| Kodak NTB | Sensitive emulsion similar to Ilford K.2 | 0·29 μm |
| Kodak NTA | Fine-grained emulsion with little background grain blackening | 0·22 μm |

### A. AUTORADIOGRAPHIC EXPOSURE

The most important precaution when an autoradiographic exposure is made is ensuring that the surface of the specimen makes contact with the emulsion and that once it is in position it is not moved until exposure is completed. The light-proof box used for storing plates during exposure should preferably be made of metal and should be kept in a cool dry place. The duration of exposure is dependent on the specific activity of the minerals present in the specimen under examination, on the sensitivity of the emulsion, and on the density of the autoradiographic image required. Sufficient alpha particles are emitted by pitchblende or uraninite for an estimate of the grade to be made by counting the track population after an exposure time of 2 min. For a mega-scopic image to be produced, however, the minimum exposure time would be about 10 min, and for an image dense enough to be reproduced on bromide paper the exposure would have to be prolonged to about 2 h.

### B. PROCESSING

There is no standard method of processing nuclear emulsions. Best results can be obtained by using somewhat different techniques depending on the type and thickness of the emulsion and on the nature of the information required. For general purposes, the metol–hydroquinone developers ID19 and D19 recommended by Ilford and Kodak, respectively, are satisfactory. A very dense image can be obtained by using a caustic–hydroquinone developer (Bowie, 1951). For emulsions of thickness 100 μm and over on which quantitative work is to be carried out, an amidol developer is most suitable. Developing and fixing time for a 100 μm emulsion with this developer is about 4½ h.

### C. QUANTITATIVE MEASUREMENTS OF RADIOACTIVITY

Autoradiographic techniques are most commonly used in mineralogy for the location of radioactive minerals in thin and polished sections (Figs. 2 and 3).

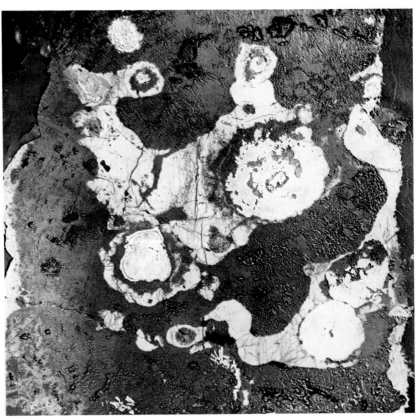

Fig. 2. Polished section showing chalcocite (white) pseudomorphous after pyrite associated with pitchblende (light grey), largely altered to metatorbernite (medium grey). Quartz (dark grey) occurs at the left- and right-hand margins. Magnification, × 140.

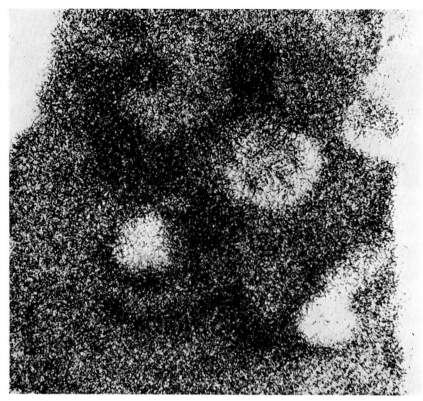

FIG. 3. Alpha-track pattern of same field as shown in Fig. 2. Note the high activity of the pitchblende compared with the metatorbernite, the pitchblende inclusions in the chalcocite and the inactive quartz. Autoradiographic negative. Magnification, ×140.

Quantitative assessments of alpha activity are normally made visually, or a microdensitometer can be used to compare the density of the images produced by an unknown and by a chemically analysed standard. An alternative method is to count the alpha-track population after a relatively short exposure.

### D. TRACK-LENGTH MEASUREMENT

The measurement of the length of alpha-particle tracks enables alpha-emitting nuclei to be identified. For accurate work it is necessary to calibrate the particular emulsion batch in order to establish the required range–energy relationship; and for results to be statistically meaningful a large number of tracks must be measured. Nevertheless, the technique is valuable for the study of small grains in thin or polished section that cannot readily be isolated for study by alpha spectrometry (Harrison and Taylor, 1966). The technique is particularly applicable to the study of weakly radioactive constituents in rocks, meteorites and deep-sea sediments. The main uncertainty of the method is in allowing for the variation in track lengths due to differing amounts of energy loss in the mineral before the alpha particles enter the emulsion. This effect can be overcome by dissolving the mineral grains and loading the emulsion with the solution obtained. Alpha-star populations produced by $^{228}$Th have been used together with track-length measurements to estimate microgram quantities of thorium in deep-sea sediments (Picciotto and Wilgain, 1954).

### III. Beta Particle Autoradiography

There is no sharp distinction between alpha- and beta-sensitive emulsion types. The faster emulsions suitable for recording alpha tracks will also record low-energy electrons, but specially sensitized emulsions are also available that are capable of recording charged particles of any energy. Beta-particle tracks are usually curved and more tenuous than alpha tracks and it is often difficult to trace them to their source. This imposes severe limitations on beta-particle autoradiography in the localization of natural beta-emitting minerals. Beta-sensitive emulsions are, however, finding a wider use in activation techniques. Thus when a thin or polished section containing manganese minerals is bombarded by slow neutrons, $^{55}$Mn yields the highly radioactive $^{56}$Mn, which emits beta particles. These will be recorded in a beta-sensitive emulsion and the track density after a known interval can be used to give a semi-quantitative indication of the manganese present if compared with standard material of known manganese content. Other elements such as barium, copper, gold, scandium and tungsten can be localized and assessed in this way.

Beta-particle emulsions suitable for mineralogical studies are given in Table 2.

TABLE 2

| Emulsion type | Characteristics | Grain diameter |
|---|---|---|
| Ilford G.5 | Sensitive to charged particles of any energy | 0·27 $\mu$m |
| Kodak NTB 3 | Similar to Ilford G.5 | 0·34 $\mu$m |
| Ilford K.2 | Slow electrons produce tracks of a few grains | 0·20 $\mu$m |
| Kodak NTB 2 | Records electrons up to 0·2 MeV | 0·29 $\mu$m |

## REFERENCES

Alexandrov, S. P. (1927). Radiographien tüjamujunscher Erze. *Z. Kristallogr. Kristallogeom.* **65**, 141.

Bowie, S. H. U. (1951). Autoradiographic techniques in geological research. *Bull. geol. Surv. Gt. Br.* **3**, 58.

Harrison, R. K. and Taylor, K. (1966). Radian anglesite from Wheal Speed, Cornwall. *Bull. geol. Surv. Gt. Br.* **25**, 41.

Mügge, O. (1909). Radioaktivität und pleochroitische Höfe. *Centbl. Miner. Geol. Paläont.* **65**, 113, 142.

Picciotto, E. and Wilgain, S. (1954). Thorium determination in deep-sea sediments. *Nature, Lond.* **173**, 632.

Stĕp, J. and Becke, F. (1904). Das Vorkommen des Uranpecherzes zu St. Joachimsthal. *Sber. Akad. Wiss. Wien, Math-naturw. Kl., Abt.I,* **113**, 585.

# CHAPTER 13

# Atomic Absorption Spectroscopy

## R. J. W. McLAUGHLIN

*Department of Geology, University of Melbourne, Australia*

## I. Introduction

The atomic-absorption method has venerable antecedents. Fraunhofer observed many dark lines in the spectrum of the sun and these were interpreted later by Bunsen and Kirchhoff (Kirchhoff, 1861a,b), who showed that the dark bands corresponded to the emission spectra of certain elements. The "Fraunhofer spectrum" is due to absorption by atoms, in the relatively cooler zones about the sun, of radiation emitted by the much hotter solar interior. The principle has been employed by astrophysicists almost since its discovery, as a powerful tool in the study of stellar material. Laboratory use has been made of the method, again for astrophysical purposes (Estabrook, 1951), but the first application to chemical analysis was more recent (Walsh, 1955; Alkemade and Milatz, 1955). Since its inception, the technique has developed with great speed, and it now offers a convenient and rapid method for the chemical analysis of various materials. The literature is already voluminous on the subject, and among the various reviews are those by Walsh (1961), Gilbert (1963), Willis (1963) and David (1964).

In emission spectroscopy the method depends upon the production of a sufficient number of atoms in the first or higher excited states. For this reason flames, while excellent for readily excited elements, e.g. alkali metals, are of little use with elements of high excitation potential. Thus zinc, even at the elevated temperatures of the electric arc or spark, does not produce intense emission of the resonance lines corresponding to the energy transition from ground to excited states. The shorter the wavelength of the spectral line corresponding to the transition, the smaller the fraction of atoms raised to the excited state by a given energy input, and so the lower the sensitivity of the emission method. If absorption, instead of emission, is used to measure the atomic concentration, greater sensitivity is obtained. The proportion of atoms in the ground state is considerably greater than that for the excited state, even in elements readily giving emission, such as caesium (Willis, 1962b).

Atoms existing in the ground state can only absorb radiation of specific wavelengths corresponding to their resonance lines in emission. Elements with complicated electronic orbitals obviously will have a greater number of such resonance lines, but even here the choice of which line to select for absorption measurement is simple. The wavelength usually selected is the most sensitive, and hence corresponds to a first-order transition. It should be noted, however, that in complex spectra the strongest emission wavelength is not necessarily always the strongest in absorption (David, 1964). The spectral width of resonance lines is very small (ca $10^{-4}$Å), and even with various broadening effects (mainly Doppler) it is still only about $0 \cdot 01$–$0 \cdot 1$Å. This very narrow band width makes absorption measurements, with a continuous background, dependent upon the use of a monochromator with considerable resolving power. Although measurements employing a continuous background are useful for exploratory work (Allan, 1962; Fassel and Mossotti, 1963), it is preferable to use a source giving a strong sharp resonance line for the element being investigated. This sharp-line source is obtained by using a spectral vapour lamp or, for a less volatile element, a hollow-cathode tube. The latter, which is now almost universally employed, consists of an anode, usually of tungsten rod, and a hollow cylindrical cathode made of, or lined with, the element whose resonance line is required. These are enclosed in a glass tube (with a fused silica end window) containing an inert gas at low pressure. In operation, bombardment of the cathode by the inert-gas ions causes atoms of the element to sputter off the cathode. These atoms are excited by collision with inert-gas atoms and give rise to a strong, concentrated, line source. The possibility of incorporating several elements in the same cathode, although attractive, is difficult to apply, owing to differing volatilities of elements (Jones and Walsh, 1960). Massmann (1962) has overcome this difficulty by using co-axial cylinders of different elements in a unidirectional stream of inert filler gas.

## II. Method and Techniques

The basic apparatus for the estimation of elements by atomic-absorption analysis (see Fig. 1) consists of the hollow-cathode emission tube, the radiation from which is passed through a flame into which the sample (in solution) is aspirated by techniques similar to those in flame photometry. The resonance beam is brought to a focus at the central portion of the flame, and the emergent beam is again focused. During its passage through the flame, ground-state atoms absorb a proportion of the radiation depending upon their concentration. The emergent beam is focused on to a slit and passed through a monochromator to filter background, and the intensity is then measured by a photomultiplier device connected to a scale or recorder. While filters suffice as monochromatic devices for some elements, e.g. alkali metals, transition elements, because of the complexity of their spectra, require a monochromator of high quality. The device outlined above is essentially a d.c. amplified

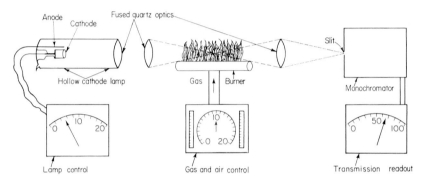

FIG. 1. Block diagram of apparatus for atomic absorption spectroscopy.

system and so can take no account of flame background at the wavelength of the absorbing resonance line. To overcome this, a.c. amplification is preferred. The hollow-cathode tube is modulated and the signal fed to an a.c. amplifier. The d.c. from the flame background is rejected, for the system is tuned to receive only the modulated signal (Box and Walsh, 1960). Double-beam instruments, in which splitting of the initial signal is effected and then the absorbed and non-absorbed beams compared, are superior because of their much greater stability, but they are more expensive than the simpler type outlined above (Russell *et al.*, 1957).

In operation the method resembles the classical colorimetric analysis, with the flame and its content of ground-state atoms simulating the liquid in the cell of the former method. Thus similar rules apply for graphs relating concentration to absorption, and at high absorption ($>70\%$) it is advisable to dilute the solution, for a small absorption increase corresponds to a large increase in concentration in this region and above. The system is confined to elements whose resonance lines are of longer wavelength than

2000 Å, because of the strong absorption by air at shorter wavelengths. This in effect precludes most of the non-metals from estimation by the technique, but some 40 elements lie in the region 2000–10,000 Å (Willis, 1963).

Besides the generally accepted technique, in which the flame is used as the source of ground-state atoms, other variants have been suggested. A method of cathodic sputtering has been described for metals (Russell and Walsh, 1959; Gatehouse and Walsh, 1960; Goleb and Brody, 1963). L'vov (1961) has utilized a furnace consisting of a carbon cylinder lined with tantalum foil, in an argon atmosphere. A small sample (ca 100 $\mu$g) is placed in the hot furnace and volatilized by a small electric arc into the hot furnace atmosphere. Very low limits of detection have been claimed with this technique ($10^{-2}$ to $10^{-4}$ $\mu$g). This method is not so rapid as the flame technique. So far only evaporated solutions have been utilized, but the method is obviously applicable to solid minerals, provided that segregation is not a problem. Somewhat similar principles have been employed by Margoshes and Scribner (1959), who replaced the flame with a plasma jet; the flash-heating technique (Nelson and Kuebler, 1963) and spark sources (Robinson, 1962; and Herrmann and Lang, 1962) have also been used. There are two recent developments that show considerable promise. The development of high-intensity hollow-cathode lamps, with increases in intensity of $10^2$ obtained by an auxiliary discharge across the open end of the cathode, will increase the sensitivity of the technique considerably. The possibility of replacing the monochromator with a resonance lamp (Russell and Walsh, 1959) will simplify the equipment and reduce instrumental costs.

The normal type of apparatus employs liquid that is aspirated into the flame. It seems likely that this will remain as the most usable technique. In the investigation of minerals, the sample must be rendered soluble and known weights contained in known volumes if quantitative data are desired. The normal methods described in most text books for decomposition of minerals nearly all involve addition of large quantities of material, with the attendant danger of introduction of impurities, e.g. sodium carbonate fusion or, worse, sodium peroxide, which is nearly always highly contaminated. For many minerals digestion in hydrofluoric and perchloric acid mixture will suffice (Hillebrand et al., 1959). Perchloric rather than sulphuric acid is preferred, because of the possibility of interference unless a very hot flame is used. Even so, there are many minerals that will not completely disintegrate with either method, and recent work has focused attention on this problem (Hoops, 1964). For these intractable materials the method of Biskupsky (1965) has been found to be of considerable value. The sample is fused with a mixture of lithium (or sodium) fluoride and boric acid. Silica is removed as the volatile fluoride and subsequent sulphuric acid digestion removes excess of boric acid and fluoride. The method is rapid and easily executed. The reagents are normally quite pure or may be readily purified. Although lithium colours the flame, this has not been found to have any serious effect in enhancement or depression of quantities estimated. For elements such as arsenic, antimony, mercury and thallium, which are relatively volatile, special refluxing techniques are necessary to prevent loss (Willis, 1963).

Standards are prepared in the usual fashion. Curves of absorbance against concentration are constructed, and these are usually straight lines or have a slight curve, concave towards the concentration axis. From these curves the concentration of an unknown may be measured from its absorption. It is only exceptionally that standards and unknowns must be of similar composition, and this is discussed later under "Interferences". A useful check on the likelihood of interference is the method of addition. In this technique, known quantities of the element being determined are added to samples of unknown and the absorptions of the mixtures are measured (Willis, 1962a). Not only may interferences be detected, but in the rare cases of such occurrence it is possible to extrapolate to the true concentration, by using a series of such additions. Great care must be exercised in the use of this technique (Fawcett and Wynn, 1961). Standards of low concentration deteriorate rapidly if stored in glass containers owing to absorption on the walls. This defect is not so serious if smooth Perspex containers are used. With the latter, no absorption has been detected for a large series of elements over a period of four months, provided that the solutions are acidified to prevent precipitation.

The sensitivity of the method may be increased by multiple-pass systems in which the resonance line from the source is passed through the flame more than once. Such methods invariably result in some loss due to scatter, and it is preferable to use an elongated flame and a single-pass system. For very hot flames the length of flame is limited, because of the danger of flash-back and explosion in the mixing chamber of pre-mix burners. Amplification of the output signal from the photo-multiplier can be carried out until the noise level becomes too great. In addition to the physical methods of increasing sensitivity, increases may be effected by methods involving liquid–liquid extraction (Dean, 1960). The use of ammonium pyrrolidine dithiocarbamate (APDTC) and various organic liquids insoluble in water, e.g. methyl n-pentyl ketone, increases of $10^2$ are quite easily and rapidly obtained (Melissa and Schöffmann, 1955; Allan, 1961a, 1961b; Willis, 1961b, 1962a). Although APDTC will complex with some 30 heavy metals and will extract over a wide pH range, in very acid solutions substituted dithiocarbamidates may be more efficient (Bode and Neumann, 1960). The method of concentration in which various selective liquid–liquid extractions are used would appear to offer quite remarkable possibilities in geochemical investigations.

The use of certain organic materials as solvents has a marked effect on sensitivity. Lower viscosity is believed to give small droplets, which thus gives a greater feed rate and a higher flame population. In addition, the more rapid dehydration in the flame gives a higher temperature and a greater opportunity for the formation of elements rather than refractory compounds (David, 1959; Willis, 1960; Robinson and Harris, 1962; Slavin and Manning, 1963).

A decrease in sensitivity is frequently necessary and, although this may be effected by dilution, it should be remembered that contamination of some elements is relatively easy at the 1 p.p.m. level (Willis, 1963). A better method of reducing the sensitivity is to choose a less sensitive resonance line. Although some workers have found that sensitivity may be reduced by turning the usual long burner through a right angle so that the path-length in the flame is short (Willis, 1960), this procedure is not recommended for rock and mineral analysis. Considerable deviation has been found for elements such as iron with this technique. The reason probably resides in the nature of the flame and concentration of the solutions.

## III. Interferences

In emission techniques, flame or arc background may cause serious difficulties, e.g. overlap of 2852·8 Å of sodium with 2852·1 Å of magnesium; $Ca(OH)_2$ bands on Na 5890 and 5896 Å; OH bands on Cu 3247 and 3271 Å. The absorption technique does not suffer from these difficulties, for measurement is of ground-state atoms, and even variation in flame temperature makes little difference to population number (David, 1960b; Willis, 1962b). Ionization interferences owing to one element suppressing another are very common in emission. In absorption the effect is considerably reduced and in hot flames is almost non-existent. In cooler flames compensation may be made, if necessary, by addition to the standards (Willis, 1960). Because of the narrow band-width of the resonance line, overlap is unlikely, and to date only a few coincidences with non-absorbing lines have been reported, e.g. nickel at 3415 Å and cobalt at 2407 Å (Robinson, 1962; Herrmann and Lang, 1962). Neither overlap poses any serious problem. A background effect, due to scattering of the resonance beam by small particles in the flame, has been noted by various workers (Allan, 1961b; David, 1961; Willis, 1962b). The more refractory the particles, the worse is the effect. Although it has not been found to be serious in biological systems, it can reach serious proportions in the analysis of rocks or minerals unless care is taken to dilute the specimens. The writer has found that a dilution of 1 g in 200 ml gives results that are almost free from this effect. At this dilution, some of the minor elements are on the limits of detection and thus the considerable value of liquid–liquid extraction, e.g. APDTC, is apparent, since the bulk of the sample is left behind by the treatment and the minor elements are effectively and efficiently concentrated.

All of the serious interferences recorded to date have been of a chemical nature and due to combination. Some elements readily form highly refractory oxides in an oxidizing flame, and so the proportion of ground state atoms is too low for detection. Examples here are silicon, aluminium, molybdenum, titanium, vanadium and tungsten. Combination of one element with another in the same solution may so lower the effective content, that sensitivity is too low or erroneous results are obtained when compared with standards. Examples here are the combination of aluminium with alkaline earths, or phosphorus or sulphur with calcium (Allan, 1958; David, 1959). This combination appears to occur in the droplets rather than in the flame itself (Alkemade and Voorhuis, 1958; Fukushima, 1959). Thus additions that lower the viscosity and cause finer droplets (organics, etc.), give incandescent particles of higher temperature, which by volatilization give atoms in the ground state, and these may be measured.

There are several means by which these chemical interferences may be eliminated or at least considerably reduced. Suppression by addition of a chemical to the unknown, removal of the interfering element, or variation of flame type, are all possible methods. The addition of a chemical to the solution works on the principle of preferential addition to the interfering element, leaving the sought-after element free. Thus in the interference of phosphorus with calcium due to phosphate formation, the addition of lanthanum or strontium, in excess, frees the calcium (Willis, 1960; David, 1960a; Willis, 1961a; Elwell and Gidley, 1961). Similarly, addition of magnesium, provided excess sulphate is present, is also effective (David, 1959). Suppression of aluminium interference on alkali earths has been effected by an excess of lanthanum, added as lanthanum chloride (Billings and Adams, 1964), or strontium (Leithe and Hofer, 1961; David, 1960a). Other additions which have been used are ethylenediaminetetra-acetic acid (Wirtschafter, 1957; Debras and Voïnovitch, 1958; Willis, 1960; West and Cooke, 1960; Herrmann and Lang, 1962); 8-quinolinol (Debras-Guédon and Voïnovitch, 1959, 1961; Wallace, 1963).

Techniques involving removal of an interfering element have not been explored as fully as methods of suppression, for the latter are much more rapid. Some application of ion-exchange removal has been reported, e.g. for phosphorus (Hinson, 1962; David, 1962).

The third method of suppression of interference, variation in flame type, shows great promise. It is known that for completely atomized metals, sensitivity is usually slightly higher in a cool flame (air–coal gas) than in a hot one (air–acetylene) (Gatehouse and Willis, 1961; Allan, 1961a). Similarly, certain elements give best results in flames rich in fuel and of a strongly reducing nature (Ca, Sr, Ba, Sn, Bi, Cr, Mo, Ru, Rh). Therefore the height in the flame at which the resonance beam passes is of utmost importance for maximum sensitivity (David, 1959, 1960c, 1961; Gatehouse and Willis, 1961). It has already been mentioned that the effect of organic additions to the solutions is to suppress interference because of higher temperature. Recently Willis (1965) has shown that by using a high-temperature flame of nitrous oxide–acetylene, many of the elements, previously difficult due to low volatility, are now relatively easily detectable. Thus aluminium, silicon and titanium amongst others can now be determined. At the same time, the chemical interferences discussed previously disappear. Other references to high-temperature flames are as follows: (Fassel and Mossotti, 1963; Slavin and Manning, 1963; Dowling et al., 1963; Chakrabarti et al., 1963; Amos and Thomas, 1965).

## IV. Applications

Although from what has been said it might be imagined that the atomic absorption method is rife with interferences, this is far from true. Interference is the exception rather than the rule, and it is on the few exceptions that so much research has centred. The advent of hot flames of long path-length seems to have solved the problem for most of these.

Most applications of atomic absorption have been in biological or agricultural fields and relatively little published work has appeared on geological material. Excellent summaries with numerous references are given by Willis (1963) and David (1964). For application to rocks or minerals the following are pertinent. Rawling et al. (1961) determined silver in lead sulphide concentrates. Fabricand et al. (1962), estimated iron, manganese, nickel, zinc and copper in sea water, and David (1962) estimated strontium in rock phosphate. Copper, nickel, iron, calcium, chromium, lead, silver, barium and sodium have been determined in oil (Barras, 1962; Sprague and Slavin, 1963). Various elements have been determined in rocks and feldspars by Billings and Adams (1964) and in rocks by Belt (1964) and Firman (1965).

An interesting possibility is the applicability of the method to isotope determination, and some work on lithium has been undertaken (Zaïdel and Korennoĭ, 1961; Slavin, 1962). If this method could be extended, it might make possible a rapid method of age determination.

REFERENCES

Alkemade, C. T. J. and Milatz, J. M. W. (1955). Double-beam method of spectral selection with flames. Appl. scient. Res. **4B**, 288.
Alkemade, C. T. J. and Voorhuis, M. H. (1958). Influence of phosphorus on the emission of calcium in the flame. Z. analyt. Chim. **163**, 91.

Allan, J. E. (1958). Atomic absorption spectrophotometry with special reference to the determination of magnesium. *Analyst, Lond.* **83**, 466.

Allan, J. E. (1961a). The determination of copper by atomic absorption spectrophotometry. *Spectrochim. Acta* **17**, 459.

Allan, J. E. (1961b). The use of organic solvents in atomic absorption spectrophotometry. *Spectrochim. Acta* **17**, 467.

Allan, J. E. (1962). Review of recent work in atomic absorption spectroscopy. *Spectrochim. Acta* **18**, 605.

Amos, M. D. and Thomas, P. E. (1965). The determination of aluminium in aqueous solution by atomic absorption spectroscopy. *Analytica chim. Acta* **32**, 139.

Barras, R. C. (1962). Atomic absorption spectroscopical estimation of chromium, iron and nickel in gas oils before cracking. Jarrel-Ash Newsletter No. 13, p. 1.

Belt, C. B., Jr. (1964). Atomic absorption spectrophotometry and the analysis of silicate rocks for copper and zinc. *Econ. Geol.* **59**, 240.

Billings, G. K. and Adams, J. A. S. (1964). The analysis of geological materials by atomic absorption spectrometry. Perkin-Elmer Newsletter No. 23, p. 1.

Biskupsky, V. S. (1965). Fast and complete decomposition of rocks, refractory silicates and minerals. *Analytica chim. Acta* **33**, 333.

Bode, H. and Neumann, F. (1960). Disubstituted dithiocarbamates. VIII. Extraction with solutions of diethylammonium diethyldithiocarbamate in organic solvents. *Z. analyt. Chim.* **172**, 1.

Box, G. F. and Walsh, A. (1960). Simple atomic absorption spectrophotometer. *Spectrochim. Acta* **16**, 255.

Chakrabarti, C. L., Lyles, G. R. and Dowling, F. B. (1963). The determination of aluminium by atomic absorption spectroscopy. *Analytica chim. Acta* **29**, 489.

David, D. J. (1959). Determination of calcium in plant material by atomic absorption spectrophotometry. *Analyst, Lond.* **84**, 536.

David, D. J. (1960a). The determination of exchangeable sodium, potassium, calcium and magnesium in soils by atomic absorption spectroscopy. *Analyst, Lond.* **85**, 495.

David, D. J. (1960b). The application of atomic absorption to chemical analysis – a review. *Analyst, Lond.* **85**, 779.

David, D. J. (1960c). Atomic spectrophotometric determination of molybdenum and strontium. *Nature, Lond.* **187**, 1109.

David, D. J. (1961). The determination of molybdenum by atomic absorption spectrophotometry. *Analyst, Lond.* **86**, 730.

David, D. J. (1962). Determination of strontium in biological materials and exchangeable strontium in soils by atomic absorption spectrophotometry. *Analyst, Lond.* **87**, 576.

David, D. J. (1964). Recent developments in atomic absorption analysis. *Spectrochim. Acta* **20**, 1185.

Dean, J. A. (1960). "Flame Photometry." McGraw-Hill, New York.

Debras, J. and Voïnovitch, I. (1958). Determination of the formation of certain strontium compounds in flame spectrophotometry. *C.r. hebd. Séanc. Acad. Sci., Paris* **247**, 2328.

Debras-Guédon, J. and Voïnovitch, I. (1959). Direct determination of aluminium in silicates by flame photometry. *C.r. hebd. Séanc. Acad. Sci., Paris* **249**, 242.

Debras-Guédon, J. and Voïnovitch, I. (1961). Analysis of iron minerals by flame spectrophotometry. *Chim. analyt.* **43**, 267.

Dowling, F. B., Chakrabarti, C. L. and Lyles, G. R. (1963). Atomic absorption spectroscopy of aluminium. *Analytica chim. Acta* **28**, 392.

Elwell, W. T. and Gidley, J. A. F. (1961). "Atomic Absorption Spectrophotometry," 1st Ed. Pergamon, Oxford.

Estabrook, F. B. (1951). Absolute oscillator strengths of chromium and nickel. *Astrophys. J.* **113**, 684.

Fabricand, B. P., Sawyer, R. R., Ungar, S. G. and Adler, S. (1962). Trace metal concentrations in the ocean by atomic absorption spectroscopy. *Geochim. cosmochim. Acta* **26**, 1023.

Fassel, V. A. and Mossotti, V. G. (1963). Atomic absorption spectra of vanadium, titanium, niobium, scandium, yttrium and rhenium. *Analyt. Chem.* **35**, 252.

Fawcett, J. K. and Wynn, V. (1961). The determination of magnesium and calcium in biological materials by flame photometry. *J. clin. Path.* **14**, 403, 463.

Firman, R. J. (1965). Interferences caused by iron and alkalis on the determination of magnesium by atomic absorption spectroscopy. *Spectrochim. Acta* **21**, 341.

Fukushima, S. (1959). Mechanism and elimination of interference in flame photometry. *Mikrochim. Acta*, 596.

Gatehouse, B. M. and Walsh, A. (1960). Analysis of metallic samples by atomic absorption spectroscopy. *Spectrochim. Acta* **16**, 602.

Gatehouse, B. M. and Willis, J. B. (1961). Performance of a simple atomic absorption spectrophotometer. *Spectrochim. Acta* **17**, 710.

Gilbert, P. T., Jr. (1963). Atomic absorption spectroscopy – a review of recent developments. Proceedings of sixth Gatlinburg conference on analytical chemistry in nuclear reactor technology. Office of Technical Services, Washington D.C., U.S.A.

Goleb, A. and Brody, J. K. (1963). Atomic absorption studies using a hollow-cathode tube as an absorption source. *Analytica chim. Acta* **28**, 457.

Herrmann, R. and Lang, W. (1962). The grinder in emission and absorption flame photometry. *Optik, Stuttg.* **19**, 208.

Hillebrand, W. F., Lundell, G. E. F., Bright, H. A. and Hoffman, J. I. (1959). "Applied Inorganic Analysis." Wiley, New York.

Hinson, W. H. (1962). Ion-exchange treatment of plant ash extracts for removal of interfering anions in the determination of calcium by atomic absorption. *Spectrochim. Acta* **18**, 427.

Hoops, G. K. (1964). The nature of the insoluble residues remaining after the HF–H$_2$SO$_4$ acid decomposition (solution B) of rocks. *Geochim. cosmochim. Acta* **28**, 405.

Jones, W. G. and Walsh, A. (1960). Hollow-cathode discharges – the construction and characteristics of sealed-off tubes for use as spectroscopic light sources. *Spectrochim. Acta* **16**, 249.

Kirchhoff, G. R. (1861a). On the chemical analysis of the solar atmosphere. *Phil. Mag.* **21**, 185.

Kirchhoff, G. R. (1861b). On a new proposition in the theory of heat. *Phil. Mag.* **21**, 241.

Leithe, W. and Hofer, A. (1961). Magnesium determinations by means of atomic absorption flame photometry. *Mikrochim. Acta* 268.

L'vov, B. V. (1961). The analytical use of atomic absorption spectra. *Spectrochim. Acta* **17**, 761.

Margoshes, M. and Scribner, B. F. (1959). The plasma jet as a spectroscopic source. *Spectrochim. Acta* **15**, 138.

Massmann, H. (1962). Abstracts of international conference on spectroscopy. *Appl. Spectros.* **16**, 56.

Melissa, H. and Schöffmann, E. (1955). Application of substituted dithiocarbamates in micro-analysis. *Mikrochim. Acta* 187.

Nelson, L. S. and Kuebler, N. A. (1963). Vaporization of elements for atomic absorption spectroscopy with capacitor discharge lamps. *Spectrochim. Acta* **19**, 781.

Rawling, B. S., Amos, M. D. and Greaves, M. C. (1961). Determination of silver in lead sulphide concentrate by atomic absorption spectroscopy. *Trans. Instn Min. Metall.* **71**, 15.

Robinson, J. W. (1962). Observations in atomic absorption spectroscopy. *Analytica chim. Acta* **27**, 465.

Robinson, J. W. and Harris, R. J. (1962). Mechanical feed burner with total consumption for flame photometry and atomic absorption spectroscopy. *Analytica chim. Acta* **26**, 439.

Russell, B. J. and Walsh, A. (1959). Resonance radiation from a hollow-cathode. *Spectrochim. Acta* **15**, 883.

Russell, B. J., Shelton, J. P. and Walsh, A. (1957). An atomic absorption spectrophotometer and its application to the analysis of solutions. *Spectrochim. Acta* **8**, 317.

Slavin, W. (1962). Atomic absorption spectrophotometry without the use of flames. Atomic absorption newsletter No. 7. Perkin-Elmer Corporation.

Slavin, W. and Manning, D. C. (1963). Atomic absorption spectrophotometry in strongly reducing oxy-acetylene flames. *Analyt. Chem.* **35**, 253.

Sprague, S. and Slavin, W. (1963). Application of atomic absorption spectroscopy to petroleum products. Atomic absorption newsletter No. 12, Perkin-Elmer Corporation.

Wallace, F. J. (1963). The determination of magnesium in aluminium alloys by atomic absorption spectroscopy. *Analyst, Lond.* **88**, 259.

Walsh, A. (1955). The application of atomic absorption spectra to chemical analysis. *Spectrochim. Acta* **7**, 108.

Walsh, A. (1961). *In* "Advances in Spectroscopy." Ed. H. W. Thompson, Vol. 2, pp. 1–22, Interscience, New York.

West, A. C. and Cooke, W. D. (1960). Estimation of anion interferences in flame spectroscopy – use of (ethylenedinitrilo) tetraacetic acid. *Analyt. Chem.* **32**, 1471.

Willis, J. B. (1960). Determination of metals in blood serum by atomic absorption spectroscopy. I Calcium, II Magnesium, III Sodium and Potassium. *Spectrochim. Acta* **16**, 259, 273, 551.

Willis, J. B. (1961a). Determination of calcium and magnesium in urine by atomic absorption spectroscopy. *Analyt. Chem.* **33**, 556.

Willis, J. B. (1961b). Determination of lead by atomic absorption spectroscopy. *Nature, Lond.* **191**, 381.

Willis, J. B. (1962a). Determination of lead and other heavy metals in urine by atomic absorption spectroscopy. *Analyt. Chem.* **34**, 614.

Willis, J. B. (1962b). Atomic absorption spectroscopy. *Proc. R. Aust. chem. Inst.* **29**, 245, 357.

Willis, J. B. (1963). *In* "Methods of Biochemical Analysis." Ed. David Glick, Vol. 11, p. 1, Interscience, New York.

Willis, J. B. (1965). The nitrous oxide-acetylene flame in atomic absorption spectroscopy. *Nature, Lond.* **207**, 715.

T

Wirtschafter, J. D. (1957). Suppression of radiation interference in flame photometry by protective chelation. *Science* **125**, 603.

Zaïdel, A. N. and Korennoï, E. P. (1961). Spectral determination of isotopic composition and concentration of lithium in solutions. *Optika Spektrosk.* **10**, 570.

# Author Index

T*

# Subject Index

The names of materials (minerals and other substances) which have been investigated by one or more of the physical methods, are given in *italics*. Principal page references are given in *italics* where they occur among a number of other page references.

## A

Abbe refractometer, 11, 52, 86, 87
Absorption
   electron probe, 216, *236–239*, 240, 244, 246
   infrared, 371–403
   microscopy: transmitted light, 33, *43–45*
   microscopy: reflected light, 112
   X-rays, 184–186, 194–196, 206, 238
   X-ray diffraction, 264, 274, 275, 280, 281, 294, 296, 299–301
*Acmite*, 80, 325, 390, 391
A.C. spark excitation, 447, 448
*Adularia*, 33, 347, 362, 363, 365, 377
Airy disc (electron microscopy), 337
*Åkermanite*, 323
*Alabandite*, 144, 146, 149
*Albite*, 252, 360, 361
$AlCl_3$, 252
*Alkali amphiboles*, 391
*Alkali feldspar*, 33, 322, 382
*Alloys*, 196, 366
*Almandine* (*almandite*), 18, 389
$Al(NO_3)_3.9H_2O$, 252
$Al_2(SO_4)_3.18H_2O$, 252
*Altaite*, 144, 147, 148
*Alumina*, 387
*α-Alumina*, 252
*β-Alumina*, 252
*γ-Alumina*, 429
Aluminium, 237
   by atomic absorption spectroscopy, 482
   for counter windows, 175
*Aluminium hydroxide*, 369
*Aluminium minerals*, 252
*Aluminium oxides*, 420, 421
*Aluminosilicates*, 320
*Alunite*, 252, 385, 425, 426
*Amblygonite*, 252
*Ammonium chloride*, 387

Ammonium dihydrogen phosphate, 173
*Ammonium illite*, 380, 381
*Ammonium nitrate*, 386
*Ammonium zeolite*, 380
*Amphiboles*, 35, 84, 321, 374, 381, 391, 411, 428, 450
Amplitude contrast (electron microscopy), 358
*Analcite*, 321
Analyzing crystals (X-ray fluorescence anal.), 172–174
*Anatase*, 388
*Andalusite*, 11, 320
*Andradite*, 389
*Anglesite*, 407, 425
*Anhydrite*, 385
Anisotropism (reflected light), 152, 153
*Ankerite*, 384
Anode excitation, 447, 448, 451
Anomalous polarization colours, 57
*Anorthite*, 252
*Antlerite*, 385
*Anthophyllite*, 428
*Antigorite*, 326, 366, 428
*Antimony*, 144, 147, 148
Antiscatter slit (X-ray), 279
*Apatite*, 386, 387, 450
*Aragonite*, 11, 321, 374, 384, 385, 423, 424
Arc excitation (emission spectrography), 446–448
*Arfvedsonite*, 321
*Argentite*, 108, 144, 146, 148
*Arsenates*, 387, 398
*Arsenic*, 144, 147, 148
*Arsenides*, 153, 156
*Arsenopyrite*, 144, 147, 150, 156, 321, 417
A.S.T.M. powder data file and index, 286–288
Asymmetric vibration (for mineral separation), 25
*Atacamite*, 387, 415, 416

501